IONIC POLYMERS

MATERIALS SCIENCE SERIES

Advisory Editors

L. HOLLIDAY,
*Brunel University,
Uxbridge, Middlesex, Great Britain*

A. KELLY, F.R.S.,
*Deputy Director, National Physical Laboratory,
Teddington, Middlesex, Great Britain*

IONIC POLYMERS

Edited by

L. HOLLIDAY

Department of Polymer Science,
Brunel University, Uxbridge, Middlesex, England

APPLIED SCIENCE PUBLISHERS LTD
LONDON

APPLIED SCIENCE PUBLISHERS LTD
RIPPLE ROAD, BARKING, ESSEX, ENGLAND

ISBN: 0 85334 603 8

WITH 68 TABLES AND 176 ILLUSTRATIONS

© APPLIED SCIENCE PUBLISHERS LTD 1975

All rights reserved. No part of this publication may be reproduced, stored in a retrieval system, or transmitted in any form or by any means, electronic, mechanical, photocopying, recording, or otherwise, without the prior written permission of the publishers, Applied Science Publishers Ltd, Ripple Road, Barking, Essex, England

Printed in Great Britain by Galliard (Printers) Ltd Great Yarmouth

LIST OF CONTRIBUTORS

S. Crisp
  Laboratory of the Government Chemist, London SE1 9NQ, England

E. W. Duck
  The International Synthetic Rubber Co., Ltd, Brunswick Place, Southampton, England

J. Economy
  Research and Development Division, The Carborundum Company, Niagara Falls, New York, USA

J. H. Elliott
  Department of Polymer Science and Technology, Brunel University, Uxbridge, Middlesex, England

L. Holliday
  Department of Polymer Science and Technology, Brunel University, Uxbridge, Middlesex, England

D. K. Jenkins
  The International Synthetic Rubber Co., Ltd, Brunswick Place, Southampton, England

Ruskin Longworth
  Plastics Department, E. I. du Pont de Nemours & Co., Inc., Wilmington, Delaware, USA

Michael J. Lysaght
  Amicon Corporation, Lexington, Massachusetts, USA

J. H. Mason
   Research and Development Division, The Carborundum Company, Niagara Falls, New York, USA

N. H. Ray
   ICI Corporate Laboratory, Runcorn, Cheshire, England

A. D. Wilson
   Laboratory of the Government Chemist, London SE1 9NQ, England

FOREWORD

For many years it has been known that some important naturally occurring polymers, such as some polysaccharide gels, contain ionic groups. In addition, polyelectrolytes containing bound ions have been used in one or two sections of industry—for example as surface active agents and in ion exchange processes. However, it is only in the last ten years or so that ionic polymers have become the subject of systematic research.

This work on ionic polymers has been of two types. On the one hand it is illustrated by the pioneering work of Brown on carboxylated rubbers, of Nielsen on the metal salts of polyacrylic acid and of the du Pont company on the ionomers. This type of research aims to use ionic groups to modify the properties of a given polymer, or to achieve a new combination of properties. The other type of research can be illustrated by the work of Eisenberg and his school at McGill University—work which has embraced both carbon polymers, and silicates and phosphates, thus bringing together organic and inorganic polymers. This type of research, which is at an early stage of development, seeks to relate the structure and the properties of ionic polymers. The present volume covers both these aspects of these materials which, despite their prior use in a few areas of application, can nevertheless be regarded as a new class of polymer with interesting properties, about which a great deal remains to be learnt.

Most polymer scientists will have a broad picture of the role of ionic bonding in polymers—that in the majority of cases its main function is to act as an easily formed cross-link. This is very much of a simplification, as the following pages show—from these a more detailed picture emerges of the various properties of ionic polymers, which will be a useful guide to future research. For example, one of the main disadvantages of the ionic bond is its hydrolytic instability, especially under acid or alkaline conditions. However, this is by no means universally true, since there are many silicate glasses which show good stability when the conditions are not too extreme. Similarly, the ionic bond may show up at a disadvantage when considering creep behaviour, but again this is not always the case. On the other hand, the fact that the ionic bond is relatively labile thermally is often an advantage. In any case, new applications are being found for these materials, as in the fields of dental cements, splints, contact lenses and other medical applications, adhesives, etc. which provide justification for more research in this field.

An important feature of the book is that it treats certain inorganic materials, e.g. borates, phosphates and silicates and the corresponding glasses (in the appropriate composition range) as ionic polymers. There are many advantages in this approach, which goes back to publications by Stevels and others, and I hope that this book will help to popularise this view, which can be very helpful in glass technology and other fields.

The book is intended for graduate students and research workers in the polymer field and related disciplines, in biochemistry, dentistry, glass and ceramics technologies and geochemistry, and could also help final year undergraduates, to supplement their normal course work.

It is self-evident that a book of this type, being edited, depends on the contribution of each author, and I should like to take this opportunity to thank them all for their co-operation.

LESLIE HOLLIDAY

Department of Polymer Science
Brunel University, Uxbridge

CONTENTS

List of Contributors v
Foreword vii
Chapter 1
CLASSIFICATION AND GENERAL PROPERTIES OF IONIC POLYMERS—*L. HOLLIDAY*

1.1 Introduction 1
1.2 Classification of Ionic Polymers 5
1.3 Some Typical Examples of Ionic Polymers 9
    1.3.1 Covalent networks and sheets containing ionic bonds . . . 9
    1.3.2 Long covalent chains containing ionic bonds 17
    1.3.3 Short covalent chains plus ionic bonds 27
1.4 Some Structural Features of Ionic Polymers 28
    1.4.1 Concentration of ionic bonds 28
    1.4.2 Aggregation of ions 30
1.5 Some Physical Properties of Ionic Polymers 35
    1.5.1 Glass transitions in ionic polymers 35
    1.5.2 Melting point 46
    1.5.3 Viscoelastic properties 50
    1.5.4 Melt behaviour 50
    1.5.5 Relaxation behaviour 54
    1.5.6 Stiffness 56
    1.5.7 Strength 62
    1.5.8 Thermal expansion 63
    1.5.9 Hydrolytic stability 64
1.6 Conclusions 66
    References 67

Chapter 2

THERMOPLASTIC IONIC POLYMERS: IONOMERS—R. LONGWORTH

2.1	Introduction	69
2.2	Preparation and Characterisation	72
2.3	Physical and Mechanical Properties	75
2.4	Electron Microscopy	79
2.5	Rheological Properties	93
	2.5.1 Steady-state viscosity at low shear stresses	93
	2.5.2 Non-Newtonian viscosities	101
	2.5.3 Viscoelastic properties	104
2.6	Relaxation Behaviour	109
	2.6.1 Dynamic mechanical properties	109
	2.6.2 Electrical properties	119
	2.6.3 Nuclear magnetic resonance studies	124
2.7	Infra-red Spectroscopy	129
2.8	X-Ray Diffraction	134
2.9	Optical Properties of Ionomers	144
2.10	Thermal Properties of Ionomers	150
2.11	The Structure of Ionomers	154
2.12	Water Absorption and Plasticisation	162
2.13	Conclusion	169
	Acknowledgments	170
	References	170

Chapter 3

CARBOXYLATED ELASTOMERS—D. K. JENKINS and E. W. DUCK

3.1	Introduction	173
3.2	Historical	173
3.3	Preparation	176
3.4	Properties	179
	3.4.1 Unvulcanised elastomer	179
	3.4.2 Vulcanisation with metal oxides	180
	3.4.3 Mixed vulcanisation	185
3.5	Nature of the Cross-Link in Oxide Vulcanisation	190
3.6	Uses of Carboxylated Rubber	194
	3.6.1 Solid and solution rubber	194
	3.6.2 Latex rubber	199
3.7	Summary	205
	References	206

Chapter 4

RIGID, HIGHLY CARBOXYLATED IONIC POLYMERS—
A. D. WILSON and S. CRISP

4.1	Ionic Polymer Cements	208
	4.1.1 Ion binding and molecular configuration	209
	4.1.2 Zinc polycarboxylate cement	212
	4.1.3 Glass ionomer cements	226
4.2	Monolithic Plastics Formed from Metal Polyacrylate Salts	237
	4.2.1 Structure and properties	237
4.3	Acrylic Soil Conditioners	242
	4.3.1 Engineering treatments	244
	4.3.2 Agricultural treatments	249
	4.3.3 The mode of action of acrylate polymers on soil properties	252
	References	257

Chapter 5

METAL DICARBOXYLATES—HALATOPOLYMERS—
J. ECONOMY and J. H. MASON

5.1	Introduction	261
5.2	Preparation	262
5.3	Structure	265
	5.3.1 Crystalline structure	265
	5.3.2 Polymeric structure	266
5.4	Properties and Uses	276
	5.4.1 Properties	276
	5.4.2 Uses	279
	References	280

Chapter 6

TECHNOLOGY OF POLYELECTROLYTE COMPLEXES—
M. J. LYSAGHT

6.1	Summary and Introduction	281
6.2	History	283
6.3	Synthesis	284
	6.3.1 Starting materials	284
	6.3.2 Fabrication of neutral resins	285
	6.3.3 Non-stoichiometric resins	285
	6.3.4 *In-situ* polymerisation	287
6.4	Properties and Physical Chemistry	288
6.5	Applications	291
	References	297

Chapter 7

CRYSTALLINE SILICATES AND PHOSPHATES AS IONIC POLYMERS—J. H. ELLIOTT

7.1	Introduction	300
7.2	Interatomic Bonding in Ionic Silicates and Phosphates	301
	7.2.1 Electronic structure	301
	7.2.2 Tetrahedral symmetry of the SiO_4 and PO_4 groups	303
	7.2.3 Silicate polymers	305
	7.2.4 Phosphate polymers	307
	7.2.5 Copolymers	308
7.3	Sources and Synthesis of Ionic Silicates and Phosphates	310
	7.3.1 Naturally occurring silicate polymers	310
	7.3.2 Synthetic silicate and phosphate polymers	312
	7.3.3 Characterisation	322
7.4	Structures of Crystalline Ionic Silicate and Phosphate Polymers	323
	7.4.1 Structures of oligomers	324
	7.4.2 Structures of infinite chain polymers	328
	7.4.3 Structures of cross-linked and branched chain polymers	332
	7.4.4 Structures of 2-D sheet polymers	337
	7.4.5 Structure of 3-D framework polymers	342
7.5	Structure/Property Relationships of Ionic Silicate and Phosphate Polymers	345
	7.5.1 Cleavage of infinite chain and 2-D sheet polymers	346
	7.5.2 The strength of asbestos	348
	7.5.3 Thermal stability	350
	7.5.4 Hydrolytic degradation of polyphosphates	353
	7.5.5 Hydrolytic stability of silicate polymers	354
	7.5.6 Hydration/dehydration properties of zeolites and bentonites	356
	7.5.7 Ion exchange	357
	7.5.8 The stability of colloidal suspensions	358
	References	361

Chapter 8

INORGANIC GLASSES AS IONIC POLYMERS—N. H. RAY

8.1	Introduction	363
	8.1.1 Polymeric nature of glasses	364
8.2	Oxide Glasses	366
8.3	Particular Oxide Glasses	379
	8.3.1 3-Connective networks—phosphate glasses	379
	8.3.2 3- to 4-Connective networks—borate glasses	386
	8.3.3 4-Connective networks—silica and silicate glasses	392
8.4	Chalcogenide glasses	400
8.5	Conclusions	404
	References	405

INDEX   407

CHAPTER 1

CLASSIFICATION AND GENERAL PROPERTIES OF IONIC POLYMERS

L. HOLLIDAY

1.1 INTRODUCTION

An ionic polymer is a polymer—either inorganic or organic—which contains both covalent and ionic bonds in its chain or network structure. A typical example is an ionomer (a du Pont trade name for a metal salt of a carboxylated polyethylene, but which is often used in a wider sense of ionic polymers) or a metal salt of polyacrylic acid. This brief definition will be amplified later, but it will serve as a beginning to indicate the field of discussion. Ionic polymers are therefore a class of materials which fall between the formal divisions of chemistry, that is between covalent solids like diamond and polyethylene on the one hand, and ionic solids like sodium chloride or sodium benzoate on the other.[1]

It will be seen later that there are many advantages in regarding glass as an ionic polymer (in fact as Dislich has pointed out, glass is the oldest thermoplastic[2]), and since glass has a history of several thousand years, it means that ionic polymers have a respectable ancestry. However, it is only in the past twenty years or so that polymer scientists have adopted this view of glass structure, or have started to synthesise some of these structures in the field of carbon polymers. Thus it is only recently that polymer scientists have begun to appreciate the importance of this new degree of freedom—of extending the type of network bonds in polymers—which forms the subject of this book, and which has recently become a fruitful field of research. It would be reasonable to say that systematic work on these materials began with the work of Brown on rubbers[11] and Nielsen on polyacrylates.[10]

By convention, ionic polymers exist in the salt or ionised form; that is the counterion—usually a cation—is not hydrogen. Thus a metal polyacrylate is an ionic polymer, but polyacrylic acid is not. By definition, the range of ionic polymers is very wide, since it includes materials as dissimilar as glass, ionomers, carboxylated rubbers, alginate gels, soluble polyelectrolytes, etc. The formula of some of these is shown below:

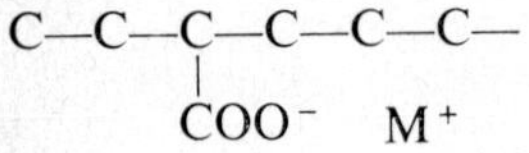

$$\begin{array}{c} \text{C—C—C—C—C—C—} \\ | \\ \text{COO}^- \quad \text{M}^+ \end{array}$$

Skeleton of ionomer or carboxylated rubber, with univalent cation.

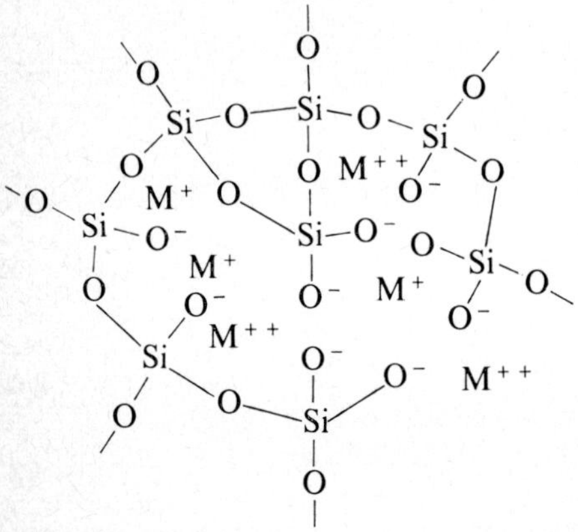

Partial 2-D representation of 3-D glass structure with silicate anion groups and univalent and divalent cations.

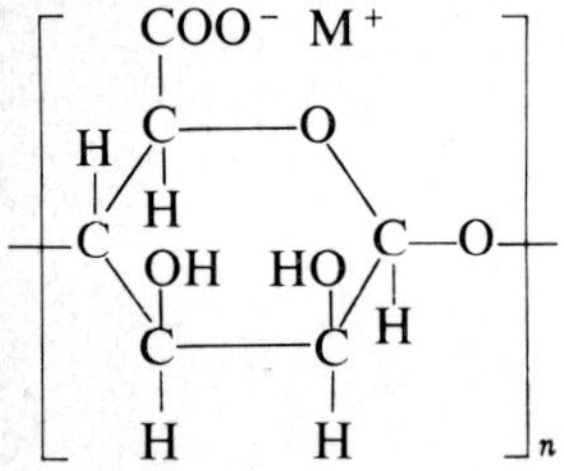

Repeat unit of metal salt of alginic acid.

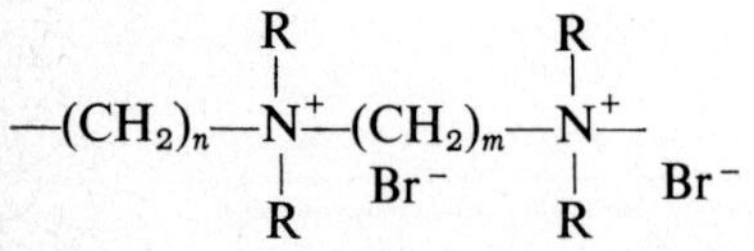

$$-(CH_2)_n-\underset{\underset{R}{|}}{\overset{\overset{R}{|}}{N^+}}-(CH_2)_m-\underset{\underset{R}{|}}{\overset{\overset{R}{|}}{N^+}}- \quad Br^- \quad Br^-$$

Simple ionene.

All the foregoing structures contain covalent polymer chains with— additionally—ionic bonds as part of the network, although it is not immediately apparent from this simple representation of, for example, a metal salt of a carboxylated rubber, how the ionic bonds act in many ways like a cross-link or network bond, as will be seen later.

It may be asked, with properties ranging from rigid infusible or high melting solids, through thermoplastics to viscous liquids at the opposite extreme, can there exist any unifying principles within such a broad class of materials? The answer which emerges is in the affirmative. It is clearly worthwhile to explore the effect of inserting different numbers and types of ionic bonds into covalent solids, and this involves covering a wide range of materials. It confers an interesting range of properties, depending on the circumstances. The ionic bond is a strong bond, like the covalent bond; however, it can under certain circumstances be thermally labile, and this may be a useful property. It may show the property of ion exchange in aqueous or molten salt systems. In some instances it may be convenient to produce a more rigid product by adding ionic bonds rather than covalent bonds. In this connection, the fact that the ionic bond can be readily formed is a great advantage.

At this point, some discussion of network bonds in a polymer is needed. The essential feature of a conventional polymer like polyethylene or polymethylmethacrylate is that the chains are made up of covalent, *i.e.* strong *directional* bonds formed by joining together atoms with a valency of two or higher, *e.g.*:

$$
\begin{array}{cc}
\mathrm{-C-C-C-C-C-} & \mathrm{-C-C-O-C-C-O-} \\
\mathrm{I} & \mathrm{II}
\end{array}
$$

In the two structures shown, I represents the backbone of a vinyl polymer like polyethylene or PVC, whilst II represents a polyether like polyethylene oxide. Those bonds which are in the chain (as opposed to pendant to the chain) are called network bonds, and these connect the network atoms, in this case C and C or C and O. The network atoms in these instances are 2-connected, whilst a network atom which represents a cross-linking point may be 3- or 4-connected. Another important variable in polymer science is the way in which a network is arranged or packed in space. In principle, there is a continuous spectrum of structures in covalent polymers from the very open, where the chains may be relatively far apart, like polyethylene (density 0·9–1·0) to the very tight, like diamond (density 3·5), or analogously from the siloxanes to silica. The two parameters, connectivity of network atoms (CN) and the relative number of covalent network bonds per unit volume (N_{cr}) have been discussed by Holliday and Holmes-Walker.[3]

In contrast to the foregoing type of structure, there exist ionic solids, made up of ionic, *i.e.* strong *non-directional* bonds, between atoms with a valency of one or higher. Since the ionic bond is non-directional, there is no network in the polymer sense, and the ions tend to cluster or pack as closely as possible. Because of this, there is no corresponding degree of freedom of atomic arrangement in a purely ionic material like sodium chloride or sodium formate, as there is in covalently bonded polymers. In fact compared with covalent polymers, ionic solids are always tightly packed aggregates of matter. Of course, ionic solids may contain covalent bonds within the cation or anion, as in sodium benzoate or a quaternary ammonium salt, but the same comments about close packing of cations and anions apply as with simpler materials like sodium chloride. This unhampered close packing of simple ionic solids partly explains the ease with which they crystallise compared with many polymers. This distinction between directional and non-directional bonds is crucial to any discussion of ionic polymers.

The foregoing assumes that there is a sharp distinction between the *directional* covalent bond and the *non-directional* ionic bond, and deliberately ignores the fact that most bonds are hybrid in character. However, the fact that the Si—O— bond in the siloxanes and inorganic glasses (to take only one example) has some ionic character does not prevent it from behaving as though it has a directional quality, and therefore quite distinct from the ionic, non-directional silicate bond Si—O^-M^+ for the purposes of classification or discussion of ionic polymers. In the majority of instances, the distinction is much sharper than in this particular example. Thus the C—C chain bond in a carbon polymer is quite distinct from the ionic metal carboxylate bond (for example in its behaviour in water) which may exist in the same polymer.

To recapitulate, ionic polymers contain both covalent bonds and ionic bonds, which together form a chain or network structure. In the vast majority of cases, the covalent bonds themselves already exist in the chain, sheet or network form which is characteristic of polymers. The ionic bonds are additional to this structure, and help to modify the properties. In certain cases, as yet unimportant but which may become of greater significance, the covalently bonded moiety is too small to be considered as a polymer (*e.g.* sebacic acid) but it acquires polymer-like properties in the salt form, *i.e.* when ionic bonds are added to the structure. In this instance, the chain or network is formed as the result of the co-operation of covalent and ionic bonds.

This distinction is important since it indicates what will happen if the ionic bond ceases to function, *e.g.* due to hydrolysis or thermal decomposition. If a covalent chain or network exists in its own right, the product will continue to show polymer-like properties even if the ionic bond dissociates.

1.2 CLASSIFICATION OF IONIC POLYMERS

At the moment, there is no accepted definition of, or method of classifying, ionic polymers. The particular field of cationic quaternary polyelectrolytes has been well reviewed by Hoover.[4] He classifies quaternary ammonium, sulphonium and phosphonium polymers in structural categories according to the position of the N, S or P heteroatom within the polymer. In a less detailed but broader review, Otocka[5] classifies ionic polymers into the two groups polyelectrolytes (subgroups homopolymers and copolyelectrolytes) and polysalts. A homopolymer polyelectrolyte would be a metal polyacrylate or polyphosphate (ionic bonds close together). A copolyelectrolyte has isolated salt groups as in a carboxylated rubber. A polysalt is a complex formed between oppositely charged polyelectrolytes. Holliday[6] draws a distinction between those materials with a continuous covalent backbone or network (divided into polyanions, polycations and polyelectrolyte complexes) and those with a discontinuous backbone, as the metal salts of dicarboxylic acids and invert glasses.†

Following on the foregoing schemes, the classification shown in Table 1 is proposed, based partly on the arrangement of the covalencies, and partly on the type of ionic bond. The references given in the table are not intended to be exhaustive, but merely representative of recent work. The diagrammatic representations which are shown, are given as a visual aid to distinguishing the various types of ionic polymer, and carry no particular implications as to the spatial arrangement of the ionic bonds. It will be seen later that the arrangement in space of the ionic groups is an important feature of these materials.

As a primary classification, ionic polymers are divided into three groups in Table 1. In the first, the covalent *network* is continuous in three (or two) dimensions; in the second, continuous covalent *chains* exist (*i.e.* they are continuous in one dimension); whilst in the third group the covalent *segments* are relatively short. As already indicated, the reason for this division is that it reflects the structure of the material when the ionic bonds

---

† It should be mentioned at this point that Eisenberg[25] uses the term counterion copolymers for ionic polymers with mixed counterions.

TABLE 1
TYPES OF IONIC POLYMERS

Chain or network type	Diagrammatic representation[a] (full lines are covalent bonds)	Ion type	Examples	Refs.
I Covalent network + ionic bonds	$M^{++}\;\|R^-$ $\qquad M^{++}\;\|R^-_{R^-}$	Polyanion	(a) Glass (b) Sheet silicates (c) Cement (d) Clay products (e) Cation exchange resins	
	$R^-\;\|M^+\;R^-$ $R^-\;\|M^+\;R^-$ $R^-\;\|M^+\;R^-$	Polycation	(a) Anion exchange resins (b) Rubbers with quaternary N cross-links	7
	$M^+\;\|R^-$ $R^-\;\|M^-$	Mixed	(a) Amphoteric ion exchangers	

CLASSIFICATION AND GENERAL PROPERTIES OF IONIC POLYMERS

				Diagram
II Long covalent chains + ionic bonds	Polyanion	(a) Linear silicates	8	
		(b) Polyphosphates	9	
		(c) Ionomers		$\overset{R^-}{\underset{M^+}{\vert}}\quad \overset{R^-}{\underset{M^+}{\vert}}\quad \overset{R^-}{\underset{M^+}{\vert}}$
		(d) Metal salts of polyacrylic acid	11	
		(e) Salts of carboxylated rubber	12	
		(f) Anionic polyelectrolytes		
		(g) Polysaccharide gels		
	Polycation	(a) Cationic polyelectrolytes	4	$\overset{M^+}{\underset{R^-}{\vert}}\quad \overset{M^+}{\underset{R^-}{\vert}}\quad \overset{M^+}{\underset{R^-}{\vert}}$
		(b) Ionenes (N^+ in main chain)	13	
	Mixed	(a) Polysalts	14	$\overset{R^-}{\underset{M_2^+}{\vert}}\ \overset{R^-}{\vert}\ \overset{R^-}{\underset{M_2^+}{\vert}}\ \overset{M_1^+}{\underset{R_2^-}{\vert}}\ \overset{R^-}{\vert}\ \overset{M_2^+}{\underset{R_2^-}{\vert}}$
		(b) Charge mosaic complexes	15	
		(c) Snake cage polyelectrolytes		
		(d) Amphoteric ion exchangers	16	
III Short covalent chains + ionic bonds	Dianion	(a) Metal salts of dicarboxylic acids		$R^-\text{———}R^-M^{++}R^-\text{———}R^-M^{++}R^-\text{———}R^-M^{++}$
		(b) Invert glasses	17	
		(c) Polysulphides	18	
	Dication	—		$M^+\text{———}M^+R^{--}M^+\text{———}M^+R^{--}M^+$

[a] These diagrams are not intended to be realistic, nor to convey any spatial aspects of the structures.

are dissociated by heat or water or other mechanism, and suggests the sort of properties the material would have under these circumstances. For example an ionic polymer with many covalent cross-links will behave as a cross-linked material even if the ionic bonds cease to be effective.

Within these groups, there is a further important distinction between organic and inorganic ionic polymers.

Organic ionic polymers

The covalent skeletal structure (whether network or chain) can be derived from virtually any normal polymer. The chain atoms are carbon, but may include other atoms such as oxygen or nitrogen (as in polyesters or polyamides). In addition the covalent skeletal structure may be made up of relatively short segments, as in the metal salts of α,ω-dicarboxylic acids. However, although there is virtually no limit to the types of structures which can be envisaged, in practice relatively few types of ionic polymer have been synthesised to date, based mostly on a few vinyl monomers or diolefins. Many interesting opportunities lie ahead. Similarly, although many alternative anions can be attached to the covalent skeleton, most work to date has involved the carboxyl group or to a lesser extent the sulphonic group. Much work therefore remains to be done with other anion forming groups such as phosphinic, acid sulphate, etc.

Turning to the polycations, by far the greatest number of examples are based on quaternary ammonium groups. These may be pendant to the chain (as in poly-N-alkyl vinylpyridinium halides) or in the chain (as in the ionenes). In the latter case, the quaternary nitrogen cation is a chain atom. Other cationic groups are sulphonium and phosphonium.

Inorganic ionic polymers

With inorganic ionic polymers, the commonest chain atoms are silicon or phosphorus, arranged alternately with oxygen thus:

$$-Si-O-Si-O-$$
$$-P-O-P-O-$$

These form the backbones of the mineral silicates, inorganic glass and the polyphosphates, to quote the most important examples. In addition to Si, P and O, many other atoms may act as chain atoms in the field of inorganic ionic polymers, such as B, Al, Zr, etc.

However, whilst there are a great variety of chain atoms, the number of *anions* in inorganic polymers is strictly limited, with silicate and phosphate being by far the most important.

Hybrid organic–inorganic ionic polymers

Just as hybrid organic–inorganic polymers of the conventional type are known, as in the siloxane copolymers, it should be possible to produce hybrid organic–inorganic ionic polymers. However, little work has been done on these materials, probably because the subject of ionic polymers is so new.

Polyelectrolytes

Polyelectrolytes are an important class of ionic polymers which deserve special mention, although they are not listed separately in Table 1. These water-soluble polymers are of great industrial importance, particularly in four main marketing areas—water treatment, paper, textiles and oil recovery. In view of this importance, they are classified separately in Table 2 (*see* Ref. 4). Non-ionic polyelectrolytes, which are not ionic polymers, are included for completeness.

TABLE 2
POLYELECTROLYTES

Non-ionic	Anionic	Cationic
Polyols	Carboxylic	Ammonium
Polyethers	Sulphonic	(1) Protonated I, II, III amines
Polyamides	Phosphonic	(2) Quaternary
Poly(N-vinyl heterocyclics)		Sulphonium
		Phosphonium

Although polyelectrolytes are often discussed separately, they differ in degree and not in kind from other ionic polymers. This difference centres around their hydrophilic nature, which depends upon the existence of a large proportion of ionic and/or hydrophilic groups in the chain.

1.3 SOME TYPICAL EXAMPLES OF IONIC POLYMERS

1.3.1 Covalent networks and sheets containing ionic bonds

Polyanions

(*i*) *Derived from silica.* This class of ionic polymer is extremely important because it contains by far the greater proportion of synthetic and natural products based on silica. This amounts to:

(a) inorganic glass (apart from invert glasses where the network is totally disrupted),

(b) sheet or network silicates,
(c) cement,
(d) clay products.

All these products can be considered to be derived from silica. As the result of the reaction of silica with metal oxides, either under anhydrous conditions (high temperature processes) or in the presence of water, the three dimensional silica lattice is disrupted and modified by the insertion of cations, by the introduction of other network forming atoms such as aluminium or boron, and in many cases by the inclusion of water. As Ray points out (*vide infra*) these cations may be regarded as chain terminators. The materials formed in this way may be crystalline or amorphous, but in general they share certain properties; they are rigid, strong, high melting and they tend to be brittle. Because they are cheap, they are of immense importance in industry, particularly in building and as ceramics.

The parent compound silica occurs in a number of polymeric forms. As pure fused silica, it consists of a random arrangement of silica tetrahedra, and is thus a glass. In the crystalline form, three basic structures quartz, tridymite and cristobalite each exist in two or three modifications depending on the temperature. Its relationship with other silicate products is seen clearly in terms of the O:Si ratio which varies from 2 in quartz and its polymorphs to 4 in the orthosilicates. Figure 1, taken from Kingery[19] illustrates this point.

The negatively charged oxygen anions are shown in the diagram—they are the oxygen atoms attached to only one silicon atom. The other oxygen atoms form bridges between the (SiO_4) tetrahedra. Thus we can consider vitreous silica, or quartz and its polymorphs to be the limiting network in which (SiO_4) tetrahedra are joined together by oxygen atoms or bridges at the four corners of each tetrahedron. As already mentioned, silica readily forms an ionic polymer by a reaction with a metal oxide, the properties of the product depending on the ratio of cations to the network atoms.

Thus

$$-O-\underset{\underset{|}{O}}{\overset{\overset{|}{O}}{Si}}-O-\underset{\underset{|}{O}}{\overset{\overset{|}{O}}{Si}}-O- + CaO \longrightarrow -O-\underset{\underset{|}{O}}{\overset{\overset{|}{O}}{Si}}-O^-Ca^{++}\ {}^-O-\underset{\underset{|}{O}}{\overset{\overset{|}{O}}{Si}}-O-$$

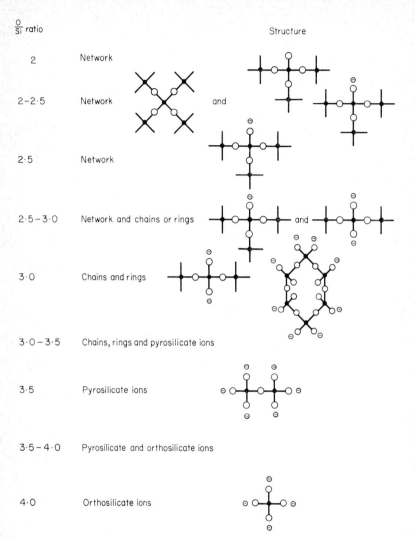

Fig. 1 *Effect of oxygen–silicon ratio on silicate network structure.* (Reprinted with permission from '*Introduction to Ceramics*' by W. D. Kingery, published by John Wiley and Sons Inc. 1960.)

To use the terminology of glass technology in this simple case, silica is called the network former and lime the network modifier. Following Stevels[20], we can describe silicate products by a useful structure parameter which can be used to correlate properties. This is the number of bridging oxygen atoms per tetrahedron.

Thus if

$X =$ average number of oxygen anions per tetrahedron;
$Y =$ average number of bridging oxygen atoms per tetrahedron;
$Z =$ average total number of oxygen atoms per tetrahedron ($= 4$ for silicates) and
$R =$ ratio of total number of oxygen atoms to total number of silicon atoms,

Then

$$X + Y = Z$$
$$X + \tfrac{1}{2}Y = R$$

Given $Z = 4$ and $R = Z$ for fused silica, it follows that $X = 0$ and $Y = 4$, that is, all the oxygen atoms are bridging oxygen atoms. For $MgOSiO_2$ $R = 3$, $Z = 4$ and therefore $X = 2$ and $Y = 2$, i.e. each (SiO_4) tetrahedron has two oxygen ions and two bridging oxygen atoms.

$$\begin{array}{c} O^- \quad\ O^- \quad\ O^- \quad\ O^- \\ |\qquad |\qquad |\qquad | \\ -O-Si-O-Si-O-Si-O-Si- \\ |\qquad |\qquad |\qquad | \\ O^- \quad\ O^- \quad\ O^- \quad\ O^- \end{array}$$

Inspection of this structure of metasilicate shows that there is one singly charged ionic bond (between a cation and an oxygen anion) for every one covalent silicon–oxygen network bond. Thus, the ratio of singly charged anions:network bonds $= X/Y$. In any discussion of ionic polymers, this ratio of bound ions to network bonds is a useful number.

Reference to Fig. 1 shows the immense variety of structures which can exist, based on the —Si—O— network bond and the —Si—O$^-$ anion. It also shows that there is no abrupt transition between network, sheet, chain and ring structures, since one class shades into another depending on the O:Si ratio.

(ii) *Cation exchange resins*. This type of ionic polymer is of considerable importance industrially. Ion-exchange resins consist of three-dimensional covalent networks based on the carbon–carbon bond, to which are attached bound ions. The covalent cross-links preserve the structural integrity of the materials in service, whilst the bound ions provide the ion-exchange sites. A cation exchanger is thus an anionic polyelectrolyte, consisting of immobilised anions fixed to the structure and mobile cations as counterions.

There is a story that the first discovery of a synthetic organic cation exchanger dates back to the chance observation of Adams and Holmes in 1935 that crushed gramophone records had ion exchange properties, stemming from the weakly acidic phenol groups in the material, which was a phenol–formaldehyde resin. Unfortunately the truth is more prosaic, since their original materials were the formaldehyde condensation products of polyhydric phenols or aromatic diamines prepared in the Chemical Research Laboratory at Teddington. It is the descendants of these materials which are now used extensively for the purification and demineralisation of water, the first commercial plant having been installed in 1937 (Fig. 2).

A typical cation exchange reaction can be represented as follows:

$$\underline{2NaX} + CaCl_2(aq) \rightleftharpoons \underline{CaX_2} + 2NaCl(aq)$$

The underlined component represents the solid phase with the —X radical standing for the bound anion. The framework of a cation exchanger (the same applies, *mutatis mutandis*, to an anion exchanger) can be regarded as a macromolecular polyanion. This can be seen as having a sponge-like structure, with counterions floating in the pores.

The structure of a typical strong acid resin is represented by the following example:

$$\begin{array}{c} -CH-CH_2--CH-CH_2- \\ | | \\ \underset{SO_3^- H^+}{\bigcirc} \underset{SO_3^- H^+}{\bigcirc} \\ -CH-CH_2- \end{array}$$

In this case styrene is copolymerised with divinyl benzene, and later sulphonated. Usually a bead polymerisation technique is employed, giving beads of 0·1–0·5 mm in diameter. For general purposes, about 8–12% divinyl benzene is used, the degree of swelling depending upon the extent of cross-linking. To produce a weak acid resin, containing carboxyl groups, one method relies upon copolymerising methyl methacrylate with divinyl benzene, and subsequently hydrolysing the ester groups after polymerisation.

Fig. 2 First commercial demineralisation plant based on ion exchange. (Built by the Permutit Co. 1937).

Anionic polyelectrolytes have also been prepared from block copolymers.[27] Sulphonic groups were introduced into the centre block of S—I—S and S—EP—S copolymers, to form materials from which tough, coherent films can be made. These swell to a considerable extent in water, in this case the block structure leading to behaviour typical of light cross-linking.

Polycations

(i) *Anion exchange resins.* These three-dimensional covalent networks based on the carbon–carbon bond, and containing bound cations, are exactly similar to the cation exchange resins discussed above. In both cases, their use as ion-exchange resins depends upon their being in the insoluble form, which requires that they should be cross-linked structures.

The starting point for a typical anion exchange resin is a cross-linked polystyrene as with the strong cation exchange resins. In this case the product is chloromethylated, and the chloromethyl group then quaternised.

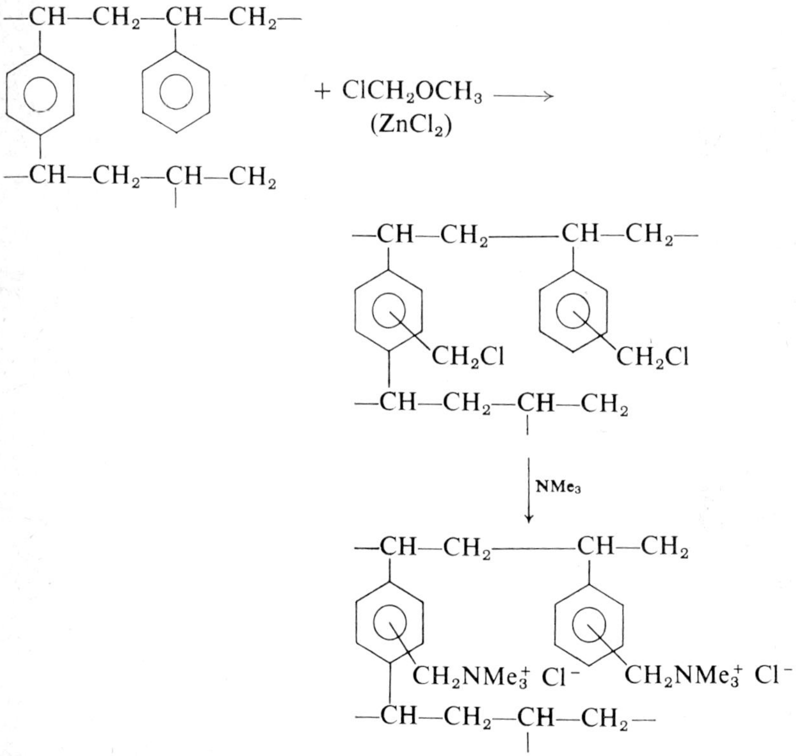

—CH—CH$_2$—CH—CH$_2$—

⬡ ⬡ + ClCH$_2$OCH$_3$ ⟶
(ZnCl$_2$)

—CH—CH$_2$—CH—CH$_2$

—CH—CH$_2$————CH—CH$_2$—

⬡—CH$_2$Cl ⬡—CH$_2$Cl

—CH—CH$_2$—CH—CH$_2$

↓ NMe$_3$

—CH—CH$_2$————CH—CH$_2$

⬡—CH$_2$NMe$_3^+$ Cl$^-$ ⬡—CH$_2$NMe$_3^+$ Cl$^-$

—CH—CH$_2$—CH—CH$_2$—

They have also been produced in block form.[27] The starting point was SBS, and the quaternary nitrogen was introduced into the butadiene centre block.

(ii) *Amphoteric ion exchangers.* Being amphoteric, these materials are difficult to classify, but it is convenient to mention them here. It is obviously possible to put bound anion and cation sites in the same three-dimensional network to produce an amphoteric resin.

$$\underset{\text{CH=CH}_2}{\bigcirc} \quad \underset{\overset{|}{\text{Cl}}}{\text{CH=CH}_2} \quad + \text{ cross-link} \longrightarrow \underset{\bigcirc}{-\text{CH}-\text{CH}_2-\text{CH}_2-\text{CH}_2-} \underset{\text{Cl}}{|}$$

$$\downarrow \text{NMe}_3$$

$$\underset{\underset{\text{SO}_3^- \text{M}^+}{\bigcirc}}{-\text{CH}-\text{CH}_2-\text{CH}-\text{CH}_2-} \underset{\text{NMe}_3^+ \text{Cl}^-}{|} \xleftarrow{\text{H}_2\text{SO}_4} \underset{\bigcirc}{-\text{CH}-\text{CH}_2-\text{CH}-\text{CH}_2-} \underset{\text{NMe}_3^+ \text{Cl}^-}{|}$$

(*iii*) *Rubbers with quaternary nitrogen cross-link points*. Compared with polyanions, there are few network or sheet structures known which contain cations bound to the polymer structure. Anion exchange resins are the most important example. Other and rather specialised examples are the rubbers which have been reported containing quaternary nitrogen cross-linking sites.[7] These have been developed using the Menschutkin reaction for polymer synthesis, of which the following is a simple example:

$$R_2N(CH_2)_n NR_2 + X(CH_2)_m X \longrightarrow \left[\underset{R}{\overset{R}{\underset{|}{\overset{|}{N}}}} \overset{X^-}{\underset{}{}} (CH_2)_n - \underset{R}{\overset{R}{\underset{|}{\overset{|}{N}}}} \overset{X^-}{\underset{}{}} (CH_2)_m \right]$$

This example shows the chain form of structure originally synthesised by Marvel.[21] The full potentionalities of this method have since been shown in the work of Rembaum on ionenes, and in the work of Dolezal et al.[7] under discussion, on liquid rubbers.

Recent work on liquid rubbers is aimed at producing pourable polymers which have molecular weights of the order of 10 000 or less, which can by a simple chemical process be chain-extended and cross-linked. If the starting polymer has active sites at the ends of the chain (in this case bromine atoms) it is possible in principle to obtain end-free networks in this way, of superior physical and mechanical properties. Chain-extension takes place by the use of bi-functional amines, as shown in the above example. Cross-linking requires the use of amines of higher functionality, such as triethylene tetramine, tris-dimethylamino phenol or hexamethylene tetramine.

The finished network, with its quaternary nitrogen cations (in the chain) and bromine anions as counterions, can be diagrammatically represented as follows:

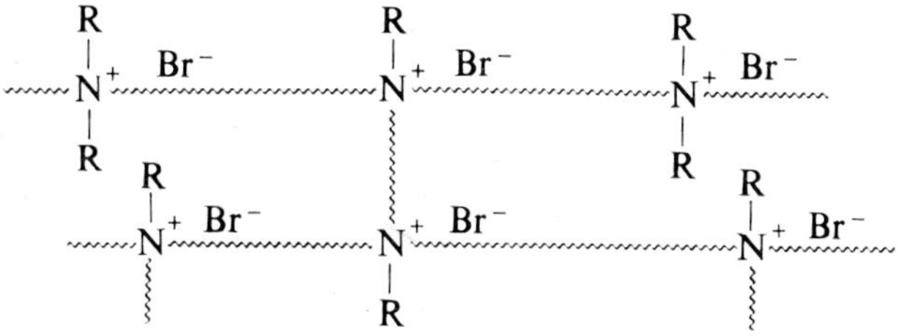

```
        R                  R                  R
        |   +   -          |   +   -          |   +   -
~~~~~~N       Br         ~N       Br         ~N       Br
        |                  |                  |
        R   R              R   R              R   R
            |   +   -          |   +   -          |   +   -
          ~~N       Br       ~~N       Br       ~~N       Br
            |                  |                  |
            R                  R                  R
```

The quaternary nitrogen cross-linking sites are mechanically and hydrolytically stable. However, these particular materials cannot be regarded as typical ionic polymers, since the bound ion sites are merely incidental to the process of chain-extension and cross-linking. They are a means, as it were, rather than an end in themselves. Nevertheless it is possible that similar structures, with a higher concentration of bound cations, might have interesting applications in biological and other areas.

1.3.2 Long covalent chains containing ionic bonds

Polyanions
Most of the important members of this class are dealt with extensively in later chapters. It is therefore only necessary to deal briefly with them here, to show how they are related to other ionic polymers.

(*i*) *Linear silicates.* Linear silicates, based on single and double strands of the type O—Si—O—Si—, are of the most widespread natural occurrence.

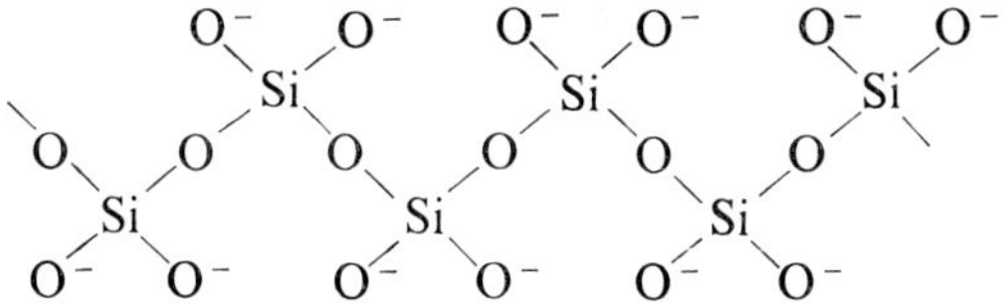

[Diagram of single and double strand linear silicates labeled "single" and "double"]

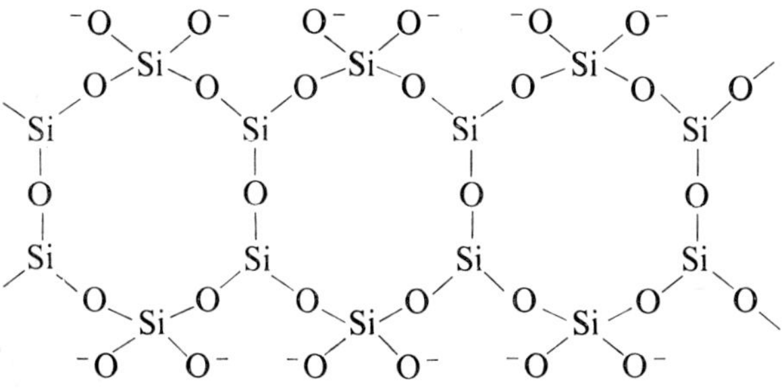

Examples of mineral silicates with single chain ions are the pyroxenes, such as enstatite ($MgSiO_3$), diopside ($CaMg(SiO_3)_2$), β-Wollastonite ($CaSiO_3$), augite, jadeite and spodumene.

Examples of silicates with double stranded chain ions are the amphiboles, which include the majority of asbestos minerals such as tremolite, actinolite, crocidolite, anthophyllite and amosite.

Many silicates and glasses contain more than one type of cation. Thus diopside contains both Ca^{++} and Mg^{++}. Where there are mixed counterions as in these examples, they can be called counterion copolymers by analogy with copolymers in carbon polymer chemistry.

The linear silicate polymers are mostly insoluble in water and are generally rigid and brittle. Crystalline sodium metasilicate (Na_2SiO_3) has a linear structure like the pyroxenes,[22] but the structure of sodium silicates in solution is more complicated, and depends on the $SiO_2:Na_2O$ ratio. The sodium silicate solutions of commerce have silica to alkali ratios varying from 1·6:1 to 3·85:1, i.e. they are rich in silica compared with the 1:1 stoichiometric sodium metasilicate. These solutions are complex mixtures of monomeric silicate ions, disilicate ions and colloidal silica micelles.

Eisenberg[23-25] has examined the viscoelastic properties of solid silicates in the composition region $M_2O:SiO_2 \approx 1$, and has compared them with the polyphosphates and other ionic polymers. His results will be discussed later.

(ii) *Polyphosphates*. Condensed phosphates are made by the dehydration of orthophosphates at elevated temperatures or by the hydration of phosphorus pentoxide. Thus

$$nNaH_2PO_4 \longrightarrow (NaPO_3)_n + nH_2O$$

They can be regarded as made up of three building blocks:

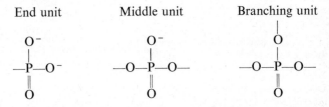

On this basis, since the branching unit is trifunctional it might be expected that cross-linked polyphosphates would occur. In fact, such branching points are readily attacked by water† and are unstable. Condensed phosphates are therefore normally encountered as rings and chains rather than sheets or networks. Condensed phosphates form soluble complexes with many metals. They are used industrially for this purpose, for example in water softening.

(*iii*) *Ionomers.* Probably the most quoted examples of ionic polymers are the ionomers of du Pont, sold under the trade name Surlyn. These are copolymers of ethylene and methacrylic acid, in which a salt or ionic linkage is formed with either a univalent or divalent cation. They can be regarded as modified low density polyethylene. Since 1965, workers in du Pont and elsewhere have published a number of interesting papers on the structure and properties of ionomers, which will be discussed in detail in Chapter 2.

Compared with normal low density polyethylene, the ionomers are characterised by the following features:

(a) High strength.
(b) Good clarity.
(c) Improved oil resistance.

(*iv*) *Metal salts of polyacrylic acid.* Amongst the earliest work in this field may be mentioned Hagedorn's,[9a] who studied the production of films, fibres and formed articles from the metal salts of polymeric carboxylic acids. More recently there has been a revival of interest in this type of material, which now forms the basis of two successful dental filling compositions as discussed later.

Much of the early work of Nielsen[10] who has worked extensively in this field dealt with a copolymer of composition 94% wt acrylic acid and 6% 2-ethylhexylacrylate. This itself is so extensively hydrogen bonded as to be infusible and therefore in order to make fully neutralised ionic polymers from this material, Nielsen adopted the techniques of powder metallurgy. The polymer and metal oxide were thoroughly mixed, and the powder moulded at 300°C under a high pressure. A large number of composites were made with these materials, some with interesting mechanical properties.

Products of this type are rich in carboxyl groups, and the following is a diagrammatic representation of the structure of the salts of polyacrylic acid formed with a bivalent cation.

† This is not true of certain phosphate glasses containing a proportion of oxide co-monomer (see *Glass Tech.*, **14**, 50, 1973).

$$
\begin{array}{cccccc}
-\mathrm{CH_2}-\mathrm{CH}-\mathrm{CH_2}-\mathrm{CH}-\mathrm{CH_2}-\mathrm{CH}- \\
| & | & | \\
\mathrm{COO^-} & \mathrm{COO^-} & \mathrm{COO^-} \\
& \mathrm{M^{++}} & \mathrm{M^{++}} & & \mathrm{M^{++}} \\
\mathrm{COO^-} & \mathrm{COO^-} & \mathrm{COO^-} \\
| & | & | \\
-\mathrm{CH_2}-\mathrm{CH}-\mathrm{CH_2}-\mathrm{CH}-\mathrm{CH_2}-\mathrm{CH}-
\end{array}
$$

Although most of this work has been based on homopolymers and copolymers of acrylic and methacrylic acids, there are many other monomers such as maleic and fumaric acids which can lead to similar interesting products, with a slightly different disposition of carboxyl groups.

(v) *Salts of carboxylated rubber.* These products, which are discussed at length later, have their origins in the work of Brown[11] to whom a patent was assigned in 1953. The first preparation of a carboxylic elastomer is recorded in a French patent of I.G. Farbenindustrie of 1933. In 1952 Miller and Reid[26] proposed the treatment of films of copolymers of butadiene and methacrylic acid with salts of polyvalent metals as a technique of hardening the film. Brown's invention specifically concerns crosslinking rubbers using ionic bonds. A wide variety of rubbers containing carboxyl groups have now been studied, including copolymers of butadiene and acrylic or methacrylic acid; butadiene, acrylonitrile and methacrylic acid; ethyl acrylate and methacrylic acid, etc., and a number are now in commercial production.

(vi) *Anionic polyelectrolytes.* The most important anionic polyelectrolytes depend for their activity either on carboxyl or sulphonic groups. Their use as ion-exchange materials has already been discussed for which purpose they are required in the solid form, mainly as beads, and whilst it is acceptable that they should swell in water it is important that they are not soluble. Their mechanical properties depend upon the degree of crosslinking and the proportion of bound anions.

Non-cross-linked anionic polyelectrolytes are widely used industrially as solutions or gels. In their anhydrous state they tend to be brittle solids, and the question of whether they will dissolve or form a gel in water is largely controlled by the proportion of bound anions in the chain. This point can be illustrated by considering the hydration behaviour of a series of copolymers containing acrylic acid with a monomer like styrene. Polystyrene is unaffected by water and polyacrylic acid is water soluble. As the proportion of acrylic acid in the copolymer increases, the behaviour of the copolymer towards an aqueous solution of sodium hydroxide changes. Copolymers containing small proportions of acrylic acid swell slightly, with larger proportions they swell extensively, and finally the point is reached where they become water soluble.

Because they are so readily available, acrylic acid and methacrylic acid are extensively used as monomers in these systems, and homopolymers and copolymers of these materials are widely used industrially. Solutions of the homopolymers are used in the acid form or as the sodium or ammonium salt in applications such as dispersing, thickening or suspending aqueous systems. Two examples will be quoted.

In the manufacture of carbon fibres, it is essential to orientate all the fibres in the same direction for maximum stiffness and strength properties when used in composites. This aligning process is normally carried out in a viscous medium, and a solution of ammonium polyacrylate is suitable for the purpose. In another field, sodium and ammonium polyacrylates are used as pigment dispersing and suspending agents in the manufacture of emulsion paints.

Related water-soluble copolymer resins are extensively used in the surface coating and other industries, such as adhesive, paper coating, etc. A typical example is a water soluble copolymer of methyl acrylate–methyl methacrylate and acrylic acid in the form of the sodium or ammonium salt.

As an example of a copolymer of lower acrylic acid concentration, the hydrogel of the sodium salt of a copolymer of methyl methacrylate and acrylic acid can be quoted. If the concentration of acrylic acid is in the region of 15% wt in the copolymer, the equilibrium water content is of the order of 60% wt in the gel, and a system of this sort has been patented for a contact lens application.

(*vii*) *Polysaccharide gels.* Many polysaccharide gels can be regarded as ionic polymers, since they contain carboxyl or sulphate groups which are capable of salt formation. Examples of such materials are the alginate gels, much used in dentistry. In the form of sodium, potassium or ammonium alginate, they form a sol of fairly thick consistency, but this can be converted into an insoluble gel form by converting it to the calcium salt (for example, by double decomposition with calcium sulphate). It is this reaction which forms the basis for their use as impression materials.

The subject of polysaccharide gels has been well reviewed recently by Rees.[12] The alginates, for example, contain carboxyl groups, attached to two types of sugar residue which are sterically closely related—β-D-mannopyranosiduronate (M) and α-L-gulopyranosiduronate (G)

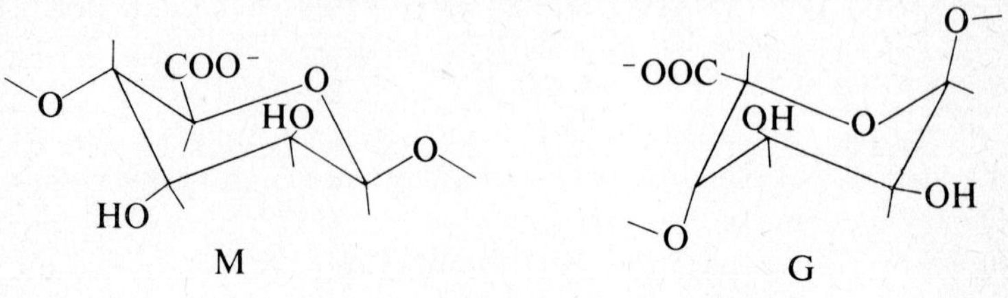

These residues are arranged in three types of sequence in the alginate molecule

$$\begin{array}{c}\text{—M—M—M—M—}\\ \text{—G—G—G—G—}\\ \text{—M—G—M—G—}\end{array}$$

In carrageenan, the sugar residues which are linked to form the polymer include galactose 2,6-disulphate (L) and galactose 6-sulphate (H):

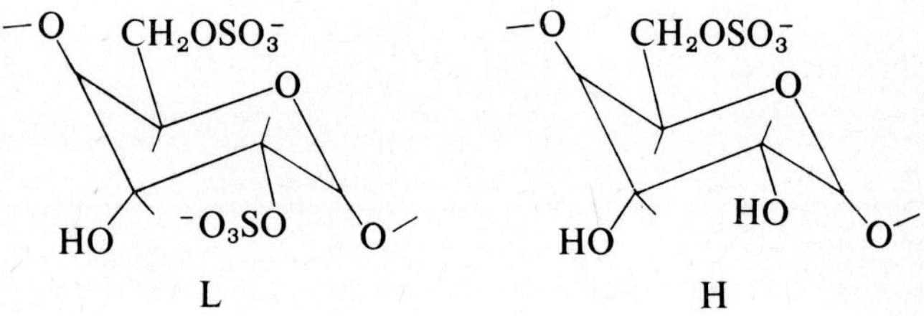

Thus in the polysaccharide gels are found six-membered carbohydrate rings containing one or two acid groups, linked by ether linkages. In certain cases they are copolymerised with rings containing no acid groups.

Polycations

(*i*) *Cationic polyelectrolytes.* The extensive literature on cationic polyelectrolytes has been reviewed by Hoover.[4] He has divided the quaternary ammonium polymers, which are the most important, into the following classes:

(a) Polymethacryloxyalkyl.
(b) Polymethacrylamidoalkyl.
(c) Polyalkenyl.
(d) Polyvinyloxy.

(e) Polyvinylbenzyl.
(f) Polydiallyl.
(g) Polyvinylpyridinium.
(h) Polyvinylimidazolinium.
(i) Polyalkylation quaternaries ('ionenes'—N^+ in polymer chain).
(j) Polycondensation quaternaries (N^+ not in polymer chain).

Of these, the ionenes will be discussed further below, as a separate group.

Cationic polyelectrolytes are used extensively throughout industry. Their use in ion-exchange resins depends upon their being available in solid form and has been discussed above. Table 3 indicates the widespread uses of cationic quaternary polyelectrolytes (Hoover[4]). It is not intended to be exhaustive.

TABLE 3

USES OF CATIONIC QUATERNARY POLYELECTROLYTES

1. Primary coagulants.
2. Flocculants.
3. Antistatic agents.
4. Soil conditioning.
5. Flame retardants.
6. Hair sprays, shampoo additives, etc.
7. Sequestering agents.
8. Grease thickening.
9. Electroconductive coating.
10. Anion-exchange resins.
11. Biocides.
12. Dye mordants.
13. Pigment retention aids.
14. Wet and dry strength additives in paper making.
15. Emulsifiers and de-emulsifiers.
16. Corrosion inhibitors.
17. Stiffening agents for fabrics and paper.
18. Sensitisers for photographic film.
19. Polyelectrolyte complexes.
20. Lube-oil additives.
21. Printing inks.
22. Adhesives.

(*ii*) *Ionenes.* These are a particular kind of polycation where the bound cations are in the main chain, as opposed to those materials where the bound positive sites are pendant to the chain. They have been extensively studied recently by Rembaum *et al.*[13,45,46] following earlier work by

Marvel, Kern and others[21,48,50] based on the Menschutkin reaction. As already mentioned, the most convenient form of the reaction involves an α,ω tertiary diamine and an α,ω dihalide—usually the bromide.

$$R_2N(CH_2)_nNR_2 + X(CH_2)_mX \longrightarrow \left[\begin{array}{cc} R & R \\ | \quad X^- & | \quad X^- \\ -N^+\!\!-\!(CH_2)_n\!-\!N^+\!\!-\!(CH_2)_m\!- \\ | & | \\ R & R \end{array} \right]$$

In this way the distance between the positive ions can be changed almost at will, and a very regular product is obtained. The result is that, generally speaking, these materials are crystalline, and up to $n = m = 10$ they are water-soluble polyelectrolytes. Following the systematic nomenclature of Rembaum, a 6-8 ionene, for example, has $n = 6$ and $m = 8$. The synthesis goes very easily, the reaction taking place at room temperature in DMF or similar solvent. Cross-linking can be achieved by using amines of higher functionality, as mentioned earlier.

Mixed ions

(*i*) *Polysalts (polyelectrolyte complexes)—cations and anions on different chains.* When two linear polyelectrolytes of different ionic charge are mixed in solution, ionically bonded network structures with interesting properties are formed. The earliest work along these lines dates back to the studies of de Jong at the University of Leiden in the 1930s, when the interaction of weakly charged natural hydrophilic polymers was investigated. In such a reaction, a separate phase is formed, illustrated by the interaction of gelatin (a polycation) and gum arabic (a polyanion). This particular example has been applied commercially in a variety of ingenious ways, following the work of B. K. Green at the National Cash Register Co., which led to the development of microencapsulation. In this process, the gel-like skin of the capsule consists of such a complex.

The interaction between weakly basic and acid functions leads to the formation of liquid or gel products called coacervates. Since the 1960s, the work of Michaels and his colleagues has led to the exploitation of the more strongly bonded precipitates which are formed when a strongly acidic polyanion reacts with a strongly basic polycation.[14] A typical example is the reaction between sodium polystyrene sulphonate and polyvinylbenzenetrimethyl ammonium chloride.

$$\underset{\substack{|\\ \text{SO}_3^- \text{ Na}^+}}{\overset{-\text{CH}-\text{CH}_2-}{\underset{|}{\bigcirc}}} \;+\; \underset{\substack{|\\ \text{CH}_2\\ |\\ \text{NMe}_3^+ \text{ Cl}^-}}{\overset{-\text{CH}-\text{CH}_2-}{\underset{|}{\bigcirc}}} \;\longrightarrow\; \begin{array}{c} -\text{CH}-\text{CH}_2-\\ |\\ \bigcirc\\ |\\ \text{SO}_3^- \text{ Na}^+\\ \text{NMe}_3^+ \text{ Cl}^-\\ |\\ \text{CH}_2\\ |\\ \bigcirc\\ |\\ -\text{CH}-\text{CH}_2- \end{array}$$

If these polyelectrolytes are mixed in dilute solution, a precipitate is formed which is almost free from the original dissolved counterions—in this case Na^+ and Cl^-. These are free to diffuse away from the polymer chains when this reaction takes place. Structures produced in this manner are hard, brittle resins when dry, and either leathery or rubber-like when wet. These materials are discussed in more detail in a later chapter, but it is relevant at this point to consider an important structural feature of these compounds, *i.e.* their stoichiometry.

A stoichiometric polyelectrolyte complex is one where the bound ions of opposite sign are equal in number, the bound ions being those which are attached by covalent bonds to the polymer chains. Thus, in the above example, a stoichiometric complex would be formed when the number of sulphonate groups in the sodium polystyrene sulphonate equals the number of quaternary nitrogen groups in the polyvinylbenzyltrimethyl ammonium chloride. If this were so, and if the groups can approach close enough, then stoichiometric poly–poly salt formation would take place, and the mobile counterions would diffuse away. Clearly, such an ideal situation is unlikely to be encountered in practice in view of the variation in molecular weight, and the packing problems involved; but it can be regarded as a limiting, if only theoretically obtainable, structure.

Non-stoichiometric complexes also exist, and their behaviour and properties are related to the extent of non-equivalence in bound anions and cations. In this case, there will be a controllable excess of bound anions or bound cations. This excess of bound ions will have the appropriate number of mobile counterions to maintain the balance of electric charges, and will therefore behave like conventional ion-exchange resins. The difference between stoichiometric and non-stoichiometric polyelectrolyte complexes shows up, for example, in their water absorption. The former may absorb 30% by weight when saturated, the latter up to ten times their dry weight. This illustrates the smaller number of ionic cross-links in the non-stoichiometric complexes. The distinction between the two types of polyion complex can be illustrated diagrammatically as follows:

Stoichiometric arrangement of bound ions

Non-stoichiometric arrangement

(*ii*) *Cations and anions on same chains.* Following the interest in charge mosaic membranes for water desalination by piezodialysis,[28] attempts have been made to produce analogous structures from block copolymers.[29,30] In one example styrene–vinyl pyridine block copolymers form the starting point. The vinyl pyridine block is quaternised and the styrene block sulphonated. In the other, a methacrylic ester is copolymerised with vinyl pyridine. By quaternising the product, followed by hydrolysis of the ester blocks, the desired structure is obtained. These two structures are shown below:

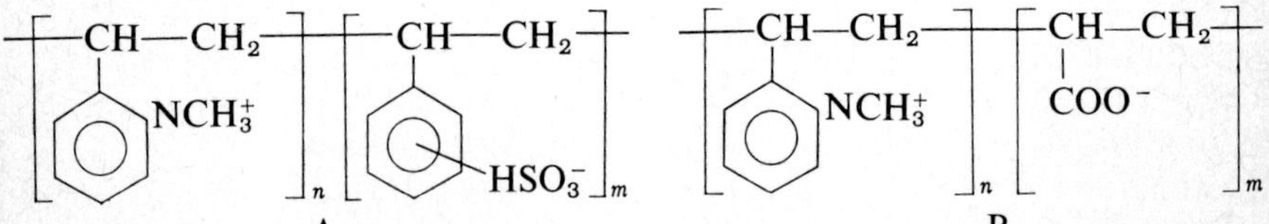

$$\left[\begin{array}{c}-CH-CH_2-\\ \vert \\ \bigcirc \\ NCH_3^+\end{array}\right]_n \left[\begin{array}{c}-CH-CH_2-\\ \vert \\ \bigcirc \\ HSO_3^-\end{array}\right]_m \qquad \left[\begin{array}{c}-CH-CH_2-\\ \vert \\ \bigcirc \\ NCH_3^+\end{array}\right]_n \left[\begin{array}{c}-CH-CH_2-\\ \vert \\ COO^-\end{array}\right]_m$$

A B

(*iii*) *Snake-cage polyelectrolytes.* These somewhat unusual structures consist of conventional cation or anion exchangers in which linear polycations or polyanions respectively have been formed by polymerisation.[31] As an example of this, a strong base ion-exchanger is converted to the acrylate form, and the acrylate anions are then polymerised within the resin. Structurally these materials are hybrids, since the original ion-exchange material is cross-linked (3-D) whilst the polymerisable counterion produces a chain (1-D) polymer.

1.3.3 Short covalent chains plus ionic bonds

Dianion type

(i) *Metal salts of dicarboxylic acids.* In 1944 Cowan and Teeter[32] published a paper which described the properties of some metal salts of dimerised fatty acids. They stated that these were a new class of resinous substances. These dicarboxylic acids were essentially dimers of linoleic acid with a molecular weight of 500–600 and a branched structure (linoleic acid, $CH_3(CH_2)_4CH\!=\!CHCH_2CH\!=\!CH(CH_2)_7COOH$ M.wt 272). The zinc salt melted at 130°C and was a transparent orange resin. The calcium and magnesium salts had melting points around 200°C. Fibres could easily be made from these materials, but were brittle. These dimers were unsaturated but no mention was made of whether any vinyl polymerisation took place.

The same class of materials has recently been investigated by Economy and his co-workers[16], and will be discussed in more detail in Chapter 5. Both aromatic and aliphatic α,ω-dicarboxylic acids were examined, and the divalent cations included Mg^{++}, Ca^{++}, Zn^{++}, Cd^{++}, Sn^{++}, Pb^{++}, and Mn^{++}. Melting points in the range 175–340°C were obtained, and some products were infusible. Thermal stabilities in air (5% wt loss at 6°C/min) ranged from 310° to 590°C. The materials are insoluble, and molecular weights were obtained from the melt viscosity.

(ii) *Invert glasses.* As the coherence of the glass network is reduced, by reducing the average number of bridging oxygen atoms per tetrahedron (*see above*), a stage is reached where discontinuous chains (or branched segments or rings) are formed (*see* Fig. 1). At this point, Y, the average number of bridging oxygens per tetrahedron is smaller than two, and there is less than 50% mole of SiO_2 in the composition. Assuming the structure to consist of chains, then the average chain length $\bar{n}$ (the average number of tetrahedra per chain) is

$$\bar{n} = 2/(2 - Y)$$

Normally glass formation ceases below this mole ratio, and crystalline products are obtained. This would happen if only one kind of cation were present. However, if two or more kinds of metal ion are used, glass formation can continue to values of $Y < 2$.[17,20] Such glasses are called invert glasses, because there is now a matrix of metal ions in which the chains are embedded (Fig. 3). Certain physical properties pass through an inversion point when $Y = 2$—for example, there is a minimum in the viscosity when plotted against Y.

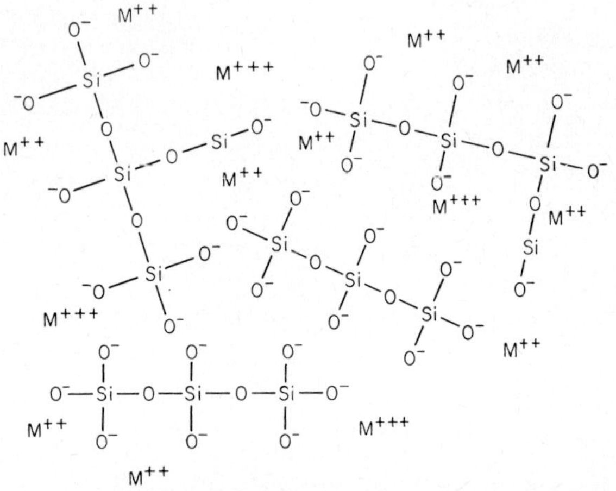

Fig. 3 *Invert glass—diagrammatic.*

(*iii*) *Polysulphides.* There appears to be a polymer-like arrangement of chains in certain polysulphides.[18] Polysulphides are formed when an alkali or alkaline earth carbonate is fused with sulphur, or when sulphur is digested with a solution of sulphide or hydrosulphide. In Cs_2S_6, for example, the ion has the form of an unbranched chain with relatively short distances between the ends of the chains. The bonds between successive S_6^{2-} chains link the ions into infinite helices similar to the helical chains in Se or Te.

1.4 SOME STRUCTURAL FEATURES OF IONIC POLYMERS

1.4.1 Concentration of ionic bonds

Ionic polymers contain bound ions, and free counterions. The bound ion is covalently bonded to the polymer network, as the carboxyl group in the polyacrylates or the quaternary nitrogen atom in the ionenes. In contrast to this, the counterion is free to move, but the actual mobility depends on the strength of the ionic bond, the temperature, the presence of liquids

CLASSIFICATION AND GENERAL PROPERTIES OF IONIC POLYMERS 29

which promote dissociation such as water, and similar factors. This distinction is easily seen by comparing ionomers with ion-exchange resins, which depend for their operation upon bound ions and mobile counterions.

Clearly, an important parameter in ionic polymers is the concentration of bound ions in the structure. In his study of the clustering of ions in organic polymers, Eisenberg[33] defines the concentration of ion pairs in terms of simple stoichiometry.

$$c = \rho N / M_i \tag{1}$$

where ρ is the density of polymer, N is the Avogadro number $= 6 \times 10^{23}$ and M_i is the average molecular weight of the chain between ion groups. By definition, this is the concentration of *bound ions* per cm³, and it is also one measure of the concentration of *ionic bonds*. In reality, the concentration of ionic bonds is more complicated than this, since the two factors of ion valency and ion coordination number must be taken into account.

In order to harmonise with previous discussions of ionic network bonds and covalent network bonds (Holliday[1,3]), one can use alternatively the concept of *relative* number of ionic and covalent network bonds per cm³, N_{ir} and N_{cr} where

$$N_{ir} = \rho / M_i \tag{2}$$

$$N_{cr} = \frac{\rho \chi (CN)}{2 M_{ru}} \tag{3}$$

where

$$CN = \frac{2a + 3b + 4c}{a + b + c} \tag{4}$$

$\chi =$ number of covalent network atoms in repeat unit, $CN =$ average connection number of network atoms and $M_{ru} =$ molecular weight of covalent repeat unit. $a =$ number of 2-connected, $b =$ number of 3-connected and $c =$ number of 4-connected atoms in the repeat unit. N_{cr} and N_{ir} are molar concentrations.

If a bound ion, of whatever charge or co-ordination number, is counted as generating a single ionic network bond, then the relative number of total network bonds is given by

$$N_r = N_{(c+i)r} = \rho (1/M_i + \chi(CN)/2M_{ru}) \tag{5}$$

Since most bound ions are univalent, this assumption is not too unreasonable.

Clearly, as already mentioned, N_{ir} is an unsophisticated measurement of the concentration of ionic bonds. For example, a high valency and coordination number may well give additional stiffness to the network, so that effectively $N_{ir} > \rho/M_i$.

This point will be discussed in more detail in the chapter on glass regarded as an ionic polymer, and will not be developed further here.

At this stage in the argument, N_{cr} and N_{ir} are being used solely as a guide to classification, for which purpose the ratio $N_{ir}/N_{(c+i)r}$ is adequate. Some approximate figures for typical ionic polymers are shown in Table 4.

TABLE 4

PROPORTION OF BOUND IONS IN IONIC POLYMERS

Polymer	$(N_{ir}/N_{cr} + N_{ir}) \times 100$
Carboxylated rubbers	0·5–2
Ionomers	2
Metal salts of dicarboxylic acids	10
Inorganic glass	~15 (covers wide range)
Polyelectrolyte complexes	~15
Metal polyacrylates	20
Polyphosphates	25
Linear silicates	33
Invert glass	> 33

This range of percentage of ionic bonds falls between the extreme values of 0% (for diamond, polyethylene, etc.) and 100% (for an ionic solid like sodium chloride).

1.4.2 Aggregation of ions

One of the most interesting features of ionic polymers is the state of aggregation of the ionic bonds. Are they uniformly distributed in space, or are they aggregated in domains? Is the state of aggregation dependent on the concentration? If there are domains, how large are they? These are some of the questions which spring to mind. These questions have already attracted some experimental work, and Eisenberg has approached the same problems from a theoretical standpoint.[33] His approach is outlined below, and is also discussed in Chapter 2.

It is known that ion association takes place in liquid media of low dielectric constant. Pettit and Bruckenstein[34] have shown that states of aggregation from ion pairs to ion sextets exist in solvents of dielectric constant between 2·27 and 7·38 for salts such as Bu_4NCl and KCl. However, the situation will clearly be somewhat different in ionic polymers, where one ion is bound to a chain, and where it is necessary that the conformation of the chain should accommodate itself to the arrangement of ions if ion multiplets are to exist. We now consider the formation of ion pairs, the formation of ion multiplets and the clustering of multiplets.

The simplest form of aggregation is the ion pair, and the probability that these will form in media of low dielectric constant rather than separated ions, is clearly established. The work required to separate an ion pair into dissociated ions for singly charged ions is

$$W = -e^2/r4\pi\varepsilon_0 K \tag{6}$$

where K is the dielectric constant ($\approx 2\cdot 3$ for polyethylene), $1/4\pi\varepsilon_0 = 1$ dyn cm^2 statcoulomb^{-2} and r = distance between the centres of positive and negative charge in the ion pair. If $r = 1.5$ Å (reasonable for —COO$^-$Na$^+$) the interaction energy is 7×10^{-12} erg per ion pair, compared with $kT = 4 \times 10^{-14}$ erg at room temperatures. The fraction of dissociated ion pairs is therefore exceedingly small.

The higher form of aggregation is the multiplet, and it is assumed at this stage that they are randomly distributed throughout the matrix. However, to simplify the calculations, it is further assumed that they are distributed on a cubic lattice, and that each multiplet consists of a spherical drop. The inside of the drop contains the ion pairs in the multiplet, whilst the outside of the drop accommodates the hydrocarbon chain which accompanies the bound ions. It is assumed that sequential ion pairs are not found in the same multiplet.

By simple arguments Eisenberg shows that

$$r_m = 3V_p/S_{ch} \tag{7}$$

where r_m = radius of multiplet, V_p = volume of an ion pair and S_{ch} = the contact surface of the chain. If, for an ethylene–sodium methacrylate copolymer, there are a relatively large number of —CH$_2$— segments between the carboxylate ions, the minimum segment to be accommodated on the surface of the drop can reasonably be assumed to be —CH$_2$—CH—CH$_2$— with the bound ion (which penetrates the drop) attached to the middle carbon atom. On this basis, with $V_p = 12$ Å^3, then $r_m = 3$ Å and V_m the volume of the multiplet = 100 Å^3. This gives a maximum value of 8 pairs for perfect volume occupation, but the number is likely to be less than this for the packing of such groups as —COO$^-$Na$^+$.

What can be said of the clustering of multiplets into higher aggregates? Bearing in mind that the multiplet drops are coated with non-ionic chain material and that cluster formation involves stretching the polymer chains so that this type of higher aggregation becomes possible, the situation is obviously more complicated. A number of different models can be considered, but whatever the model chosen, the following factors will be involved.

(1) Upon cluster formation, work is done to stretch the segments of polymer chain between ionic groups from the distance corresponding to random dispersed multiplets to the distance corresponding to higher clusters, which will be further apart.

(2) Electrostatic energy is released when multiplets aggregate.

(3) The cluster is not infinitely stable, and above some temperature T_c the cluster decomposes. At this temperature electrostatic and elastic forces balance.

(4) Some ring formation will take place between sequential ion pairs incorporated in the same cluster.

A plausible picture of a cluster might show it to be based on a central ionic drop (containing up to 8 ion pairs, coated with a non-ionic skin made up of 8 three-carbon chain segments) surrounded by ion pairs and multiplets. These are attracted by electrostatic forces acting through the non-ionic skin of the droplet. The total number of ion pairs in such a cluster is calculated to be of the order of 100, for an ethylene–sodium methacrylate copolymer containing 4·5% mole of the salt (equivalent to 1 bound ion per 45 chain carbon atoms). The ionic phase in this case will occupy around 30% of the total volume, and will itself contain non-ionic material which accompanies the bound ions in the clusters. There are many assumptions in these calculations of Eisenberg, which are necessarily very approximate, but they provide a starting point for visualising the fine structure of an ionic polymer. Only experimental work will show whether and at what ion concentration clustering will begin, as the ion concentration is increased. At low ion concentrations, as the spacing between the ionic groups increases, the stage will be reached when there will be more physical entanglements than ionic groups. Since these entanglements act as cross-links, they will oppose the aggregation of multiplets into clusters. The nature of the polymer chain—for example its stiffness or the presence of bulky side groups—will also play a large part. Similarly at high ion concentrations, the stage may be reached where extensive cluster formation is impeded by the highly entangled and hindered nature of the structure. Thus the most orderly arrangement of the ions may be found at some intermediate range of concentrations.

At the moment there is a paucity of experimental evidence relating to the aggregation and clustering of ions, and unfortunately some of the evidence is conflicting. However, the picture which is beginning to emerge is broadly in favour of Eisenberg's theory of primary (multiplet) and secondary (cluster) types of ion aggregation.

The most direct evidence on ion aggregation comes from X-ray studies, as might be expected. There is no doubt that these studies confirm the existence of ionic heterogeneities in the materials which have been examined, but these cover only a narrow range of compositions—crystalline and amorphous copolymers of carboxylic acids containing a relatively small proportion of ionic groups. This is the range of compositions which comprises the materials which are generically termed ionomers (up to 8–10% mole salt groups).

In the case of the ethylene ionomers, the original du Pont work[37] suggested that ionic domains exist. The X-ray peak which is observed with the neutralised copolymer (whether the cation is mono- or divalent) persists above the melting point—in fact up to 300°C. Small amounts of water intensify the peak, which virtually disappears however if the sample is saturated with water. From this evidence it was inferred that this was a three-phase system, with amorphous and crystalline polyethylene regions, and ionic domains of diameter ~ 100 Å. This ionic phase consists of an arrangement of carboxylate groups co-ordinated around cations. The estimated domain size is of the same order of magnitude as the calculations of Eisenberg. Later work is beginning to refine this picture.† Although the work of Roe[52] cast doubt on this proposed morphology, the most recent studies tend to support it. Roe examined the caesium salt of a similar material to the du Pont workers (5% mole acrylic acid: 75% neutralised) and found no evidence of cluster formation, but strong evidence for the existence of dimers and quadruplets. Very recently three other investigations have been reported. MacKnight[75] has examined a typical ethylene ionomer (4·1% mole methacrylic acid: various counterions) and concluded that the salt groups are largely concentrated in spherical clusters of diameter 16–20 Å located in the amorphous phase. Marx et al.[76] investigated an ethylene ionomer and a rubber copolymer of butadiene and methacrylic acid and concluded that the aggregation of the scattering sites corresponded to multiplet formation—from dimers to tetramers and above,

† Discussed in detail in Chapter 2.

depending on the salt concentration. Finally Eisenberg[79] has examined a number of styrene ionomers and found two peaks in the X-ray pattern at ~ 22 Å and 60–80 Å which he has tentatively assigned to multiplet and cluster formation respectively. Presumably these can co-exist in the same material. Thus it will be seen that the latest work confirms the existence of primary aggregates or multiplets, but leaves the existence of clusters 'at the level of a working hypothesis'[79].

Other evidence of the aggregation of ions comes from various directions. There have been a number of viscoelastic studies of ionomers and carboxylated rubbers which bear on this problem. The work of Tobolsky[35,36] supports the idea of ionic domains of unspecified size, and therefore of unknown degree of aggregation. The work on the polyethylene ionomers is complicated by the presence of crystallinity, but the results obtained with the amorphous carboxylate rubbers are simpler to interpret. The high strength of these materials is ascribed to hard ionic clusters which give rise to a two-phase reinforced structure. In a separate study of a rubbery copolymer of butadiene–lithium methacrylate,[43] Otocka and Eirich postulated the existence of quadrupolar, *i.e.* multiplet links.

The study of water uptake by styrene and ethylene ionomers[79] has also thrown some light on this problem. In the case of the copolymers of styrene–sodium methacrylate, 1 water molecule is absorbed per ion pair up to 6% mole sodium methacrylate in the material. In the range 7–10% mole sodium methacrylate, this figure increases to 3–5 water molecules per ion pair. This strongly suggests that a new structural feature appears in the system around 6% mole of the salt. This may correspond to a change from multiplet to cluster formation. In contrast to this mode of behaviour, ethylene ionomers behave differently. Even at low concentrations of ions, these materials absorb more than 1 molecule of water per ion pair. Eisenberg suggests on these grounds that the structural feature which appears with styrene ionomers at 6% mole of salt, are already present at lower concentrations with ethylene copolymers.

Information on ion aggregation in other ionic polymers can be briefly summarised as follows. The case of polyelectrolyte complexes is complicated by the presence of water as an essential third component, since anhydrous polyanion–polycation complexes exist but are of little practical interest. The presence of dissolved, mobile counterions further complicates the situation. Based on studies of mechanical properties,[38] it has been concluded that a polyelectrolyte complex is made up of regions of varying intensity of polyion–polyion interaction. This can be rationalised by a Takayanagi type of model of regions of different stiffness in series and in parallel (as applied previously to composite materials and crystalline polymers), but this throws no light on whether and to what extent aggregation occurs. In the case of the single-phase amorphous silicate glasses, and also the polyphosphates[79] there appears to be no evidence of ion aggregation.

This picture of ion aggregation lends strong support to the suggestion that the dielectric constant of the medium in which the salt groups are immersed is central to the whole question. With a high dielectric constant as in the polyphosphates, no special driving force exists for domain formation. Furthermore, the higher the concentration of salt groups, the more polar the system. For that reason in many systems, such as the copolymers of acrylic and related acids, there will be a lower concentration below which there will be no aggregation, because of the distances between salt groups and other steric factors, and an upper limit above which there will be no phase separation for reasons of dielectric constant.

It is clear from this, as Eisenberg has emphasised, that the problem of ion aggregation in ionic polymers requires much more work in view of its importance and the fragmentary nature of our present knowledge.

Figure 4 attempts to summarise pictorially some of the foregoing information. It includes the interpretation of Bonotto and Bonner[39] of ethylene–acrylic acid salt copolymers which has been widely used, and which is based on a study of mechanical properties.

1.5 SOME PHYSICAL PROPERTIES OF IONIC POLYMERS

Although it is difficult at this stage to make wide ranging generalisations about ionic polymers, enough systematic work has been done in a few cases to permit comment on certain physical properties. For some systems, more detailed information will be found in later chapters.

1.5.1 Glass transitions in ionic polymers

This is a subject which has been investigated by Eisenberg, Otocka, Ray and others. The transition from a glassy to a rubbery consistency is a characteristic of many polymeric solids (the chief exception being highly cross-linked materials), and depends upon the thermal energy of the molecules. At some stage as the temperature increases, the segmental mobility increases to the point where rubber-like deformation under a small applied stress becomes possible. It can be expected that this segmental mobility will depend upon the nature of the inter-chain forces, and in

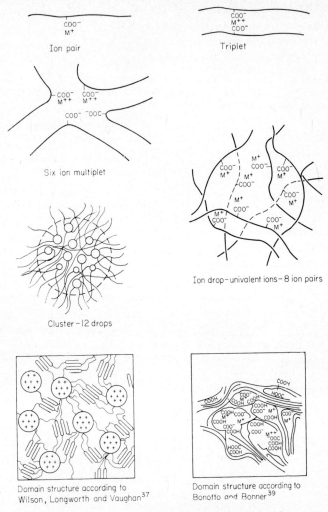

Fig. 4 Some representations of ionic polymers. (Reprinted with permission from Macromolecules 1, 514, 1968. Copyright by the American Chemical Society.)

the case of ionic polymers upon the degree of ionisation and the presence of ionic bonds. It should also depend upon the nature of the cation and anion. Experimental work confirms this and shows that the electrostatic forces between the bound ions in the polymer chain, and the counterions reduce the segmental mobility and increase T_g. This increase in T_g can be large in certain systems. Unfortunately the interpretation of the data in the literature is not consistent, and this raises certain difficulties in discussing the magnitude of ΔT_g in relation to structure.

The main difficulty revolves around the datum line against which the increase is to be measured. This can be illustrated by considering the system styrene–methacrylic acid studied by Nielsen,[10] from which the following data is taken:

TABLE 5

GLASS TRANSITIONS IN SALTS OF COPOLYMERS OF STYRENE–METHACRYLIC ACID

Mole % S	Mole % MA	$T_g °C$ (A) polymer	$T_g °C$ (B) sodium salt	ΔT_g (B)–(A)	ΔT_g (B)–T_g of polystyrene
100	0	100	—	—	—
98[a]	2	117	120	+3	20
94[a]	6	122	120	−2	20
90[a]	10	132	143	+11	43
60[a]	40	175	185	+10	85

[a] T_g measured from the maximum in the mechanical damping peak at ½ c/s.

In the penultimate column, the increase in T_g is shown in going from the acid to the salt form. This is quite small. In the last column, the increase in T_g is shown in going from polystyrene to the salt form of the copolymer. This is large, but it is also evident that the difference in T_g between the acid form of the copolymer and polystyrene is almost as large. It is clear that the increase in T_g caused by hydrogen bonding in this instance is almost as great as the increase caused by ionic bonding.

This leaves us with a choice as to what should be regarded as the host polymer for the ionic bonds—in this example polystyrene is one possibility, the acid form of the copolymers is the other. It is an arbitrary choice, and the literature favours the former. Thus Eisenberg[25] and Otocka[43] adopt this convention, and for the sake of uniformity it will be followed here also.

Before focusing attention on ionic polymers and the effect of the ionic bond on the glass transition temperature, it is enlightening to consider the analogous effect of covalent cross-links. In very highly cross-linked carbon polymers there is no indication of a glass transition below the decomposition temperature of the polymer. At lower concentrations of cross-links, it has been found that the glass temperature increases with the degree of cross-linking. This phenomenon has been studied by Fox and Loshaek[65] and Gibbs and Di Marzio.[66] On the Gibbs–Di Marzio theory, the criterion of glass formation is that the configurational entropy, which is temperature-dependent, becomes zero. Cross-linking decreases the configurational entropy and therefore raises the glass temperature. On this basis, they calculate

$$\frac{T_g - T_g^\circ}{T_g} = \frac{Kx}{1 - Kx} \qquad (8)$$

where T_g is the glass transition temperature of the cross-linked polymer, T_g° is the glass temperature of the uncross-linked material, x is the cross-link density and K is a constant to a first approximation independent of the material. This extremely useful relationship has been applied to inorganic as well as organic polymers. For example, Ray and Lewis[67] have shown that it describes the behaviour of a phosphate glass in which the extent of covalent P—O—P cross-linking was varied. If this relation were to apply to ionic polymers, and the ions acted as cross-linking sites, a concave upwards relationship between T_g and ion concentration would emerge.

(*i*) *Homopolymers—high concentration of ionic bonds.* At this point it is convenient to discuss ionic polymers in the two categories of homopolymers and copolymers. Homopolymers such as polyacrylic acid salts and polyphosphates have a high concentration of ionic bonds when fully neutralised. In many cases they have one ionic bond for every two backbone or chain atoms. Copolymers will generally have less, depending on the concentration of the monomer which introduces the bound ion. Although this is an arbitrary distinction, since one category merges into the other, it is useful in discussion. In practice, many of the commercially important solid ionic polymers have a relatively low concentration of bound ions.

Eisenberg has done a considerable amount of work on this topic,[25] and has compared the behaviour of linear silicates, polyphosphates and polyacrylates. These are all polyanions, two being inorganic and one organic. Despite the wide differences in structure, the behaviour of these materials is very similar. For this limited range of materials, the glass transition temperature can be represented approximately by an equation of the following general form for the fully neutralised polymers:

$$T_g = A(q/a) + B \qquad (9)$$

The individual equations are as follows:

$T_g = 625(q/a) - 12$ polyphosphates—partly or fully neutralised
$T_g = 635(q/a) + 132$ linear silicates ($M_2O:SiO_2 = 1$)
$T_g = 730(q/a) - 67$ polyacrylates—fully neutralised only

These are also shown in Fig. 5. In these relations, q is the counterion charge in units of 1 electron and a is the internuclear distance in angstrom units between the cation and anion at closest approach. A similar relation applies for ionenes which have only been tested over a limited range and which, being copolymers, have their bound ions fairly widely spaced. For counterion copolymers, i.e. systems with mixed counterions such as a polyphosphate with two different cations, the equations are also valid if a number average value for q/a is taken. This means in effect that the glass temperature is a linear function of the molar proportions of the components. Unfortunately this only applies for simple systems, since in multicomponent phosphate glasses, for example, it is not valid.

It is interesting that univalent cations with their single positive charge are effective in raising T_g. To a first approximation, the difference in going from a univalent to a divalent cation is adequately accounted for by the difference in the q/a parameter. However, there is a significant difference to be seen in the behaviour of individual ions which is overlooked in the above simple picture. For example, the T_g of calcium phosphate is considerably higher than that of cadmium phosphate (520°C and 450°C respectively) for the same value of q/a. Similarly the difference in T_g between barium and lead phosphates is bigger than would be expected (*see* Table 4, Chapter 8). Other exceptions will be noted when inorganic glasses are discussed.

The similarity of slopes for these polymers is interesting, but further work will be required to show the significance of this if any. The important part played by the parameter q/a is clearly significant. It has been explained by Eisenberg in terms of the work involved in removing a bound anion from the co-ordination sphere of a cation, or vice versa. At T_g, kT is just large enough for this to take place. Thus kT_g will be proportional to the electrostatic work involved, i.e.

$$T_g \propto W_{el} \propto \int F_{el} da$$

where F_{el} represents the electrostatic interaction and $a =$ internuclear distance. From this it can be seen that

$$T_g \propto q_a q_c / a$$

and since the bound ion charge is fixed

$$T_g \propto q/a \tag{10}$$

This is in line with the behaviour shown in Fig. 5. The line for the polyphosphates represents both the partly and the fully neutralised salts, if the parameter q/a is corrected for the concentration of counterion in the former case. In this case, with q and a fixed, T_g is proportional to the ion concentration. For the polyacrylates, however, the behaviour of the partly neutralised salt cannot be extrapolated from the line of the fully neutralised salt. The glass temperature is higher than would be expected from the extent of salt formation. Thus the extrapolated T_g for polyacrylic acid from the q/a plot of the metal polyacrylates is $-67°C$, but the actual value is

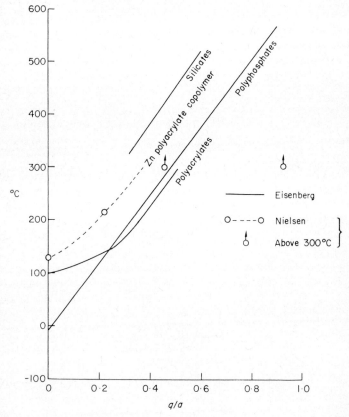

Fig. 5 Glass transition temperature as a function of q/a parameter.[25,68]

105°C. This very large difference can be explained by the very strong effect of hydrogen bonding on T_g in the case of the partly neutralised materials.

The results of Fields and Nielsen[68] are also shown for the partly neutralised zinc salt of their 94/6 acrylic acid/2-ethylhexylacrylate copolymer. The agreement with Eisenberg's data is only fair, but there are significant differences in experimental methods which might explain part of the difference in results.

(ii) Copolymers—lower concentration of ionic bonds. Although the work of Moacanin and Cuddihy[51] does not deal specifically with ionic polymers, it deals with a related problem. They examined glass transitions in the system lithium perchlorate–polypropylene glycol. Although this is not an ionic polymer, the solution of the salt in the polymer gives strong ion–dipole interactions, and an increase in T_g was observed with increasing salt concentration, denoting a lowering of segmental mobility.

A series of ionic polymers based on butadiene have been examined by Otocka and Eirich.[43] These included polyanions (Li counterion), polycations (iodide counterion) and polysalts of the two with the counterions left in. Their exact compositions were as follows:

TABLE 6

COMPOSITION OF POLYANIONS AND POLYCATIONS EXAMINED BY OTOCKA AND EIRICH[43]

	Butadiene–lithium methacrylate copolymers		
Polymer	Acid content eq/g	Carboxylate content eq/g (90% conversion acid groups)	Mole fraction salt
RA-1 Li	8.08×10^{-4}	7.29×10^{-4}	0.042
RA-2 Li	12.6×10^{-4}	11.34×10^{-4}	0.069
RA-3 Li	18.22×10^{-4}	16.45×10^{-4}	0.104

Butadiene methyl (2-methyl-5-vinyl) pyridinium iodide copolymers	
Polymer	Mole fraction pyridinium ion
RB-1Q	0.017
RB-2Q	0.060
RB-3Q	0.117

Stoichiometric mixtures of the first two members of each series were also prepared.

The effect of ion concentration on increasing T_g is shown in Fig. 6a where it is seen that the increase is greatest with the polycations, least with the polyanions whilst the polysalts occupy an intermediate position.

A number of ethylene–metal acrylate copolymers have been studied by Otocka and Kwei.[44] Sodium and magnesium were used as the counterions, and the carboxylate concentration varied from 0·66–2·78 COO$^-$ per 100

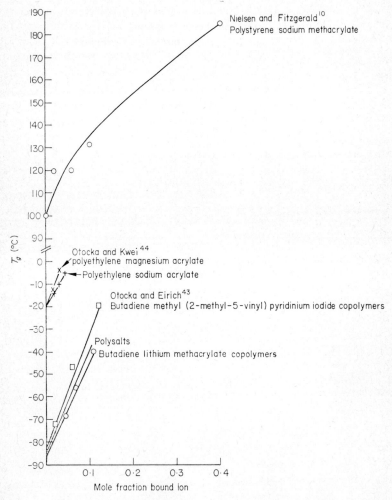

Fig. 6a *Glass transition temperature of copolymers versus ion concentration.*

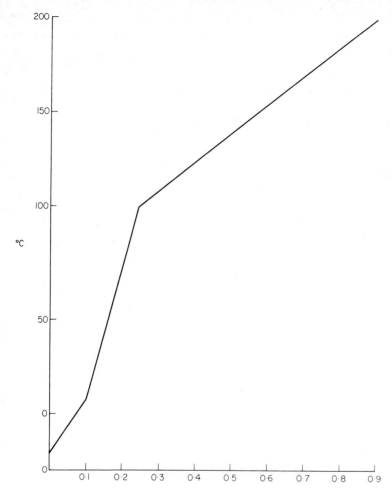

Fig. 6b *Glass transition temperature of copolymers versus ion concentration. (Reprinted with permission from Polymer Preprints* **14**, *871, 1973. Copyright by the American Chemical Society.)*

CH_2 groups. The results of this investigation are also shown in Fig. 6a. The data were obtained by dynamic mechanical tests, and the glass temperature identified with the β transition. The glass temperature follows the copolymerisation equation, being proportional to the ion concentration:

$$T_g = n_1 T_{g_1} + n_2 T_{g_2} \tag{11}$$

where n is the mole fraction of the component.

As already seen in Table 5, Fitzgerald and Nielsen found an increase in T_g in the system styrene–sodium methacrylate copolymers with increasing ionic bond concentration, these results being also shown in Fig. 6a.

Examination of Fig. 6a shows that T_g increases sharply with increasing ionic bond concentration in all cases, but there is not enough data to draw any quantitative conclusions, since dT_g/dc varies from 3–5·5°C mole %$^{-1}$.

Some very recent work[79] on the sodium salts of copolymers of ethyl acrylate and acrylic acid, over a very wide range of ion concentrations, has revealed some interesting features of this system. The results are shown in Fig. 6b, from which it will be seen that significant changes of slope occur at around 10%, and again at 25% mole of salt. From 0–10% mole of sodium salt, the increase is linear and can be described by the simplified copolymer equation. The rate of increase is 2·7°C mole%$^{-1}$. Above 10%, however, the rate of increase more than doubles to 5·8°C mole%$^{-1}$, and above 25% it decreases again to about 2°C mole%$^{-1}$. As Eisenberg suggests, this increase in slope may indicate the onset of clustering at around 10% mole of the salt. This effect is not evident in the system polystyrene–sodium methacrylate shown in Fig. 6a (which, however, was not studied in great detail), and further work on this subject is obviously needed.

The work on homopolymers already discussed indicates that the cation valency (through the parameter q/a) has an important influence on T_g. Unfortunately there is little information from the copolymer systems described which bears on this point, although it is reasonable to assume that cation valency should have a similar effect in these cases. There is ample confirmation from Fig. 6a that a univalent cation (Na^+ and Li^+) has a large effect on the glass temperature, as has a univalent anion (I^-). There is some indication from the behaviour of polyethylene magnesium acrylate that the magnesium ion is more effective than the sodium in raising T_g, but the evidence for the moment is slight. This is clearly a useful area for future research.

It appears that the ionic bond, based on electrostatic forces, behaves similarly to a covalent bond in increasing T_g, and the increase is of the same order of magnitude. However, the effect is more complex because of the role played by ion valency, as well as by a possible effect on chain stiffening,[40] and on ion aggregation effects.[79]

(iii) Inorganic glasses. It is convenient to discuss inorganic glasses separately, since their behaviour is more complicated than the behaviour of the simple systems dealt with above. There is no reason to believe that it

will not ultimately be possible to rationalise the behaviour of all ionic polymers in terms of a small number of basic parameters, but that time has not yet arrived, Meanwhile differences in behaviour are found when comparing carbon-based ionic polymers with inorganic glasses. In some cases, a close examination shows that these differences are more apparent than real. For example, the transformation temperatures (equivalent to T_g in plastics) of alkali phosphates increase in the order K < Na < Li, which is in the order of increasing q/a, whilst the transformation temperatures of alkali silicates and borates increase in the order Li < Na < K. In the former case the smaller the cation, the higher the transformation temperature. In the latter two cases the reverse is true. This is explained by Ray in Chapter 8 in terms of the relative importance of network packing, and cation co-ordination number which differs in these two examples. Another difference pointed out by Ray is that all alkali cations lower the T_g of silica and raise the T_g of B_2O_3, whilst some cations raise the T_g of P_2O_5 and some lower it.

Before seeking for explanations of some of these differences, the following should be borne in mind.

(1) It is very important to make an allowance for the structure of the starting material against which the comparison is being made. Does the introduction of ionic bonds increase or decrease the degree of connectivity of the system? As already seen, the addition of ionic bonds to a linear butadiene, styrene or ethylene copolymer increases the connectivity of the network by introducing a form of cross-link, and this increases T_g. In a similar way, the addition of cations to boric oxide converts part of the boron to the 4-valent state, and hence makes the system more cross-linked.

On the other hand, silica is a three-dimensional network structure which is highly cross-linked to begin with. The addition of cations may reduce the connectivity of the network, depending on ion valency, radius and co-ordination number.

Thus, in order to facilitate comparisons between ionic polymers of different origin such as polyacrylates and polyphosphates, it is helpful to begin with analogous structures, in this case polyacrylic acid and linear polyphosphoric acid as the conceptual starting points. It has been seen that there are many similarities between the behaviour of the salt forms of these materials which are readily rationalised in terms of the original polyacrylic acid with a T_g of 102°C and polyphosphoric acid with a T_g of −10°C. To the glass technologist, however, it may be more sensible to compare the behaviour of a polyphosphate with phosphorus pentoxide, with a T_g of 270°C.

(2) In the case of inorganic oxide glasses, having network bonds of the type Si—O, B—O, P—O which themselves are thermally labile (and the properties of which are affected by the presence of neighbouring ionic bonds if any), the ionic bonds and network bonds may be of comparable stability in the region of the transformation temperature. In such circumstances, the nature of the cation becomes important. With temperature labile covalent network bonds, bond switching may occur at higher temperatures or under stress. Ray has calculated[53] an energy parameter E' which is measured in temperature units, and which represents the internal energy required for bond switching to occur. This energy will be reflected in the transformation temperature, and figures for some typical bonds are given in the following table. It is evident from this, that some cations produce bonds which are more thermally labile than the original network, whilst with others the reverse is true.

TABLE 7

ENERGY PARAMETERS FOR BOND SWITCHING[53]

Original network bonds	E' (°K)	Network with ionic bonds	E' (°K)
Si—O—Si	544	Si—O$^-$Na$^+$	394
		Si—O$^-$Ca$^+$	1174
P—O—P	600	P—O$^-$K$^+$	490
		P—O$^-$Pb$^+$	786
B—O—B	540	Na$^+$ $^-$B—O	1550
		K$^+$ $^-$B—O	2900

The reader is referred to the original paper and to the chapter by Ray, in which it is shown that the transformation temperature can be calculated with considerable accuracy using this approach, based on the assumption that at the transformation temperature, a sufficient proportion of covalent and ionic bonds are switching to loosen the network. This thermal lability of the original network bonds is the biggest difference between inorganic glasses and carbon-based ionic polymers, and tends to obscure the role of the ionic bond in the material.

1.5.2 Melting point

The process of melting is a reversible one, and this means that bonds which are broken on melting must be capable of reforming on cooling. It also means that there must be enough thermally labile bonds to disrupt the polymer structure sufficiently, so that flow becomes possible, *i.e.* it must dissociate into 1-D fragments. The effect of ionic bonds on the softening or melting behaviour of a polymer (the precise behaviour depending on its crystallinity if any), depends on the nature and strength of all the cohesive forces which bond the material together. Thus, in addition to the ionic bonds which will be present in an ionic polymer, there will also be a 1-D or 3-D continuous arrangement of covalent bonds, as well as van der Waals' bonds and possibly hydrogen bonds as well.

In order to clarify the effect of ionic bonds on melting behaviour, it is necessary to bear the following facts in mind about the behaviour of polymers which do not contain ionic bonds:

(1) Carbon polymers melt if their structure is 1-D (linear or branched), and if the cohesive forces between their chains is not too great. Thus cellulose acetate melts, but cellulose itself decomposes on heating (decomposes in air at 270°C; out of air 350°C). Carbon polymers with 2-D or 3-D structures, exemplified by graphite, thermosets and diamond do not melt reversibly on heating since this would require the reversible breakdown of C—C bonds.

Clearly it is a necessary but not sufficient condition that if an ionic polymer is to melt, it should be capable of dissociating into 1-D covalently bound chains by the reversible breakdown of ionic bonds.

(2) Some inorganic oxide polymers such as silica melt reversibly on heating, despite the fact that they have a 3-D structure. In this case, the network bonds themselves such as Si—O are capable of reversible bond switching on heating. This is an important distinction with carbon polymers.

The next point to consider in discussing the effect of ionic bonds on the melting behaviour of ionic polymers is whether the ionic bonds themselves are thermally labile at a temperature below the decomposition temperature of the polymer. Thus in a carbon polymer containing metal carboxylate bonds acting as cross-links, will these bonds dissociate reversibly below the decomposition temperature of the rest of the structure? In simple ionic solids, reversible melting behaviour is widely encountered, although there are many important exceptions to this. When ionic solids do melt, the melting point depends upon the ratio of the heat of fusion to the entropy of fusion ($T_f = \Delta H_f / \Delta S_f$), and it may vary over a very wide range as the following examples show. Whether an ionic solid decomposes or melts on heating depends on the energetics of the rival processes.

TABLE 8

MELTING/DECOMPOSITION BEHAVIOUR OF SOME IONIC SOLIDS

Melt	°C	Decompose
NaF	988	Na oxalate
NaCl	807	Na palmitate
NaI	651	
Na acetate	324	
Na formate	253	
CaF_2	1360	$CaSO_4$
$CaCl_2$	772	Ca acetate
CaI_2	740	Ca formate
BaF_2	1280	Ba oxalate
$BaCl_2$	963	Ba propionate
BaI_2	740	Ba palmitate
Ba laurate	260	

In the case of the materials which decompose rather than melt, it will be seen that they all contain polyatomic anions, although it will also be seen that some salts containing polyatomic anions (*e.g.* sodium acetate) melt rather than decompose. Clearly the strength of the chemical bonding within the anion is a matter of the first importance in deciding how the material will behave on heating.

Turning to ionic polymers, both types of behaviour are encountered. Examples of ionic polymers which melt, and which decompose on heating are shown in Table 9.

TABLE 9

MELTING/DECOMPOSITION BEHAVIOUR OF SOME IONIC POLYMERS

	Melt	Decompose
Organic	Ionomers (carboxylated polyethylene salts)	Metal polyacrylates
	Carboxylated rubber salts	Ionenes
	Styrene–sodium methylacrylate copolymers (styrene rich)	
Inorganic	Glass Silicates (chain)	Silicates (fibrous and layer)

There is unfortunately a shortage of data on the melting points of ionic carbon polymer systems, so that it is not possible to draw any general conclusions at this stage, as to whether the materials are thermoplastic, and if so, what is the effect of ion valency and concentration on the melting point. For example, both univalent and divalent metal polyacrylates decompose before they melt. These have a high concentration of carboxylate bonds. On the other hand, styrene or ethylene copolymers containing up to 10% or higher molar concentration of sodium methacrylate units are fusible and thermoplastic.[10,44,56] At some intermediate concentration therefore, there is a changeover from a range of products which melt, to a situation where the products are infusible. Products with a low concentration of ionic bonds melt, whilst those with a high concentration decompose on heating. This may denote a tendency of ionic bonds to raise the melting point in carbon polymers, or may reflect the higher degree of cross-linking.

Before reaching any general conclusion, it is necessary to consider the special case of crystallisable ionic polymers such as ionomers, where over a limited range the inclusion of ionic bonds lowers the melting point, perhaps as the result of reducing the crystallinity. This is illustrated by the data of Otocka and Kwei[44] on the behaviour of the sodium and magnesium salts of copolymers of acrylic acid with ethylene.

TABLE 10

MELTING POINTS AND CRYSTALLINITIES OF ETHYLENE–METAL ACRYLATE COPOLYMERS[44]

$COOH/100CH_2$	Molar %	Acid copolymer M.Pt °C	Na salt		Mg salt	
			M.Pt °C	Crystallinity	M.Pt °C	Crystallinity
0·66	1·32	107	99	24	95	18
1·57	3·14	101	95	18	92	<5
2·26	4·52	98	93	11	infusible	
2·78	5·56	96	91	<5	infusible	

The melting points of the sodium and magnesium salts are lower than the original acid copolymer and lower than the 'original' polyethylene. The melting point falls with increasing cation concentration over the range studied, as does the crystallinity. The effect of the magnesium ion is greater than that of the sodium ion, although at higher magnesium concentrations the materials become infusible, whilst at the same concentrations the sodium salts melt. The data can be represented by Flory's copolymer crystallisation equation:

$$\frac{1}{T_m} - \frac{1}{T_m^\circ} = -\frac{R}{\Delta H_0} \ln N \qquad (12)$$

where T_m is the melting point of the copolymer, T_m° of the homopolymer, ΔH_0 is the heat of fusion of the homopolymer crystals and N is the mole fraction of crystallisable units.

Turning to inorganic ionic polymers, the situation is different because in the cases about which we have most knowledge, the main chain bonds are themselves thermally labile, as Si—O, so that the derived ionic polymers are thermally labile. Whether the melting point is higher or lower than the parent material depends primarily on whether the ionic bonds make the structure more or less connected.

1.5.3 Viscoelastic properties

As with the glass transition and melting point, the viscoelastic properties of ionic polymers—for example flow, stress-relaxation and creep—involve a number of features which are unique to these materials. These include some factors which have already been mentioned:

(a) the lability of the ionic bond or cross-link.
(b) the relative lability of the main chain bonds.
(c) the formation of ion multiplets, clusters or other microaggregates.
(d) the effect of ion aggregation on crystallisation behaviour, when crystallisable main chains are present.

It has already been seen that the ionic bond acts as kind of cross-link, and the high viscosity of ionic polymer melts provides evidence that its effect persists above the melting point. In the case of carbon polymers, the ionic bond is more labile than the covalent cross-link with which it is compared, in its response to heat or long-term stress. The situation is not so clear cut in the case of silicate polymers, where reversible main chain scission or bond switching readily takes place at high temperatures.

1.5.4 Melt behaviour

The poorer flow properties, as evidenced for example by a drop in melt index, show that the presence of ionic bonds leads to a greater inter-chain attraction. As might be expected, over the range studied, the viscosity increases as the concentration of ions increases (with one exception which will be discussed later). However it is difficult to speak confidently about the magnitude of the effect of ion type and valency on flow, as the following examples show.

At a descriptive level, Brown[11] found that the monovalent salts of carboxylic elastomers containing 0.1 ephr of carboxyl groups become plastic when warmed, and mill and sheet normally, in contrast to the zinc salts which crumble at the same temperature. At a higher temperature, however, the zinc salts also become somewhat plastic.

Cooper[54] looked only at the flow behaviour of the divalent metal salts of a low molecular weight copolymer of butadiene and methacrylic acid. His work showed the following:

(1) E_{vis}, the activation energy for viscous flow, increases with the acid content of the copolymer, and hence the number of ionic cross-linkages.

(2) It is much higher than that of the hydrocarbon polymer segments.

(3) It depends on the nature of the cation: Zn, Ca, Sr, Pb being approximately equal, but Mg and Cd were about twice as great. E_{vis} varied within the range 12–30 kcal mole^{-1} for the materials studied.

The monovalent metal salts (in this instance Li) of similar but higher molecular weight materials have been examined by Otocka and Eirich.[43] They studied their stress-relaxation behaviour, and above 20°C obtained a value of E_{vis} of about 20 kcal mole^{-1}. However it is not possible to compare this too closely with Cooper's work because of differences in molecular weight and polymer composition.

Turning to ethylene copolymers of the ionomer type, it has been shown[39] that there is little difference in the flow behaviour of mono- and divalent metal salts at the same molar concentration, as measured by the melt index at 190°C. The results obtained by Bonotto and Bonner are shown in Table 11. The great effect of salt formation on flow is clearly shown by these figures. The same paper shows that there is also little difference between mono- and divalent metal ions in the viscosity versus shear rate behaviour. This subject is discussed in detail in Chapter 2.

Recently a study has been made of a series of terpolymers of styrene-*n*-butyl methacrylate–methacrylic acid in the form of their potassium salt.[77] Over the range of ion concentrations studied, that is up to around 8% mole, an increase of viscosity of up to five orders of magnitude was observed. However, there was also a significant effect to be seen of polymer composition (styrene content) and molecular weight which complicates the picture.

TABLE 11
FLOW BEHAVIOUR OF ETHYLENE–ACRYLIC ACID COPOLYMER SALTS

	Control	Low conversion to salt				High conversion to salt			
		Na^+	K^+	Li^+	Ca^{++}	Mg^{++}	Na^+	K^+	Li^+	Ca^{++}	Mg^{++}
% ionised	0	30	25	28	30	37	66	63	67	63	64
M.I. (*ASTM* D1238) 190°C	67·0	3·8	4·5	5·2	3·1	2·5	0·3	0·6	0·2	0·1	0·2

It is very interesting to note that the formation of amine salts has been shown [78] to have only a very small effect on flow properties, as measured by the melt index at 190°C. In this case the salts were formed by aliphatic diamines in an ethylene ionomer, and the negligible reduction in flow properties can be contrasted with the large effect found with mono- and divalent cations with same material. It appears from this that the diamine salts dissociate completely at 190°C, and the system behaves like a solution of the diamine in the acid copolymer (in contrast to this, the same study showed that diamine salt formation has a significant effect on stiffness).

There are differences between the flow properties of silicates and silicate glasses on the one hand, and carbon polymers such as the metal salts of ethylene–acrylic acid copolymers discussed above. In some ways, these differences reflect the fact that far more is known about glass and the crystalline silicates, and also a far wider range of temperatures is accessible for study. In other ways, these differences depend upon the starting point under discussion—whether one is comparing silica with the silicates and polyethylene with the ionomers, where the trend of connectivities is in the opposite direction. The effect of introducing ionic bonds is to reduce the viscosity of silica and to increase the viscosity of polyethylene. Figure 7 shows how the viscosity of a multicomponent glass falls as the number of oxygen ions per tetrahedron increases up to the value of 2.[20]

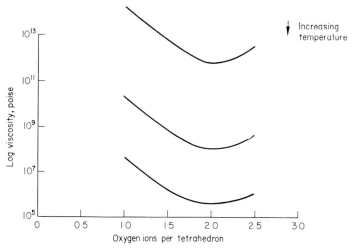

Fig. 7 Viscosity of a multicomponent silicate glass as a function of the number of oxygen ions per tetrahedron. (After Stevels.[20])

An increase in the number of oxygen ions per tetrahedron is synonymous with a reduction in the number of Si—O network bonds (the increase in viscosity above two oxygen ions per tetrahedron corresponds to the region of invert glasses). Apart from this broad picture, it is not possible to summarise briefly the effect of ion valency, radius and type, and also the effect of temperature, on the viscosity of glasses and silicate melts. For further information, the reader is referred to the chapter by Ray, and to the literature.

1.5.5 Relaxation behaviour

A discussion of relaxation behaviour involves the region bounded by the solid phase and the melt, and is concerned mainly with creep, stress-relaxation and dynamic mechanical properties. Studies of these phenomena have thrown light on the behaviour both of ionic and covalent bonds under the effect of stress and/or temperature. In stress-relaxation measurements, a sample is placed between two clamps and stretched rapidly. The stress at constant length is then measured as a function of time, and the data is usually presented as a plot of the logarithm of Young's modulus $E_r(t)$ against log time over a range of temperatures, usually covering from below to above the glass transition temperature.

When there is no bond interchange as with polystyrene or polymethylmethacrylate, the principle of time–temperature superposition applies. That is to say, a single master curve will fit all the data when log $E_r(t)$ is plotted against log t, if the curves are shifted horizontally along the time axis on to one reference curve. This is explained by postulating that the relaxation mechanism above T_g consists of an irreversible movement of chain segments, a process which speeds up as the temperature is increased (for a fuller discussion see Eisenberg[23]). At the stresses which are encountered, no scission of the polymer chains occurs.

The phenomenon of time–temperature superposition is not generally observed with ionic polymers (except when the ionic bonds are sparse[56]), nor is it observed when covalent bond interchange occurs. This can be explained by bond switching, as illustrated in Fig. 8. There is ample evidence that the switching of ionic bonds is normal behaviour under the effect of stress or temperature or both. On the other hand, the switching or interchange of covalent bonds is infrequently encountered, and is restricted to certain types. For example, it is highly unusual with carbon bonds and common at elevated temperatures with the network bonds which exist in glasses or in polymeric sulphur.

Ionic bond interchange at normal temperatures forms the basis of ion

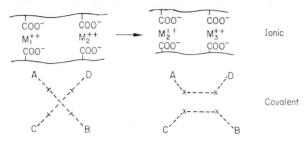

Fig. 8   Bond interchange in ionic and certain covalent systems. (After Cooper[54] and Eisenberg.[23] Reprinted with permission from Macromolecules **1**, *514, 1968. Copyright by the American Chemical Society.)*

exchange with aqueous electrolytes, between the counterion of the ionic polymer, and the corresponding ion in solution. It was also shown to occur by Cooper[54,55] in an anhydrous carboxylated rubber system. When the zinc salt of a liquid butadiene–acrylic acid copolymer was heated with excess finely divided cadmium acetate in the presence of zinc oxide, the polymer after purification contained cadmium. The following results were obtained:

TABLE 12

REPLACEMENT OF BOUND ZINC BY CADMIUM[55]

Time of heating mixture at 120°C hours	Ratio Cd/Zn
0	0·25
1	0·67
5	0·90
9	0·90

Magnesium and lead acetates were also found to take part in this type of exchange reaction. On the other hand, when the corresponding oxides were used, very little exchange occurred.

Evidence of the same phenomenon has been published by Bonotto and Bonner[39] in their extensive study of ethylene–acrylic acid copolymer salts. Also the process of ion exchange is very clearly shown up in glasses, where it can be used as a criterion to distinguish between network-forming atoms, and cations. If a glass containing sodium ions is immersed in molten potassium nitrate, an exchange of cations takes place, so that potassium ions enter the glass (despite their greater size), and sodium ions enter the melt.

Ionic bond interchange has been postulated to explain the stress-relaxation or creep behaviour of the following systems:

Butadiene–lithium methacrylate copolymers A	Otocka and Eirich
Butadiene–2-methyl-5-vinylpyridinium iodide copolymers B	Otocka and Eirich
Polyelectrolyte complexes of A and B	Otocka and Eirich
Styrene–sodium methacrylate copolymers	Eisenberg and Navratil
Butadiene–metal acrylate copolymers	Cooper

The polycation B above has more stable ionic linkages than the polyanion A, but in both cases the bonds are more labile than normal covalent cross-links above a transition temperature (which in the case of the polycation is above T_g). Whilst it appears that ionic bonds act as cross-links with a finite life-time under stress, it also appears[56] that above a critical ion concentration additional relaxation mechanisms such as ion clustering may play a part.

The rate of exchange depends upon the valency of the ion, and upon other factors such as ion charge and electropositive nature. Thus the work of Fitzgerald and Nielsen[10] on styrene–metal methacrylate copolymers shows that the rate of stress-relaxation is reduced in the following order in the temperature range 115–175°C: styrene–methacrylic acid copolymer > styrene–sodium methacrylate copolymer > styrene–barium methacrylate copolymer. The effect of ion valency is marked in this case. With the ions which he used, Cooper[54] found the rate to fall in the following order Pb > Zn > Mg. Similarly the creep rate decreased in the following order Pb > Zn > Ca > Mg.

In a work devoted to ionic polymers, a discussion of covalent bond interchange can find no place except for special cases, such as when discussing inorganic glasses, where the process rivals and complements ionic bond exchange. This is discussed by Ray in a later chapter. The interested reader is also referred to the review by Eisenberg[23] of the viscoelasticity of inorganic polymers.

1.5.6 Stiffness

Since this is one of the most important and informative properties of a material, it is of great interest to know how it is affected by the inclusion of ionic bonds. It is well known that the normal effect of cross-linking is to increase the modulus, so that it might be expected that—if ionic bonds increase the connectivity of a material—the stiffness will increase. That in fact is the general experience, as shown by the following examples. A related effect can also be seen in the retention of stiffness at higher temperatures compared with the un-ionised form.

In the case of rubbers, the pioneer work of Brown[11] on carboxylic elastomers showed that the salt form has a higher modulus than the acid form, and he also showed that a divalent ion (in this case Zn^{++}) is more effective in raising the modulus than a monovalent ion—for this system. Other workers,[36] using a terpolymer of butadiene-acronitrile-methacrylic acid, found that the zinc salt has a modulus three times that of the un-neutralised material at ambient temperature. They also found that the modulus of the zinc salt approximated to that of the sulphur-cured material. Unfortunately this work throws no light on the effect of ion valency on modulus, since only zinc was studied. Otocka and Eirich[43] examined two series of butadiene copolymers, one a polyanion the other a polycation (in each case with a univalent counterion) and found a considerable increase in modulus on salt formation.

Bonotto and Bonner[39] have made a thorough study of the bulk physical properties of salts of ethylene–acrylic acid copolymers (6·3% mole acrylic acid). The salt form showed a modulus increase of six times, which is already attained at a conversion of the acid groups of ~30%. This work only showed minor differences when equivalent quantities of univalent or divalent ions were used (*i.e.* 1 mole of M^+ or 0·5 mole of M^{++}). This increase in stiffness is even more marked at higher temperatures (80°C) but this only applies to the fully neutralised material. Similar results were found by Ward and Tobolsky.[35]

Using a similar type of ionomer (in this case a copolymer of ethylene and methacrylic acid[78]), it has been found that aliphatic diamines also produce salt-like cross-links and hence increase the stiffness of the acid copolymer. However, these materials are not as effective as metal ions, since maximum stiffness is not reached until 100% neutralisation (*cf.* 30% with mono- and divalent metals). This may reflect the formation of a greater proportion of intra- as opposed to intermolecular bonds.

Considering next materials of higher modulus, the metal salts of polyacrylic and polymethacrylic acids have been investigated by Nielsen[10,57] as well as copolymers of lower carboxyl content. The flexural modulus of zinc polyacrylate is $2\frac{1}{2}$ times that of polyacrylic acid at normal temperatures. Furthermore, whilst the modulus of polyacrylic acid falls off sharply in its transition region of ~90°C, the modulus of the zinc salt remains high up to 300°C. These results are illustrated in Fig. 9, and discussed in more detail in Chapter 5. In this work on zinc polyacrylate, the samples were prepared by the techniques of powder metallurgy, that is high temperature sintering of powdered oxide and polyacrylic acid. Attention is also drawn to some earlier work on the mechanical properties of divalent metal polyacrylates.[80] These were prepared by the aqueous polymerisation of solutions of the corresponding metal acrylate monomers, and the modulus of calcium polyacrylate was measured over the range of water contents 25–50%, where the water acts as a plasticiser. Extrapolation of this data to zero water content would give results roughly in accord with the data of Nielsen. It is noteworthy that the modulus of plasticised calcium polyacrylate *increases* with temperature over the range studied, but no explanation has been advanced for this phenomenon.

Other data of Nielsen on the metal salts of copolymers of styrene and methacrylic acid with a lower carboxyl content shows that the zinc salts have a higher modulus than the sodium salts over a range of temperatures

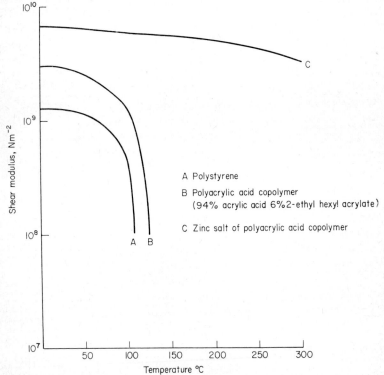

Fig. 9 Shear modulus from dynamic mechanical test (about 1 cps). (After Nielsen. Reprinted by kind permission of the Society of Plastics Engineers, Inc. from Polymer Eng. Science, **9,** *357, 1969.)*

as seen in Fig. 10. Since Figs. 9 and 10 deal with different polymers, direct comparison is not possible, but the higher modulus and retention of modulus of the zinc salt of polyacrylic acid compared with the corresponding salt of the styrene–methacrylic acid copolymer presumably reflects the greater number of ionic bonds in the former material (there are more than twice as many.)

The foregoing information is summarised in Table 13, which also includes data on strength which is discussed later.

Although it is not possible to draw detailed conclusions from the work to date, which is at an early stage, the following picture emerges from the above data:

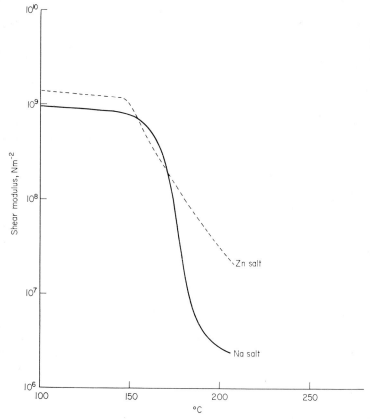

Fig. 10   Shear modulus of metal salts of styrene (60%)–methacrylic acid (40%) copolymer.[10]

TABLE 13
STIFFNESS AND STRENGTH OF SOME IONIC POLYMERS[a]

Ionic polymer system	Composition	Unneutralised modulus (E) Nm^{-2}	Univalent ion: modulus (E) Nm^{-2}	Divalent ion: modulus (E) Nm^{-2}	Unneutralised strength Nm^{-2}	Univalent ion: strength Nm^{-2}	Divalent ion: strength Nm^{-2}	References
Rubbers	SBR–MA 0·7% mole MA	—	$1·3 \times 10^6$ Na$^+$	$1·8$–$4·4 \times 10^6$ wide range of divalent oxides	—	$6·2 \times 10^6$ Na$^+$ (875%)	$1·3$–$3·9 \times 10^7$ wide range of divalent oxides (740–890%)	Dolgoplosk[69]
	B–ACN–MA 4% mole MA	—	—	—	2–3×10^6	—	8×10^6 (0°C) Zn^{++}	Tobolsky, Lyons and Hata[36]
	B–MA 6·7% mole MA	$\ll 7 \times 10^5$	$>1·4 \times 10^6$ Na$^+$	$>1·0 \times 10^7$ Zn^{++}	$<7 \times 10^5$ (>1600% ext)	$1·2 \times 10^7$ Na$^+$ (900%)	$4·1 \times 10^7$ Zn^{++} (400%)	Brown[11]
	B–MA 9% mole MA	6×10^6	$2·2 \times 10^7$ Li$^+$	—	—	—	—	Otocka and Eirich[43]
	B–pyridinium (P)[b] 9% mole P	3×10^6	$5·5 \times 10^7$ I$^-$	—	—	—	—	Otocka and Eirich[43]
Polyolefins	Ethylene–AA 6·3% mole AA	$4·8 \times 10^7$	$3·4 \times 10^8$	$2·8 \times 10^8$	$1·5 \times 10^7$ (470% ext)	$2·7 \times 10^7$ (360%)	$2·5 \times 10^7$ (365%)	Bonotto and Bonner[39] ~30% conversion
		$4·8 \times 10^7$	$2·8 \times 10^8$	$2·5 \times 10^8$	$1·5 \times 10^7$ (470% ext)	$3·3 \times 10^7$ (300%)	$3·1 \times 10^7$ (210%)	Bonotto and Bonner[39] ~65% conversion
	Ethylene–AA 8% mole AA	$4·1 \times 10^7$	$4·0 \times 10^8$ Na$^+$	$3·2 \times 10^8$ Ca^{++}	$1·4 \times 10^7$ (490% ext)	$3·4 \times 10^7$ Na$^+$ (320%)	$3·5 \times 10^7$ Ca^{++} (230%)	A range of ions used Ward and Tobolsky[35] 47% conversion
Polyacrylic acid copolymers	Styrene–MA 40% mole MA	—	$\sim 1 \times 10^9$ Na$^+$ (shear)	$\sim 2 \times 10^9$ Zn^{++} (shear)	—	—	—	Nielsen[10,57]
Polyacrylic acid	AA–acrylate[c] 94% mole AA	7×10^9 (flex)	—	$1·8 \times 10^{10}$ Zn^{++} (flex)	7×10^7 (flex) $1·7 \times 10^8$ (comp)	—	$6·8 \times 10^7$ (flex) $3·7 \times 10^8$ (comp.)	Nielsen[10,57]

KEY SBR = styrene–butadiene rubber; B = butadiene; MA = methacrylic acid; AA = acrylic acid.

[a] The table does not include data on plasticised calcium polyacrylate[80] or diamine neutralised ionomers[7] for simplicity.
[b] The polycation investigated by Otocka and Eirich was a copolymer of butadiene and 2-methyl-5-vinyl pyridine quaternised with methyl iodide.
[c] The mechanical properties of some metal polyacrylate composites are discussed in Chapter 5.

(1) The stiffening effect of ionic bonds is greater in the rubbery region, where large elastic strains are possible.

(2) The use of univalent ions leads to an increase in modulus. This can be substantial, since it varies by a factor of 4–18 in the examples quoted.

(3) The use of divalent ions leads to a greater increase in modulus than univalent ions. This effect is most marked with elastomers. There is an exception to this in the case of polyethylene where univalent and divalent ions have approximately the same effect.

Polyethylene is a crystalline polymer, and the introduction of ionic bonds changes the morphology, by reducing the crystallinity. Further work may show that this is a general phenomenon with crystalline polymers. The increase in modulus found with polyethylene containing ionic bonds is a compound of the effects of reducing the crystallinity (lowering the modulus) and introducing ionic bonds (increasing the modulus).

Clearly further work is required to understand the effect of ion valency and ion type on modulus.

(4) Where an excess of metal oxide is used in preparing the salt form, as in some of the work on rubbers, this will act as an inert filler and will increase the modulus slightly in accordance with normal composite experience.[59]

(5) Where ion clustering occurs, to form a macroscopic ionic phase, this may also increase the modulus by the mechanism mentioned in (4) above, i.e. the structure may behave as a composite. This effect would operate over and above the cross-linking effect.

The foregoing discussion has dealt with carbon polymers. The case of silicate glass is more complicated and is dealt with later by Ray. It is more difficult to elucidate the part played by the cations on stiffness but it appears that the effect is not so great as with carbon polymers as the following table shows:

TABLE 14
STIFFNESS OF SOME SILICATE GLASSES

	$E\ (N\ m^{-2})$
Fused silica	7×10^{10}
'E' glass	7×10^{10}
'A' glass	$7 \cdot 4 \times 10^{10}$
High modulus glass	11×10^{10}
	(up to 15×10^{10} has been reported[58])
High strength glass	$8 \cdot 7 \times 10^{10}$

1.5.7 Strength

Strength is a more complex property than stiffness, since it is greatly affected by specimen imperfections, and involves larger strains. It is also more affected by rate of strain. Although it is difficult to make broad generalisations about strength, it is usually true to say that two routes to higher strength are to increase stiffness and/or ductility. Since the introduction of ionic bonds increases stiffness, it would be expected that it would also increase strength. This in fact is what happens, despite a concomitant reduction in strain to failure. Table 13 shows the sort of increase in strength which is encountered. Usually the increase in strength is less than the increase in stiffness. The anomalous results obtained with the metal polyacrylates in flexure can probably be explained by imperfections in the specimens, which are very difficult to fabricate.[10]

In the case of rubbers, it appears that a divalent ion has a greater effect on strength than a univalent ion. This is in line with the experience with stiffness. In the case of polyethylene, ion valency appears to have little or no effect. This is also in line with experience with stiffness. Otherwise there is not enough data available to draw general conclusions on the effect of ion valency.

The early work of Brown[11] on rubbers showed the importance of the amount of base added in relation to the stoichiometric quantity. Experience with carboxylated polyethylene is quite different. In the case of the butadiene–acrylonitrile–methacrylic acid copolymer, maximum strength is not achieved until twice the theoretical quantity of zinc oxide is added, whilst to reach maximum modulus, over three times the theoretical quantity is needed. In the case of the ethylene–acrylic acid copolymer, both maximum stiffness and strength are reached with less than the stoichiometric quantity of base. The explanation for these differences awaits further research.

Turning to the glasses, the behaviour of oxide glasses is interesting since the tensile strength appears to be practically independent of composition, being generally determined by the surface condition. The maximum strength of flaw-free fibres at room temperatures may approach 7×10^9 Nm^{-2} whether made of silica or a cation-containing glass, or even asbestos, which is an anisotropic mineral silicate.[60,61] In view of the difficulty of making accurate measurements of this kind, and also in view of the paucity of information, it would be premature to conclude definitely at this stage that strength is independent of composition.

1.5.8 Thermal expansion

Generally speaking thermal expansion and mechanical properties are related, and therefore it is to be expected that the insertion of ionic bonds into a polymer network will affect the expansion behaviour. For example Barker[62,63] has shown that the following very approximate relationship is valid for a large number of materials:

$$E\alpha^2 \simeq 15 \text{ N m}^{-2}{}^\circ\text{C}^{-2} \tag{13}$$

where E is Young's modulus and α is the linear coefficient of expansion. Since the introduction of ionic bonds into polymers increases E, it can be expected that it will reduce the coefficient of expansion. Unfortunately information on this subject is sparse, but the following figures obtained by Nielsen[57] for metal polyacrylates are in general agreement with this prediction:

TABLE 15
EFFECT OF IONIC BONDS ON THERMAL EXPANSION

Material	Coefficient of linear expansion ($^\circ C^{-1} \times 10^5$)
Polystyrene	8·0
Polyacrylic acid	5·5
Calcium polyacrylate	2·3
Zinc polyacrylate	1·4

In this example, the zinc salt has a thermal expansion coefficient which is only one quarter of the acid form. This is greater than the reduction which would be expected from the change in modulus, based on Barker's relationship.

There is little available information on the effect of ion valency on thermal expansion, but some results of Eisenberg[79] indicate that Ca^{++} reduces the thermal expansion of the polyphosphates to a greater extent than Na^+:

TABLE 16
THERMAL EXPANSION AND CATION VALENCY

Material	Coefficient of linear expansion ($^\circ C^{-1} \times 10^5$)
$NaPO_3$	3·4
$Ca(PO_3)_2$	1·0

1.5.9 Hydrolytic stability

The picture which has emerged up to this point is that the ionic bond in ionic polymers behaves in certain ways like a normal cross-link. However, in the carboxylated rubbers and polyethylenes, properties such as creep, stress-relaxation, flow and ion interchange indicate that it is less stable to temperature and stress. The divalent metal polyacrylates, however, with their higher concentration of ionic bonds, appear to be much more stable.

What of the chemical stability of the ionic bond, in particular the hydrolytic stability, and resistance to aqueous acids and alkalis? From the information available, it can be concluded that ionic polymers cover the whole range of hydrolytic stabilities, from those which disintegrate in water, to those which swell but remain coherent, to those which are completely stable. To illustrate the extremes of behaviour, water has little or no effect at room temperature on some ionomers or some salts of carboxylated rubbers, whilst it has a considerable effect on polyelectrolyte complexes and the divalent metal salts of α,ω dicarboxylic acids.

Bonotto and Bonner[39] studied the water absorption of ethylene–acrylic acid copolymer salts as a function of conversion, with five cations. Their results are shown in Fig. 11, from which the great difference between potassium and sodium on the one hand, and lithium, magnesium and calcium on the other can be seen. A higher water absorption was accompanied by a reduction in stiffness, indicating that water has a plasticising effect. These results show that at 60% conversion to the salt form, approximately 2 molecules of water are absorbed per ion pair for the potassium salt, 1 molecule for the sodium salt, and 0·2 molecules per ion pair for Li, Ca and Mg. It is probable that equilibrium water uptake at 100% conversion will considerably exceed these figures in the case of the potassium and sodium salts, based on the shape of these curves. It has already been mentioned that Eisenberg[79] has made use of the water absorption data for styrene–sodium methacrylate ionomers to support the argument that the increase from 1 water molecule per ion pair to 3–5 molecules per ion pair as the ion content increases reflects a change in the degree of aggregation of the ions. In view of the data of Bonotto and Bonner, it is likely that there are at least two separate effects involved in controlling water uptake, *i.e.* ion type and ion aggregation. Clearly the nature of the chain will also be an important variable.

The effect of ion type shows up also in the case of the metal polyacrylates. Sodium polyacrylate is water soluble, whilst divalent metal polyacrylates are insoluble in water, although they may be more or less hydrophilic depending on the method of preparation. For example barium, zinc,

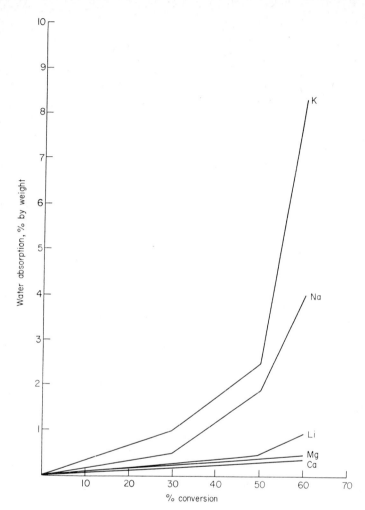

Fig. 11 *Effect of cation on water absorption of salts of ethylene–acrylic acid copolymers.*[39]

calcium and magnesium polyacrylates prepared by polymerising aqueous solutions of the corresponding metal acrylates, contain 50–70% by weight of water at equilibrium.[80] In this form they are stiff hydrogels. On the other hand, when prepared by the method of Nielsen[57] by reacting the finely powdered metal oxide with powdered anhydrous acrylic acid at high temperatures and pressures (in the region of 300°C and 10 000 psi), zinc polyacrylate is almost unaffected by water. Nielsen studied the effect of water at 22°C and 100°C on zinc polyacrylate prepared by this high temperature method, over periods of 90 days and 7 days respectively. Ninety days immersion in water at room temperature gave a water content of 0·3%, and 7 days at 100°C gave a value of 1·6%. There was little effect on strength or stiffness in either case. Clearly these figures, which indicate excellent hydrolytic stability, suggest a structural change resulting from the method of preparation.

Zinc polyacrylate forms the basis of a novel dental cement developed by Smith.[70] This and the alumino-silicate glass–polyacrylic acid dental cement developed by Wilson[71] are discussed fully in Chapter 5. The alumino-silicate glass acts in this case as a source of calcium and aluminium cations. These cements are completely stable in water, as is to be expected from their use in dentistry. They are made by reacting an aqueous solution of polyacrylic acid (approximately 50% weight) with zinc oxide or powdered glass of special composition. The cement matrix phase therefore has a substantial water content at equilibrium.

A wide range of polyacrylates has been investigated more recently.[72] The following multivalent metal oxides form hydrolytically stable cements —zinc, copper, lead, mercuric, yttrium, cadmium, bismuth. Other metal oxides form cements which either absorb water extensively or disintegrate completely. The same work has also shown that polyacrylate cements are not stable towards strong acids or alkalies.

There is a very extensive literature in the fields of geochemistry and glass technology on the hydrolytic stability of silicates and glasses. It is beyond the scope of this chapter to deal with this subject even briefly, and the reader is referred to the appropriate literature for further information on this subject. Generally speaking, these materials have excellent hydrolytic stabilities at all temperatures under neutral conditions, although they are subject to attack by concentrated acids and alkalis. A systematic study of the relationship of silicate structures to strong mineral acids has been made by Murata.[73]

1.6 CONCLUSIONS

An attempt has been made in the foregoing pages to classify ionic polymers, and to deal briefly with the relationship between some important physical properties and their structure. Silicates, glasses and polyphosphates have been included in the discussion, despite the wide gulf which separates the chemistry of these inorganic polymers from synthetic carbon polymers. Even on a formal descriptive level it is instructive to compare these materials since interesting analogies can be seen, and it is probable that lessons can be learnt from silicate polymers—about which we know a great deal—which can be applied to organic ionic polymers. We are at a very early stage of development with ionic polymers, and it is clear that this subject will be an important area for research in the future.

REFERENCES

1. Holliday, L. (1970). *Inorg. Macromol. Rev.*, **1**, 3.
2. Dislich, H. (1971). *Ang. Chemie Int. Edn.*, **10**, 363.
3. Holliday, L. and Holmes-Walker, W. A. (1972). *J. App. Polym. Sci.*, **16**, 139.
4. Hoover, M. F. (1969). *Polymer Preprints*, **10**, 908.
5. Otocka, E. P. (1971). *J. Macromol. Sci. Revs. Macromol. Chem.*, **C5(2)**, 275.
6. Holliday, L. (1972). *Chemistry and Industry*, **23**, 921.
7. Dolezal, T., Edwards, D. C. and Wunder, R. H. (1968). *Rubber World*, **158**, 46.
8. van Wazer, J. R. (1958). *Phosphorus and its Compounds*, Interscience, New York.
9. Rees, R. W. and Vaughan, D. J. (1965). *Polymer Preprints*, **6**, 287.
9a. Hagedorn, M. (1936). US Patent 2,045,080.
10. Fitzgerald, W. E. and Nielsen, L. E. (1964). *Proc. Roy. Soc.*, **A282**, 137.
11. Brown, H. P. (1957). *Rubber Chem. Tech.*, **30**, 1747.
12. Rees, D. A. *Chemistry and Industry*, 19 August 1972, p. 630.
13. Rembaum, A., Baumgartner, W. and Eisenberg, A. (1968). *J. Polym. Sci.*, **B66**, 159.
14. Michaels, A. S. (1965). *Ind. Eng. Chem.*, **57**, 32.
15. Schindler, A. and Williams, J. L. (1969). *Polymer Preprints*, **10**, 832.
16. Economy, J. E., Mason, J. H. and Wohrer, L. C. (1966). *Polymer Preprints*, **7**, 596
17. Trap, H. J. L. and Stevels, J. M. (1959). *Glasstech. Ber.*, **6**, 31.
18. Wells, A. F. (1962). *Structural Inorganic Chemistry*, O.U.P., 3rd edn., p. 421.
19. Kingery, W. D. (1960). *Introduction to Ceramics*, Wiley, New York, p. 152.
20. Stevels, J. M. (1960–61). *Philips Technical Review*, **22**, 300.
21. Gibbs, C. F. and Marvel, C. S. (1934). *JACS*, **56**, 725.
22. Grund, A. and Pizy, M. (1952). *Act. Cryst.*, **5**, 837.
23. Eisenberg, A. (1970). *Inorg. Macr. Rev.*, **1**, 75.
24. Eisenberg, A. and Takahashi, K. (1970). *J. Non-Crystalline Solids*, **3**, 279.
25. Eisenberg, A. (1971). *Macromolecules*, **4**, 125.
26. US Patent 2,604,668 (29 July, 1952).
27. Lopatin, G. and Newey, H. A. (1971). *Polymer Preprints*, **12(2)**, 230.
28. Weinstein, J. N. and Caplan, S. R. (1968). *Science*, **161**, 70.
29. Schindler, A. and Williams, J. L. (1969). *Polymer Preprints*, **10**, 832.
30. Stille, J. K., Kamachi, M. and Kurihara, M. (1971). *Polymer Preprints*, **12(2)**, 223.
31. Hatch, M. J., Dillon, J. A. and Smith, H. B. (1957). *Ind. Eng. Chem.* **49**, 1812.
32. Cowan, J. C. and Teeter, H. M. (1944). *Ind. Eng. Chem.*, **36**, 148.
33. Eisenberg, A. (1970). *Macromolecules*, **3**, 147.
34. Pettit, L. D. and Bruckenstein, S. (1966). *J.A.C.S.*, **88**, 4783.
35. Ward, T. C. and Tobolsky, A. V. (1967). *J. App. Polym. Sci.*, **11**, 2403.
36. Tobolsky, A. V., Lyons, P. F. and Hata, N. (1968). *Macromolecules*, **1**, 515.
37. Wilson, F. C., Longworth, R. and Vaughan, D. J. (1968). *Polymer Preprints*, **9**, 505.

38. Hoffman, A. S., Lewis, R. W. and Michaels, A. S. (1969). *Polymer Preprints*, **10**, 916.
39. Bonotto, S. and Bonner, E. F. (1968). *Macromolecules*, **1**, 514.
40. Eisenberg, A., Farb, H. and Cool, L. G. (1966). *J. Polym. Sci.*, **A-2**, **4**, 855.
41. Eisenberg, A., Matsuura, H. and Yokoyama, T. (1971). *J. Polym. Sci.*, **A-2**, **9**, 2131.
42. Eisenberg, A., Matsuura, H. and Yokoyama, T. (1971). *Polymer J.*, **2**, 117.
43. Otocka, E. P. and Eirich, F. R. (1968). *J. Polym. Sci.*, **A-2**, **6**, 921; **A-2**, **6**, 933.
44. Otocka, E. P. and Kwei, T. K. (1968). *Macromolecules*, **1**, 401.
45. Noguchi, H. and Rembaum, A. (1969). *Polymer Preprints*, **10**, 718.
46. Hadek, V., Noguchi, H. and Rembaum, A. (1971). *Polymer Preprints*, **12**, 90.
47. Lehman, M. R., Thompson, C. D. and Marvel, C. S. (1933). *JACS*, **55**, 1977.
48. Kern, W. and Brenneisen, E. (1941). *J. Prakt. Chem.*, **159**, 193.
49. Berlin, A. A., Zherebtsova, L. V. and Razvodovskii, Y. F. (1964). *Polym. Sci. USSR*, **6**, 67.
50. Razvodskii, Y. F., Neksavov, A. K., and Yenikolopyan, N. S. (1971). *Polym. Sci. USSR*, **13**, 2226.
51. Moacanin, J. and Cuddihy, E. F. (1966). *J. Polym. Sci.*, **C**, **14**, 313.
52. Roe, R. J. (1971). *Polymer Preprints*, **12(2)**, 730.
53. Ray, N. H. (1971). IXth International Conference on Glass, **A1**, **5**, 633.
54. Cooper, W. (1958). *J. Polym. Sci.*, **28**, 195.
55. Cooper, W. (1958). *J. Polym. Sci.*, **28**, 628.
56. Eisenberg, A. and Navratil, M. (1972). *Polymer Letters*, **10**, 537.
57. Nielsen, L. E. (1969). *Polymer Engineering and Science*, **9**, 356.
58. Wilson, M. L. and Scott, G. E. (1970). *Glass Technology*, **11**, 76.
59. Holliday, L. (1966). In *Composite Materials*, Ed. L. Holliday, Elsevier, Chapter 1.
60. Kelly, A. (1966). *Strong Solids*, O.U.P., pp. 52 et seq.
61. Zukowski, R. and Gaze, R. (1959). *Nature*, **183**, 35.
62. Barker, R. E. Jr. (1963). *J. Appl. Phys.*, **34**, 107.
63. Barker, R. E. Jr. (1967). *J. Appl. Phys.*, **38**, 4234.
64. Eisenberg, A. and King M. (1971). *Macromolecules*, **4**, 204.
65. Fox, T. G. and Loshaek, S. (1955). *J. Polym. Sci.*, **15**, 371.
66. a. Gibbs, J. H. and DiMarzio, E. A. (1958). *J. Chem. Phys.*, **28**, 373.
    b. DiMarzio, E. A. (1964). *J. Res. Nat. Bur. Standards*, **68A**, 6, 611.
67. Ray, N. H. and Lewis, C. J. (1972). *J. Mat. Sci.*, **7**, 47.
68. Fields, J. E. and Nielsen, L. E. (1968). *J. App. Polym. Sci.*, **12**, 1041.
69. Dolgoplosk, B. A., Tinyakova, E. I., Reikh, V. N., Zhuravleva, T. V. and Belonovskaya, G. P. (1959). *Rubber Chem. and Tech.*, **32**, 321/328.
70. Smith, D. C. (1968). *Brit. Dent. J.*, **125**, 381.
71. Wilson, A. D. and Kent, B. E. (1971). *J. App. Chem. Biotech.*, **21**, 313.
72. Elliott, J., Holliday, L. and Hornsby, P., unpublished work.
73. Murata, K. J. (1945). *The American Mineralogist*, **28**, 545.
74. MacKnight, W. J. (1973). *Polymer Preprints*, **14**, 813.
75. MacKnight, W. J., Taggart, W. P. and Stein, R. S. (1973). *Polymer Preprints*, **14**, 880.
76. Marx, C. L., Caulfield, D. F. and Cooper, S. L. (1973). *Polymer Preprints*, **14**, 890.
77. Erhardt, P. F., O'Reilly, J. M., Richards, W. C. and Williams, M. W. (1973). *Polymer Preprints*, **14**, 902.
78. Rees, R. W. (1973). *Polymer Preprints*, **14**, 796.
79. Eisenberg, A. (1973). *Polymer Preprints*, **14**, 871.
80. Hopkins, R. P. (1955). *Ind. Eng. Chem.*, **47**, 2258.

CHAPTER 2

THERMOPLASTIC IONIC POLYMERS: IONOMERS

RUSKIN LONGWORTH

2.1 INTRODUCTION

The word 'ionomer' is used to describe generally ionic polymers consisting of a hydrocarbon backbone and pendant carboxylic acid groups which are neutralised either partially or completely with metal or quaternary ammonium ions. An excellent general review of ionic polymers has recently appeared.[1]

This chapter is particularly concerned with the ionomers as originally described,[2,3] that is, homogeneous random copolymers of ethylene and methacrylic acid which are neutralised either partially or completely with metals, particularly sodium and zinc. The range of compositions is such that they can be fabricated by conventional plastics processing equipment into finished products. This consideration applies equally well to ionomers made from ethylene copolymers with either methacrylic or acrylic acids. Ionomers prepared with lithium or ammonium derived cations are processable but have not as yet been commercialised. Because their structural features are of interest, reference will be made to ionomers made from styrene–methacrylic acid copolymers, but these materials are difficult to fabricate except by simple compression moulding techniques and are not sold commercially. Although usable thermoplastic ionomers have been made from copolymers based on a polyoxymethylene backbone,[80] they, too, are not available commercially.

The only ionomers commercially available at the present time are sodium and zinc ionomers introduced in 1964 by the Du Pont Company under the trade mark Surlyn®[3] and since then have become increasingly important plastics. Ionomers come between the two extremes of intractable, ionic polymers described elsewhere in this book on the one hand and tractable, low density polyethylene on the other. Although the weight fraction of metal ion in the commercial ionomers is only a few percent, this should be more than enough to form cross-linked gel if suitably combined; and it is remarkable that the ionomers show no tendency to gel, either in the solid state or in solution. They remain truly thermoplastic even after exposure to the elevated temperatures encountered in processing.

Although this chapter is concerned with scientific aspects of ionomers, it is generally the case that the incentive to study such aspects comes from a growing commercial interest, and so the features which make ionomers commercially important will be briefly described.

In the first place, ionomers are flexible, tough and truly thermoplastic. At melt temperatures similar to those used for low density polyethylene (150°C to 250°C) they can be extruded into films by blow moulding (lay flat tubing, foamed sheet, rod, tubes, etc.). They can be injection moulded into solid objects and blow moulded into bottles. The films and sheets can be postformed by various methods into skin and blister packages.

The most striking feature of the ionomers is their clarity in contrast to low density polyethylene, which is the parent hydrocarbon polymer. This clarity is illustrated in Fig. 1 in which the transparency of 0·120 in. thick slabs of an ionomer, a methacrylic acid copolymer and low density polyethylene are compared. The translucency of the copolymer and the ethylene homopolymer is due, of course, to their spherulitic crystallinity. The transparency of the ionomers led earlier workers to suppose that similar crystallinity was either absent or present to a small degree in these

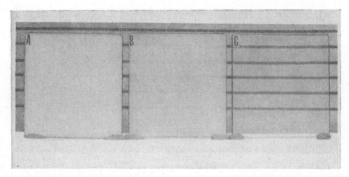

Fig. 1 Optical characteristics of polyolefins. (a) Low density polyethylene; (b) Ethylene–methacrylic acid copolymer (3·5 mole percent acid); (c) Sodium ionomer of (b). Slabs 6 × 6 × 0·120 in (Du Pont Company).

materials. However, this is not the case; ionomers are almost as crystalline as low density polyethylene. The explanation for the transparency will be dealt with below.

There are other, less visible qualities that ionomers have which are of great importance in commercial applications; in particular, toughness and melt strength are notable. Both of these are illustrated in Fig. 2. The hot, molten film of the ionomer was drawn down by vacuum over the sharp edges of the metal nail. The melt strength is necessary to allow the plastic film to be drawn down without tearing; the toughness guarantees resistance of the film to puncturing. The melt strength of a polymer is not determined solely by the viscosity of the material (which in turn varies with weight average molecular weight [4]) or by the elasticity of the molten resin (which varies with higher molecular weight averages [5]) but reflects also a quality in the chemical composition. The distinction between ionomers and polyethylene will be made clear in due course. Another packaging application

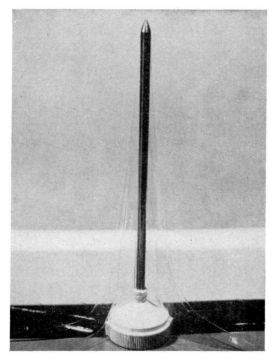

Fig. 2 Sample of ionomer film vacuum drawn over 4 in. nail (Du Pont Company).

where this melt strength is of importance is in the sealing together of plastic films under heat and pressure to form a pouch. Polyethylene is inadequate in many instances because the tension between the two layers causes the seal to tear apart before the fused polyethylene has had time to crystallise. The high melt strength of the ionomer reduces this hazard and its broader melting range compared to polyethylene permits more flexibility in the heating and cooling cycle.

An important recent application where the toughness and clarity are of importance is as a surface coating for glass bottles. A thin layer of the ionomer (0.004 in) is applied to the hot glass bottle from a fluidised bed of the powdered ionomer. The film of plastic forms a continuous bag around the bottle. This bag has the remarkable property of retaining more than 95 percent of the shards and fragments when the bottle is dropped and broken.

Compared to other packaging films, ionomers have a unique balance of oil and grease resistance and permeability which has made them important in the meat packaging industry.

2.2 PREPARATION AND CHARACTERISATION

Ionomers are generally prepared from copolymers having either methacrylic or acrylic acid as one component. At the present time, the other component has been mostly either ethylene[2,3,5,9,14,25,54] or styrene.[17,75,82]

However, other ionomers have been made by specific methods. Phillips and MacKnight[83] prepared an ethylene–phosphonic acid copolymer by treating polyethylene with phosphorus trichloride:[58,59,83]

$$-RCH_2CH_2- + PCl_3 + \tfrac{1}{2}O_2 \longrightarrow -RCHCH_2- + HCl$$
$$\underset{POCl_2}{|}$$

$$\xrightarrow{H_2O} -RCHCH_2- + HCl$$
$$\underset{PO(OH)_2}{|}$$

This was treated with dilute base (CsOH) to form the ionomer. An ionomer based on the polyoxymethylene backbone has been described by Wissbrun.[80] A copolymer of trioxane and epichlorhydrin was treated with disodium thioglycolate. Excess thioglycolate was used to avoid hydrolysis of the chlorine.

$$-CH_2-O-CH-CH_2-O-CH_2-O- \; + \; -SCH_2CO_2^- Na^+$$
$$|$$
$$CH_2Cl$$
$$\rightarrow -CH_2-O-CH-CH_2-O-CH_2-O-$$
$$|$$
$$SCH_2CO_2^- Na^+$$

The principal concern here is with ionomers based on the polyethylene backbone. These copolymers can be made by a substantially similar process to that used to manufacture low density polyethylene. At the levels of methacrylic acid typical of the commercial ionomers, *i.e.* up to about six mole percent acid, the copolymers resemble low density polyethylene in many respects. They are flexible, tough and exhibit spherulitic crystallinity typical of low density polyethylene, but with a melting point several degrees lower than that of polyethylene, *i.e.* about 105°C. Consequently, there is present in these copolymers short and long chain branching to about the same degree as in low density polyethylene. Butyl short chain branches are present to the same extent as the pendant carboxyl groups and the long chain branch points are ten times less frequent. Another consequence is that the ionomers have very broad molecular weight distributions, weight average to number average ratios of 10 to 20 are characteristic.

The composition of the copolymer is an important consideration, as will become apparent below in discussing the structure of the ionomers. There are two aspects to the question, the uniformity of the composition of the copolymer between chains and the distribution of methacrylic acid groups along the chains.

Methacrylic acid is a much more reactive monomer than ethylene. Q and e for the acid are 2·34 and 0·65, and 0·015 and $-0·20$ for ethylene.[7] Therefore, even at the low (10 percent) conversions typical of the high pressure, continuous processes, it might be expected that the product would consist of a mixture of a copolymer rich in methacrylic acid with ethylene homopolymer. However, numerous attempts to fractionate the copolymers as made commercially have succeeded only in fractionating by molecular weight and no variation in the chemical composition of the fractions has been found.

The copolymers are initially, at low levels of methacrylic acid, flexible materials resembling low density polyethylene. As the acid content is increased, they become increasingly transparent (around six mole percent acid) and around eight to ten mole percent acid are quite transparent and leathery in appearance. Above ten mole percent they are brittle, glassy solids.

Ionomers made from copolymers containing more than six mole percent acid are not available commercially at the present time.

Although a description of the commercial synthesis of ionomers is not available, Rees[3] has described various procedures. However, they can conveniently be prepared in the laboratory by several methods. The copolymer is fluxed on a heated two roll rubber mill between 150°C and 200°C and then, in the case of alkali metals, a pre-calculated amount of the hydroxide is added carefully dropwise as a ten percent solution in water. Care must be exercised to avoid spattering of the strong alkali solution (safety goggles are essential) otherwise loss of metal will occur. It is found that the alkali is readily absorbed into the copolymer, which becomes cloudy, and there are steam bubbles in the melt. A rapid increase in viscosity is observed. When all of the water has been driven off, the molten plastic suddenly becomes quite clear and can be stripped off the rolls and sheeted out to form a transparent, tough, flexible sheet.

This method is particularly convenient for the preparation of ionomers differing in degrees of neutralisation and also when the product is to be moulded in bars or sheets. If a mill is not available, then the method of MacKnight, *et al.* is satisfactory.[9a] The starting material was an ionomer of undetermined degree of neutralisation. The plastic was ground to a fine powder and refluxed in tetrahydrofuran with dilute hydrochloric acid to produce the un-ionised copolymer. (The time was not given, but 12 hours would be adequate.) The copolymer was then refluxed in tetrahydrofuran containing various amounts of sodium hydroxide for various times until different degrees of neutralisation were obtained. The mixture was then added to excess methanol–water mixture (proportions not given) and washed and dried *in vacuo* at 100°C. The degree of neutralisation was was determined by infrared analysis using the 1700 cm^{-1} carbonyl band.[52]

One of the difficulties in characterising ionomers is their insolubility in many common organic solvents.[2b] The parent copolymers of low acid content (less than 3 mole percent) are soluble in solvents typical of polyethylene, such as xylene, toluene, 1-chloronaphthalene. With increasing acid content, more polar compounds are necessary, such as diglyme (diethylene glycol dimethyl ether) and dimethyl formamide. Generally satisfactory are alcohols such as butanol or amyl alcohol. Suitable solvents for the commercial ionomers are 60/40 mixtures of toluene/isobutyl alcohol or methyl isobutyl ketone and perchloroethylene/*n*-butyl alcohol. These will maintain the ionomer in solution at temperatures above 70°C. A mixed solvent free from hydroxyl groups, suitable for infra-red spectroscopy is decahydronaphthalene/diethylene glycol dimethyl ether/dimethyl sulphoxide in the ratio 70/20/10 by volume.

In suitable solvents the acid groups can be titrated.[10,54] Thus, with hot 75/25 toluene/butanol as a solvent, and starting with a partially neutralised copolymer, the remaining acid groups can be titrated with sodium hydroxide using phenolphthalein as indicator. When completely neutralised, the total acid content can then be determined by titration with hydrochloric acid using thymol blue as indicator.

This method allows one to characterise the ionomer with respect to metal content and degree of neutralisation. Determination of molecular weight is not straightforward with the ionomer itself. A satisfactory procedure would be to remove the metal as described by refluxing with hydrochloric acid and then to treat the copolymer by methods suitable for low density polyethylene, *i.e.* xylene or perchloroethylene may be used as a solvent for the determination of intrinsic viscosity, for example.[11]

Ionomers of styrene copolymers can be prepared by similar techniques.[38,70] Solutions of the copolymer in benzene or benzene–methanol mixtures could be titrated with sodium hydroxide in methanol with phenolphthalein as indicator.

2.3 PHYSICAL AND MECHANICAL PROPERTIES

The changes in physical and mechanical properties on ionisation of ethylene based acid copolymers have been presented by Rees and Vaughan[2b] and Bonotto *et al.*[6a,b,c]

TABLE 1
PROPERTIES OF ETHYLENE–METHACRYLIC ACID COPOLYMERS[2b]

% Acid (molal)	Stiffness (psi)	Yield point (psi)	Ultimate Tensile strength (psi)	Ultimate elongation %
1·4	10 000	950	2 560	610
3·5	10 000	890	3 400	550
5·9	16 500	1 400	5 000	500
13·0	161 000	5 500	5 000	260

TABLE 2

PROPERTIES OF ETHYLENE–ACRYLIC ACID COPOLYMER SALTS. MONOVALENT IONS

	% neutralised[a]	Melt index (43·25 psi), dg min^{-1}	Melt index (432·5 psi), dg min^{-1}	Secant modulus 1% extension, psi[c]	Tensile strength (ultimate), psi[c]	% elongation at break	Density, g/cc
Control–starting material 14·8% acrylic acid wt	0	67	2·570	7 000	2 150	470	0·949
Sodium salt	12·0	12·2	256	33 400	3 150	420	0·9568
	30·0	3·9	92	48 600	4 000	330	0·9586
	47·5	1·0	30	42 500	4 600	310	0·9603
	66·0	0·3	7·6	39 700	4 800	280	0·9633
Potassium salt	8·0	16·3	360	29 300	3 050	470	0·9588
	25·0	4·5	110	52 600	3 700	410	0·9626
	51·0	2·7	49	49 500	4 450	370	0·9684
	63·0	0·57	15	44 800	5 000	390	0·9750
Lithium salt	12·0	18·7	442	26 300	3 150	410	0·9516
	28·5	5·2	116	48 900	3 850	350	0·9510
	52·5	1·4	38	48 500	4 100	260	0·9493
	67·5	0·2	5·4	36 900	4 600	250	0·9446

[a] Infra-red measurements.
[b] Density 23°/4° (ASTM-D-1505-57T).
[c] 1 psi = 6·9 × 10^3 N m^{-2}.

PROPERTIES OF AN IONOMER, POLYETHYLENES AND PLASTICISED VINYLS

Properties	Ionomer HXQD-2137	Polyethylene		Plasticised PVC	
		Low density	High density	Low Durometer[a]	Medium Durometer[b]
Melt index, 44 psi (190°C) (D1238-52T)	1·6	1·5	5·0	0·2	0·6
Density, 23°/4°C. (D1505-57T), g/cm³	0·955	0·920	0·960	1·273	1·32
Hardness, Durometer 'D' (D1484-57T)	61	45	65	47	74
Secant modulus of elasticity at 1% strain (D882-61T), psi	50 000	23 000	150 000	5 000	118 000
Tensile strength (D882-61T), psi	3 600	1 800	4 600	3 300	3 700
Yield strength (D882-61T), psi	1 925[c]	1 200	4 200	None	None
Elongation (D882-61T), %	425	600	15% @ yld.	210	120
Brittleness index, 80% OK (D746-55T)	−105°C	−80°C	−70°C	−20°C	+10°C
Tensile impact (23°C) (D1822-61T), ft-lb/in³	413	388	54	425	68
Stress-crack resistance (50°C, Igepal) (D1693-60T)	No failures, 500 hr	50% failures, 24 hr	—	No failures, 500 hr	—
Light transmission, 20-mil plaque (D1003-61), %	88·7	84·7	66·9	85·1	86·1
Haze, 20-mil plaque (D1003-61), %	5·3	56·7	99·5	5·8	11·6
Abrasion resistance, (Taber H-22 wheel), g lost/1000 cycles	0·011	0·018	0·006	0·032	0·053
Grease resistance[d] (JAN. B-121-12; 1953)	No failures, 24 hr	100% fail, 24 hr	No failures, 24 hr	100% fail, 24 hr	—
Cold shrinkage (D955-51), %	1·0	1–3·5	2·0–4·0	1·0–5·0	—
Specific heat, cal/°C/g	0·61	0·63	0·54	—	—

[a] Union Carbide QYTZ, 26% dioctyl phthalate plus stabilisers.
[b] Union Carbide QYST, 16% dioctyl phthalate plus stabilisers.
[c] Not a true yield but point of inflection.
[d] Time required for a solution of turpentine, grease, and red dye to go through a 5 to 10 mil thick specimen at 23°C.

The changes in modulus (stiffness) with increasing acid content are given in Table 1[2b] for a series of ethylene–methacrylic acid copolymers. These polymers were all of approximately the same molecular weight and since the crystallinity decreases with increasing copolymerisation, the increase in stiffness is due to an increase in the amorphous T_g. There is a substantial increase in tensile strength but a decrease in elongation.

The effect of degree of neutralisation is shown in Table 2[6c] for copolymers of ethylene with acrylic acid neutralised with various alkali metals. The modulus increases with degree of neutralisation up to about 30% and then a plateau is reached; and in some cases, the toughness decreases somewhat.

A useful comparison between an ionomer, low and high density polyethylene, and some plasticised PVC compounds is given in Table 3[6a] for mechanical properties and in Table 4[6a] for electrical properties.

TABLE 4

ELECTRICAL PROPERTIES OF AN IONOMER, LOW DENSITY POLYETHYLENE, AND PLASTICISED PVC

Properties	Ionomer HXQD-2137	Low density polyethylene	Plasticised PVC
Dielectric constant (D150-54T)			
60 cycles	2·56	2·33	5·85
10^6 cycles	2·48	2·33	3·25
Dissipation factor (D150-54T)			
60 cycles	0·002 5	0·000 4	0·100 0
10^6 cycles	0·004 0	0·000 2	0·059 0
Volume resistivity at 23°C, 50% RH (D257-58), megohm-cm.	$4·070 \times 10^{10}$	$17·2 \times 10^{10}$	$0·002\ 9 \times 10^{10}$
Dielectric strength[a] (D149-55T), short time, v./mil	510	550	300–400

[a] Measured on ⅛-in. thick plaque.

These differences in the mechanical properties of ionomers, coupled with the transparency, stimulated investigations of the structural changes accompanying neutralisation. The nature of the crystallinity present, the breakdown of spherulitic structure and the location of the metal ions turned out to be problems requiring a variety of techniques for their solution. The molecular structures which are consistent with the evidence have turned out to be much more complicated than was anticipated, and in the succeeding sections the results obtained with these techniques will be examined in some detail and the various solutions to the structural problems examined and compared.

2.4 ELECTRON MICROSCOPY

As shown in Fig. 1, ionomers exhibit a high degree of clarity compared to polyethylene. Preliminary examination of the surfaces of an ionomer and the parent copolymer showed that this clarity was due to an absence of spherulite formation rather than crystallinity of the polyethylene chains.[2] A more extensive examination was made by Davis et al.[15] and a summary of their findings follows.

The absence of spherulitic structure is demonstrated most clearly by comparing surfaces of the copolymer and the ionomer derived from it. The surface structure was developed by first swelling in xylene vapour and then drying and annealing. The surfaces were replicated by condensing chromium vapour on to the surface at a 30° angle and stripping off with polyacrylic acid. The surface of the copolymer is typical of polyethylene (Fig. 3), the edges of lamellae can clearly be seen and these lamellae run down into larger, globular spherulites. By contrast (Fig. 4), the ionomer is seen to be free from spherulitic structure of any kind. There is no regularity on the surface, only a random, grainy structure can be distinguished.

A better insight into the morphology of ionomers was gained from an examination of thin films prepared by casting from solution.

The technique used was to prepare a dilute (0·1%) solution of the ionomer or the copolymer and to place a few drops of this solution on to the surface of either hot water (80°C) or dilute (2%) base. The base was chosen from rubidium, caesium or sodium hydroxides. The caesium or rubidium hydroxides were used to increase the contrast in the electron microscope. The thin film (0·1 micron) was picked off the surface of the water or base directly on to the grid and put into the microscope and examined by transmission.

The cast films exhibited structure typical of copolymer or ionomer depending on whether they had been cast on hot water or base, regardless of whether the starting solution was a copolymer in xylene or an ionomer in one of the mixed solvents described above. Hence, because the solutions of the copolymers were more readily prepared, the cast ionomer films were prepared by adding solutions of the copolymers to hot solutions of base.

(a)

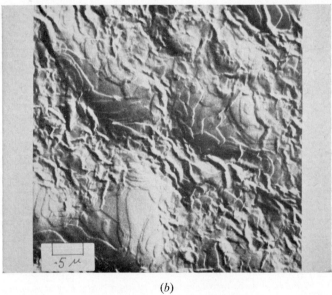

(b)

Fig. 3 Surface of ethylene–methacrylic acid copolymer (3·5 mole % acid) at low (a) and high (b) magnification.[15]

Fig. 4   Surface of a sodium ionomer, copolymer as in Fig. 3.[15]

Fig. 5   Film of low density polyethylene, thickness 0.1 μ.[15]

Fig. 6   Film (0·1 μ) of ethylene–methacrylic acid copolymer (0·5 mole % acid) cast from solution in xylene on to water.[15]

One series of experiments was designed to show the effect of variation in the acid content of the copolymer on the structure of the copolymer and the corresponding ionomer.

Figure 5 shows the structure of a film of low density polyethylene cast on to dilute rubidium hydroxide solution from a 0·1% solution in xylene. The initiating sites for the growth of two-dimensional spherulites can clearly be seen. The fine structure of the spherulites is also visible, rather more clearly than usual; ribbons of lamellae can be seen radiating out from the core of each spherulite to give the characteristic 'wheatsheaf' pattern.

In Figs. 6 and 7 are contrasted the behaviour of an ethylene–methacrylic acid copolymer containing 0·5 mole percent acid when cast on to either water or dilute rubidium hydroxide. The structure of the film cast on to water is essentially similar to the film of polyethylene, exhibiting spherulitic structure. In the case of the film cast on to base, there can still be distinguished the boundaries of spherulitic growth, but the ribbons of lamellae are poorly formed. There also appear dark regions, away from the centre of the spherulites, indicative of some sort of agglomeration.

The distinction between the copolymer film and the ionomer is more marked with a two mole percent methacrylic acid copolymer, as shown in Figs. 8 and 9. The typical spherulitic development is seen for the copolymer

THERMOPLASTIC IONIC POLYMERS: IONOMERS

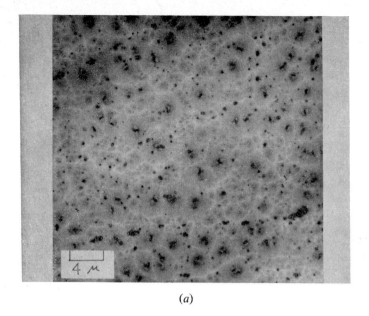

(a)

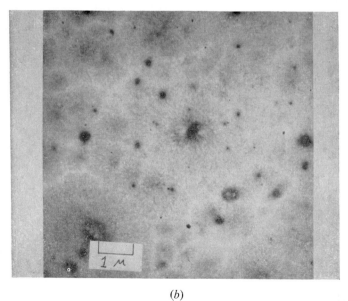

(b)

Fig. 7 Film similar to Fig. 6 cast on to dilute RbOH solution. (a) low; (b) high magnification.[15]

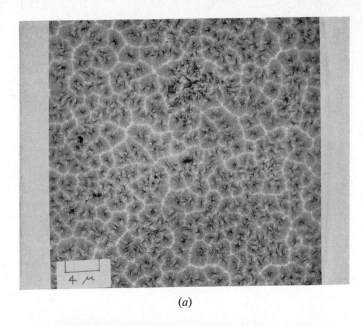

(a)

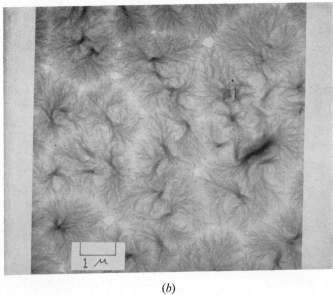

(b)

Fig. 8   Film (0·1 μ) of ethylene–methacrylic acid copolymer (2·0 mole % acid) cast on to water at (a) low and (b) high magnification.[15]

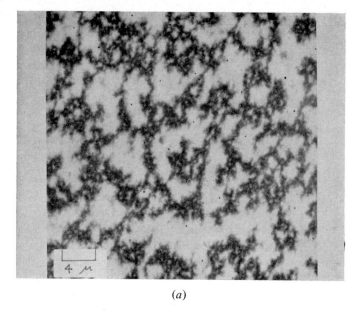

(a)

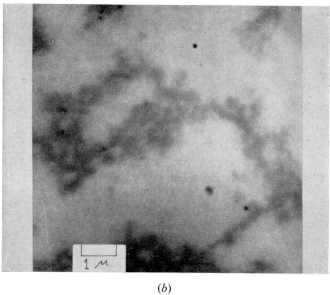

(b)

Fig. 9   Film similar to Fig. 8, cast on to RbOH solution at (a) low and (b) high magnification.[15]

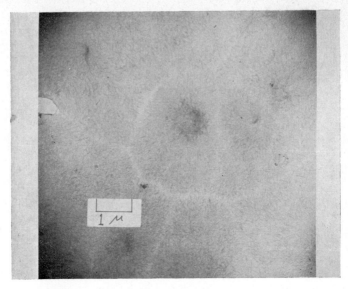

Fig. 10 Film (0·1 μ) of ethylene–methacrylic acid copolymer (3·5 mole % acid) cast on to water.[15]

but the ionomer film is characterised by more or less randomly arranged fragments of lamellae with irregular lumps of aggregated material, in which are located the ions. Similarly, for a copolymer containing 3·5 mole percent methacrylic acid the contrasting behaviour in Figs. 10 and 11 is found. In the case of the copolymer, the spherulitic structure is less well-developed; but in the ionomer film, there is no trace of spherulitic or any long-range morphological feature characteristic of polyethylene present. There is present, instead, an irregular, granular structure, the grains being about 150 Å in diameter. The visibility of the grains is due to fluctuations in the electron density throughout the film; and hence, represents regions of high concentration of ions.

Finally, in this series, Figs. 12 and 13 show the results for a copolymer and ionomer containing six mole percent methacrylic acid. In spite of the rather high level of comonomer, there are still lamellar and spherulitic structures, but the ionomer film is seen to consist of agglomerates of smaller regions or grains of ionic material.

The overall impression from this series of experiments is that for the ethylene–methacrylic acid copolymers, spherulitic morphology persists to quite high levels of copolymerised acid, but that when this acid is converted to the salt form, the ionic centres along the chains aggregate to form grains

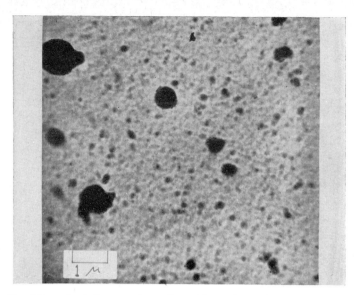

Fig. 11 Film of ethylene–methacrylic acid copolymer (3·5 mole % acid) cast on to RbOH solution.[15]

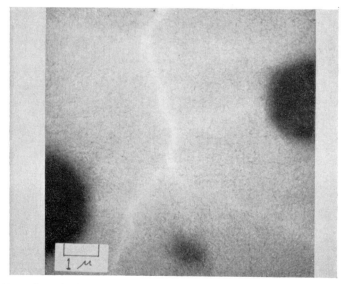

Fig. 12 Film of ethylene–methacrylic acid copolymer (6·0 mole % acid) cast on to water.[15]

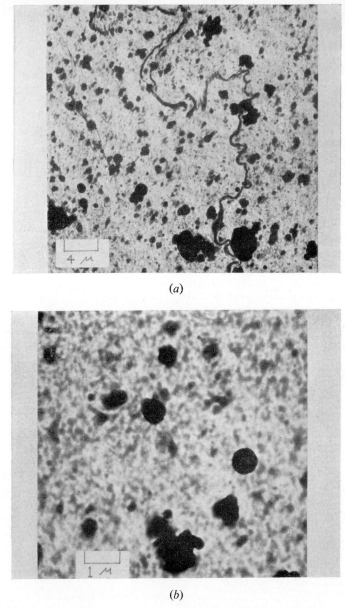

Fig. 13 Film of ethylene–methacrylic acid (6·0 mole % acid) cast on to RbOH solution at (a) low; (b) high magnification.[15]

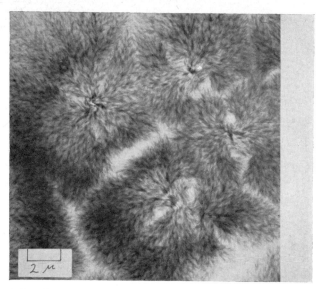

Fig. 14 Film of an ionomer, sodium salt of ethylene–methacrylic acid copolymer (1·68 mole % acid) cast from solution on to water.[15]

or regions about 100–1000 Å in diameter, thus destroying the polyethylene-like structure. The conversion to the ionomer type of structure is complete with about 3 mole percent copolymerised acid.

In the preceding experiments, the films were prepared by casting from solutions of the ethylene–methacrylic acid copolymer. In Figs 14 through 17, the results are shown of casting from solutions of sodium ionomers in DMSO/tetralin/diglyme mixture. Figures 14 and 15 are of film made from a solution of a sodium ionomer made from a 1·68 mole percent methacrylic acid copolymer by casting on to water and dilute sodium hydroxide respectively. Similarly, Figs. 16 and 17 are for the ionomer from a 5·4 percent acid copolymer. When cast on to water, the ionic structure is removed and the film is typical of the un-ionised copolymer.

The method of preparation of these films, by evaporation of a dilute solution, raises doubts that the observed granular structure of the ionomers is typical of the bulk material. To answer this question, thin sections were cut from solid material and examined in various ways.

Figure 18a shows a section of a copolymer containing 3·5 mole percent methacrylic acid which was neutralised by immersion in dilute rubidium hydroxide solution for 3 hours at 70°C. This is a low magnification view, which shows tearing and formation of holes along the spherulite boun-

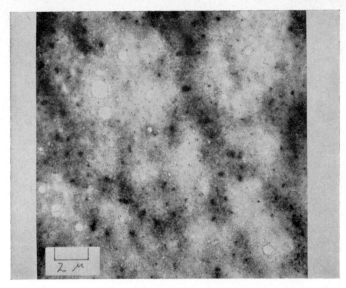

Fig. 15 Film of an ionomer, sodium salt of ethylene–methacrylic acid copolymer (1·68 mole % acid) cast from solution on to dilute NaOH.[15]

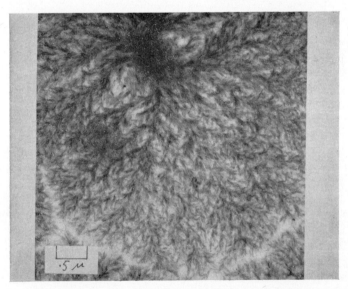

Fig. 16 Film of an ionomer, sodium salt of an ethylene–methacrylic acid copolymer (5·4 mole % acid) cast from solution on to water.[15]

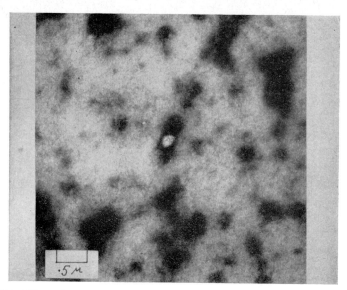

Fig. 17 Film of an ionomer, sodium salt of an ethylene–methacrylic acid copolymer (5·4 mole % acid) cast from solution on to dilute NaOH.[15]

daries. Within the spherulites themselves, the typical structure is almost completely destroyed, and converted to the granular structure which is shown at high magnification in Fig. 18b.

Because of the rather low stiffness of the copolymer, the above section was prepared by embedding the copolymer in a casting syrup of 50/50 methyl methacrylate/styrene. With the ionomers even this was rather difficult, but a section could be cut from a carefully dried ionomer prepared from a 1·75 mole percent acid copolymer. This is shown at fairly high magnification in Fig. 19. Thus, the granular surface shown in Fig. 4 is representative of the structure of the interior of the sample.

This electron microscopic study shows clear differences in morphology between low density polyethylene and acid copolymers on the one hand and ionomers on the other. It leaves open the question of the distribution of the ions within the ionic granules or domains, but there is little doubt as to their formation. Another question which these results do not answer is the location of the carboxyl groups of the copolymer. Even when the copolymer contains several mole percent acid, it still resembles low density polyethylene and there is no evidence that shows the carboxyl groups to be segregated outside of the lamellae. And yet, under the influence of the ions, below the melting point of the crystals, such segregation can take place.

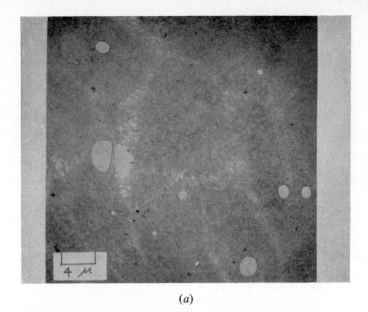

(a)

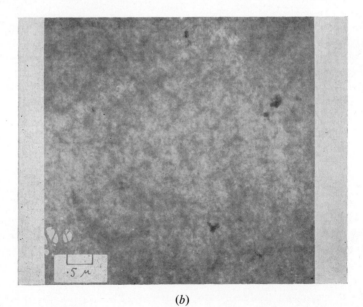

(b)

Fig. 18 Section of an ethylene–methacrylic acid copolymer (3·5 mole % acid) treated with dilute RbOH solution (a) low, (b) high magnification.[15]

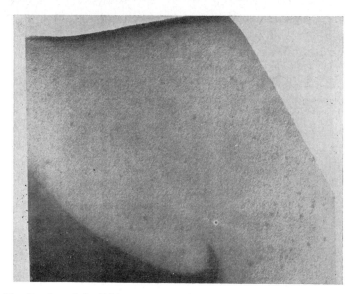

Fig. 19   Section of the lithium ionomer of an ethylene–methacrylic acid copolymer (1·75 mole % acid).

Other electron microscopic studies of ionomeric polymers have not been as detailed, but the formation of granular ionic material has been observed. Films of ethylene type ionomers containing phosphonic acid groups were prepared in a similar fashion to the above by treating sections of the polymer with caesium acetate solution.[59]

An examination of ionomers from butadiene–methacrylic acid copolymers was made by Marx *et al.*[48] The granular size was found to be considerably smaller than for the ethylene ionomers; ionic domain diameters were about 26 Å and 13 Å.

2.5 RHEOLOGICAL PROPERTIES

2.5.1 Steady-state viscosity at low shear stresses

This section is concerned with the effect of ionisation on the low shear stress, or Newtonian, viscosity.

It is first necessary to consider the effect of hydrogen bonding, carboxyl–carboxyl association particularly, on the melt viscosity of hydrocarbon polymers. The question which is difficult to resolve is whether the hydrogen bonds can best be described as time-dependent cross-links which are regarded as extra entanglements or as decreasing the segmental mobility which can be described by free volume considerations. In the case of polyamides, for example, the viscosity of 66 nylon at 280°C is about three times that of linear polyethylene of similar weight average molecular weight. It is known that polyamides are extensively hydrogen bonded even at these elevated temperatures[21] and it seems unlikely that the polyamide chain is stiff enough to account for the difference. Longworth and Morawetz[16] examined the melt viscosities of copolymers of styrene and methacrylic acid and a typical result was that the viscosity of a copolymer containing about ten mole percent acid was about twice that of the corresponding homopolymer of styrene. They interpreted their results by

$$\begin{array}{c} | \\ \text{C} \\ \text{O} \diagup \diagdown \text{O} \\ | \quad\quad | \\ \text{H} \quad\quad \text{H} \\ | \quad\quad | \\ \text{O} \diagdown \diagup \text{O} \\ \text{C} \\ | \end{array}$$

saying that a carboxyl dimer functioned as a temporary cross-link and that segmental flow occurred only via those segments which were not hydrogen bonded at any given moment. This conclusion was disagreed with by Fitzgerald and Nielsen[17] who argued that, since the viscosities were measured between 120°C and 160°C, then the increased viscosities were due to the decreased free volume caused by the increase in glass temperature (the glass temperature, T_g, of polystyrene is about 100°C; that of poly(methacrylic acid) 200°C).

The effect of hydrogen bonding on the viscosities of copolymers of ethylene with acrylic and methacrylic acids was made by Blyler and Haas.[18] These authors were not able to compare the viscosities of the copolymers (containing up to eight mole percent acid) with polyethylenes of similar molecular weight, but showed that there was an increase in the activation energy for viscous flow (at constant stress) from about 12 kcal mole^{-1} (typical of branched polyethylene) to about 17 kcal mole^{-1} when the amount of acid was eight mole percent. Since the increase activation energy for viscous flow of branched polyethylene over linear polyethylene (6–7 kcal mole^{-1}) is thought to be caused by the presence of long chain branches, then they argued that the effect of the carboxyl groups was simply to increase the number of such branches. At higher stresses, the cross-links are no longer effective and there is a decrease in activation energy.

Convincing evidence that the hydrogen bonds can function as cross-links was provided by an experiment in which the carboxyl groups of a copolymer containing 5·3 mole percent acrylic acid was esterified by diazomethane. The viscosities of the two copolymers are shown in Fig. 20

$$\sim\sim\sim\underset{\text{COOH}}{|} + CH_2N_2 \longrightarrow \sim\sim\sim\underset{\text{COOCH}_3}{|} + N_2$$

at 130°C and 160°C. The control sample has a viscosity at low shear rates which is about three times that of the esterified sample. In addition, the activation energy for viscous flow of the esterified sample is only 10 kcal mole^{-1}, similar to that of branched polyethylene.

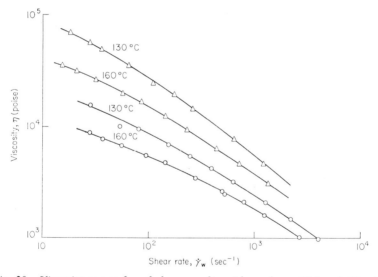

Fig. 20 Viscosity curves for ethylene–acrylic acid copolymer (5·3 mole % acid) and 85% methyl ester at 130°C and 160°C: (△) EAA—5·3%, control; (○) EAA—5·3%, 85% ester.[18]

The melt viscosities at 190°C of a series of ethylene–methacrylic acid copolymers which included several ethylene homopolymers, and up to fifteen mole percent methacrylic acid which was synthesised as nearly as possible under similar conditions were measured by Pieski.[19] The weight average ($\overline{M}_w$) molecular weights of the homo- and copolymers were measured by light scattering. A multiple regression analysis of the viscosities with respect to molecular weight and methacrylic acid content gave the equation

$$\log \eta_0 = 3 \cdot 23 \log \overline{M}_w + 0 \cdot 033 \, (\% \text{ MAA}) - 15 \cdot 5$$

where the acid content is in weight percent. The influence of acid on the melt viscosity is small compared to the influence of molecular weight but nonetheless significant.

The low shear stress viscosities of ionomers were measured by Longworth and Vaughan.[20] A series of ionomers was prepared by partially or completely neutralising ethylene–methacrylic acid copolymers containing up to eighteen mole percent acid. The viscosities were measured at 120°C and 140°C at shear stresses of $4 \cdot 0 \times 10^3$ dynes/cm^2 ($4 \cdot 0 \times 10^2$ Nm^{-2}) in a cone and plate rheometer. Since the viscosity of the partially neutralised copolymer included a contribution from the free acid, the measured viscosity of the ionomer was corrected for this. The values of the relative viscosity, η_r, which is the ratio of the viscosity of the ionomer to that of the ethylene homopolymer, are shown in Fig. 21 as a function of the mole

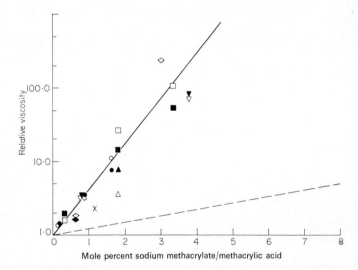

Fig. 21 Increase in viscosity relative to polyethylene of ethylene–methacrylic acid ionomers (Na). Mole % MAA: (○) 1·7; (◇) 3·1; (□) 3·5; (×) 5·85; (▽) 7·5; (△) 17·8. Solid symbols, 120°C; open symbols, 140°C; – – – line for un-ionised copolymers.[20]

percent of the sodium salt of methacrylic acid. The curve for the relative viscosities of the un-ionised copolymers is shown for comparison. This figure illustrates the dramatic effect that rather small amounts of neutralisation can have on the melt viscosity. Thus, two mole percent methacrylic acid will increase the viscosity by about half, whereas a similar percentage of the sodium salt results in about a twentyfold increase. Recollecting that these materials are truly thermoplastic, soluble linear polymers, this great increase in viscosity is a measure of the powerful intermolecular attraction between the ionised residues. This type of behaviour can be characterised as an 'internal viscosity' and a more quantitative measure is presented in the discussion of viscoelastic properties.

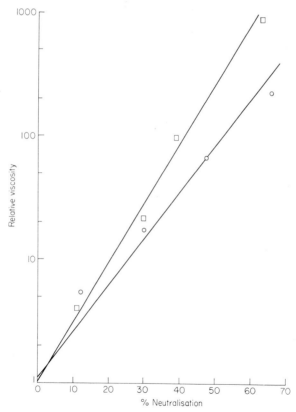

Fig. 22 Relative increase in viscosity with increasing neutralisation of ethylene–acrylic acid copolymers at 190°C. (○) sodium salts; (□) calcium salts; (Data from Ref. 6c).

Similar results were obtained by Bonnotto and Bonner[60] as shown in Fig. 22, which illustrates the relative change in melt viscosity with increasing neutralisation of a 14·8 percent (by weight) ethylene–acrylic acid copolymer with both sodium and calcium. In this case the viscosity of the parent homopolymer was not given, but the increase in viscosity for sodium salts is in line with the increase reported in Fig. 21. Calcium is rather more effective than sodium, and even in this case the ions evidently do not form permanent cross-links. However, this result must be accepted with caution. A problem in the preparation of ionomers from bases other than the alkali metals is the low solubility of the hydroxides of other metals, particularly alkaline earth metals. The hot copolymer can absorb a considerable amount of water during the preparation of the ionomer, but it can readily occur that the metal is present as an insoluble hydroxide or oxide and consequently will be present as a sort of partially reacted filler. In this case it is difficult to be sure that the metal has reacted stoichiometrically with the carboxyl groups of the copolymer. If the reaction is incomplete then the situation in the final product is that there are microscopic particles or agglomerates of metal hydroxide partially reacted with the acid groups of the polymer. These are much less effective at restricting flow than the molecularly dispersed intermolecular links formed by the alkali metal ions. Since the salts of divalent metals form gels in other systems (*e.g.* calcium methacrylate copolymerised with methyl methacrylate[84]) one might expect that the zero stress viscosity of the ionomer would become too high to be measurable. That this is not the case leads to the conclusion that the metal salts are not molecularly dispersed throughout the hydrocarbon matrix. Some unpublished results by the author for barium ionomers support this conclusion for calcium ionomers (barium is convenient for the preparation of ionomers from divalent metals because its hydroxide is appreciably more soluble in water than calcium).

The temperature dependence of the viscosities of the polymers shown in Fig. 21 was measured between 120°C and 180°C. Values of the activation energy for viscous flow, ΔE_{visc}, were calculated from the equation

$$\Delta E_{\text{visc}} = d \ln \eta_T / d(1/T)$$

with T in °K, and are shown in Fig. 23 for copolymers and fully neutralised sodium ionomers. These activation energies are similar to those reported by Rees and Vaughan,[2b] Blyler and Haas[18] and Sakamoto *et al.*[13] It is somewhat surprising that there is very little increase on neutralising the individual copolymers, at least in the temperature range studied, in view of

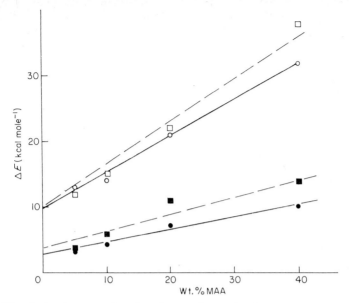

Fig. 23 Activation energies for viscous flow of ethylene–methacrylic acid copolymers and fully neutralised ionomers. (○) copolymers; (□) ionomers. Open symbols, 10^0 sec^{-1}; closed symbols 10^4 sec^{-1}.[20]

the much stronger attractions between the chains of the ionomers than the copolymers.

This observation led to the consideration that the temperature dependence of the viscosity of these materials at these temperatures is governed by changes in the free volume rather than by changes in specific chemical interactions. The data were treated by the method of reduced variables, otherwise known as the WLF (Williams, Landel and Ferry) method.[22] The method consists in calculating a 'reduced' viscosity, a_T, given by

$$a_T = \frac{\eta_T}{\eta_S} \cdot \frac{T_S}{T}$$

where η_T is the viscosity at temperature T and η_S is the viscosity at an arbitrary reference temperature T_S. The ratio T/T_S, in °K, is an entropy correction and is usually rather small (for a discussion of the theoretical background, see Ref. 23). The theory then gives the temperature dependence of a_T as

$$\log a_T = \frac{-A(T - T_S)}{B + (T - T_S)}$$

where A and B are constants and one material differs in its behaviour from another only in the choice of T_S, the values of A and B being the same. Furthermore,

$$T_S = T_g + T_0$$

where T_g is the glass temperature and T_0 will be the same for the group of materials under examination. This equation predicts a very large, non-linear, increase in the activation energy with decreasing temperature as T_g is approached.

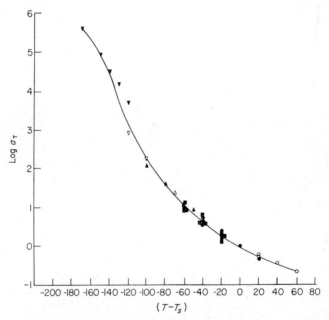

Fig. 24 WLF plot for ethylene–methacrylic acid copolymers and sodium ionomers at temperatures from 120°C to 180°C. Mole % acid: (○) 1·7; (□) 3·5; (△) 7·5; (▽) 17·8. Open symbols, copolymers; closed symbols, ionomers.[20]

The results of applying the equation to the viscosity data on the copolymers and ionomers of Fig. 21 are shown in Fig. 24.[20] The reference conditions were those of a 1·7 mole percent acid copolymer at 120°C (T_S). The values of A and B were found to be $A = 3\cdot33$, $B = 243\cdot0$. The values of T_S

TABLE 5
REFERENCE TEMPERATURES (T_s) OF ETHYLENE–METHACRYLIC ACID COPOLYMERS AND IONOMERS

% MAA		T_S (°C) at				
wt	mole	0	10	50	95	% Neutralisation
5	1·7	120	160	160	180	
10	3·5	160	180	180	180	
20	7·5	190	250	220		
40	17·5	240	290	280		

are given in Table 5. The changes in T_S with composition are quite substantial. The T_g of low density polyethylene is about -40°C (the temperature of the β transition[36]), and the effect of copolymerisation with about 2 mole percent methacrylic acid will be to raise this by about ten to twenty degrees, and so T_0 (i.e. $T_S - T_g$) is about 140°C. Thus, the T_g of a 3·5 mole percent copolymer is around room temperature, and copolymers in this composition range are rather leathery in feel. Similarly, for a 7·5 mole percent acid copolymer a glass temperature of about 50°C is predicted, corresponding to a brittle glassy polymer and such is found to be the case. The effect of neutralisation is to raise the value of T_S by about 40°C. This is about what would be expected, since the T_g of polyacrylic acid is raised by 200°C on neutralisation,[24,54] and a similar increase is to be expected for polymethacrylic acid and these data only cover compositions up to 18 mole percent. Thus the small differences between the activation energies for viscous flow of copolymers and ionomers can be accounted for by the small differences in T_g.

2.5.2 Non-Newtonian viscosities

As the shear stress is increased, the viscosity of an ionomer decreases, in common with all high polymers. Typical behaviour is shown in Fig. 25, which shows the change in viscosity with increase in degree of neutralisation and shear rate.[20] From the data points the parameters of the empirical equation

$$\log (\eta_{(\dot\gamma)}/\eta_0) = Q(\log \dot\gamma\tau)^2 \quad (\dot\gamma \geq 1/\tau)$$

were calculated.[26] η_0 is the zero stress, or Newtonian, viscosity; $\eta_{(\dot\gamma)}$ is the shear-rate dependent non-Newtonian viscosity at shear rate $\dot\gamma$; τ has the dimensions of time and for $\dot\gamma < 1/\tau$, $\eta_{(\dot\gamma)} = \eta_0$. The parameters were fitted

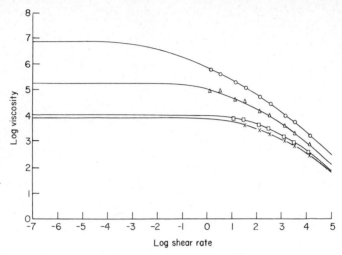

Fig. 25 Shear rate dependent viscosities of an ethylene–methacrylic acid copolymer (3·5 mole % acid) at various degrees of neutralisation, 160°C. % neutralisation: (×) 0; (□) 10; (△) 50; (○) 100.[20]

by linear multiple regression analysis. The value of η_0 is readily obtained by differentiation. The parameters were used to calculate the curves shown in the figure. As discussed above, there is a dramatic increase in the low shear rate viscosity with increasing degree of neutralisation, but the effect is much less at higher shear rates, as reported earlier by Rees and Vaughan.[2b] This suggests that there is a breakdown in some flow unit in the ionomer at high shear rates so that the behaviour corresponds more nearly to that of the un-ionised copolymer.

Similar results were found by Sakamoto *et al.*[13] Figures 26 and 27 show the results for a 4·1 mole percent ethylene–methacrylic acid copolymer and the sodium salt. These results also include the effect of change of temperature. In these figures are included for comparison the values of the dynamic viscosity. An empirical correlation has been found between dynamic and steady-state viscosities at corresponding frequencies and shear rates.[27] This interesting result has been justified theoretically by Strella.[28] The physical significance is that the flow processes are similar in the two types of experiment. Viscous flow of high polymers can be separated into two types of behaviour, segmental motion of short lengths of chain, independently of molecular weight and co-operative motion of these segments leading to transport of the molecule as a whole. At low shear rates and frequencies, both types of motion take place. As the shear rate and frequency are

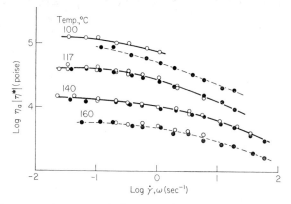

Fig. 26 Apparent viscosity η_a (○) versus shear rate $\dot{\gamma}$ and complex viscosity $|\eta^*|$ (●) versus angular frequency ω for an ethylene–methacrylic acid copolymer (4·1 mole % acid).[13]

increased, however, the processes are somewhat different. For steady-state flow, as the shear rate is increased, the rate of deformation of the whole molecule becomes faster than the longest relaxation time and so the molecule absorbs progressively less energy, and the viscosity decreases as viscous energy is dissipated by the lower molecular weight chains;[23] there

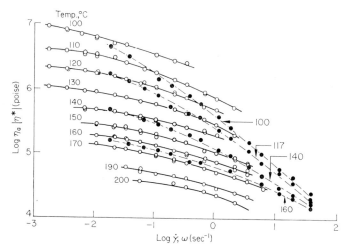

Fig. 27 Apparent viscosity η_a (○) versus shear rate $\dot{\gamma}$ and complex viscosity $|\eta^*|$ (●) versus angular frequency ω for the sodium salt of an ethylene–methacrylic acid copolymer (4·1 mole % acid).[13]

is no longer segmental motion of the higher molecular weight chains. For the dynamic case, as the frequency is increased, the viscous energy is dissipated by progressively shorter segments of all the chains.

In the case of the ethylene–methacrylic acid copolymer (Fig. 26), the two curves overlap at the higher temperatures; but at the lowest temperature, 100°C, there are some crystalline fragments present and these have the effect of sharply increasing the steady-flow viscosity. The dynamic viscosity does not increase to the same extent because the segments which are in crystallites no longer contribute to the viscosity, behaving essentially as rigid filler particles.

The ionomer behaves quite differently and the dynamic viscosity is lower than the steady-flow viscosity over almost the entire range of frequency and shear rate. This implies the existence of more or less rigid domains which are associated with the ions. Actually, the flow behaviour of the ionomers is quite complicated, because even though the dynamic viscosity is lower than the corresponding steady-flow viscosity, it is nonetheless considerably higher than the viscosity of the parent copolymer. The sets of curves do tend to converge at very high shear rates and frequencies, implying that the limiting segment is the same, *i.e.* a short length of hydrocarbon chain.

The behaviour of the calcium ionomer was similar to the sodium. The curves are parallel, with the viscosities of the calcium salt being about two to three times greater.

2.5.3 Viscoelastic properties

In this section, the discussion of viscoelastic properties will be confined to the behaviour at higher temperatures, above the crystalline melting point of polyethylene.

The dynamic viscosity and modulus have been reported by Sakamoto *et al.*[13] Results for a 4·1 mole percent copolymer of ethylene and methacrylic acid and the sodium derivative (59% ionised) are shown in Figs. 28 and 29. The results for a calcium derivative are similar to those for the sodium. This particular un-ionised copolymer is above its melting point in the temperature region studied, 100–160°C, and so the flow response is relatively uncomplicated and results can satisfactorily be treated by the time–temperature superposition method. The relaxation spectra were calculated according to the first approximation method,[29] with the results shown in Figs. 30 and 31. The un-ionised copolymer shows only a flow or terminal region whereas the ionised copolymer exhibits a rubbery plateau which tends to decrease with increasing temperature. This result indicates the existence of structure in the molten copolymer above the crystalline melting point, which is associated with the ions. Evidently the ions are able

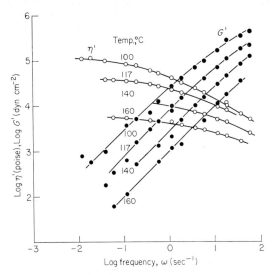

Fig. 28 Dynamic viscosity η' (○) and modulus G' (●) of an ethylene–methacrylic acid copolymer (4·1 mole % acid) as a function of frequency and temperature.[13]

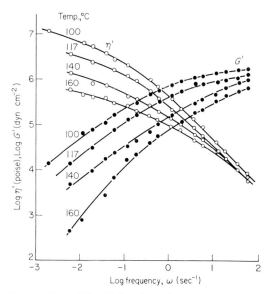

Fig. 29 Similar results to Fig. 28 for the sodium ionomer (59% neutralised).[13]

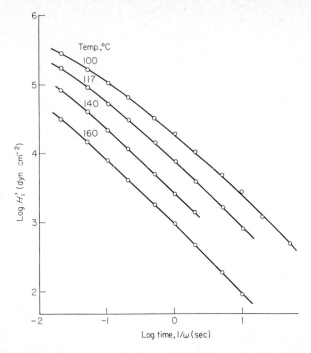

Fig. 30   Relaxation spectra of ethylene–methacrylic acid copolymer (4·1 mole % acid).[13]

to move from one domain to another during flow; this motion becomes easier with increasing temperature and more difficult with increasing frequency, or shorter time scale. The rubbery plateau region is shifted to shorter times with increasing frequency, and this reflects a loosening of the ionic structure, or domain. (However, X-ray diffraction studies described below indicate that these structures do not disappear even at very high temperatures (300°C)). The calcium derivative exhibited similar behaviour to the sodium except for the presence of a maximum in the plateau region, but this may be a consequence of using the first approximation method.

Longworth and Vaughan[20] measured the elastic recovery of polymer melts following steady flow for a series of ionomers of increasing degree of neutralisation. The results were analysed graphically to give the parameters of a series of retarded elastic, or Voigt, elements represented by the equation

$$\gamma/\sigma = \sum_i J_i \exp(1 - t/\tau_i) = J_\infty$$

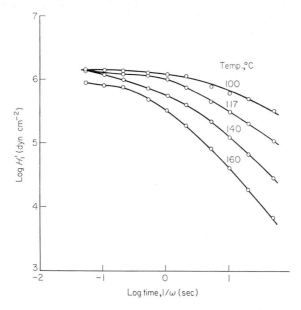

Fig. 31 Relaxation spectra of a sodium ionomer (4·1 mole % methacrylic acid, 59% neutralised).[13]

where J_i is the compliance and τ_i is a retardation time, equal to the product $J_i\eta_i$ where η_i is an 'internal viscosity,'[30] reflecting the resistance to flow experienced by a single segment. J_i is molecular weight dependent, but η_i is not. J_∞ is the total elastic compliance.

The base copolymer contained 5·4 mole percent methacrylic acid and the recovery after steady flow was measured at 180°C with a cone and plate rheometer. The shortest time for which the recovery could be measured was about one second (which is much too long for measurement of the instantaneous elastic recovery). The curve could be analysed into a series of three Voigt elements of retardation times 5, 28 and 79 seconds. Similar recovery curves were obtained for ionomers of increasing degree of neutralisation with sodium. It was found that, as the degree of neutralisation was increased, elements with longer retardation times had to be added to describe the behaviour. In other words, the relaxation processes at the shortest times were common to all samples, the effect of the ions was to create relaxation processes at longer times. To compare the behaviour, Fig. 32 shows the values of the parameters of the retarded elastic elements with the longest time constants.

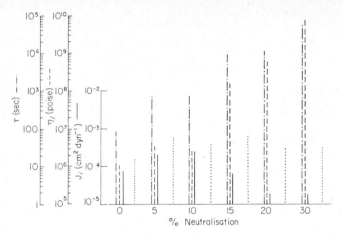

Fig. 32 Parameters of the Voigt elements (τ_i, J_i, η_i) of the elements of longest τ for an ethylene–methacrylic acid copolymer (5·4 mole % acid) with increasing degree of neutralisation. (. . . J_∞, total elastic compliance of each sample).[20]

The most striking feature is the enormous increase in retardation time from less than 100 seconds to almost 10^5 seconds. This increase is virtually entirely due to an increase in η_i from 10^6 poise to 10^9 poise. At the same time, there is very little increase in the elastic components. The retardation for the ionomer at 30% neutralisation is almost entirely viscous in character, and the total elasticity (J_∞) changes hardly at all. Since the elasticity is strongly dependent on chain length, this result shows clearly that the effect of ions is to greatly increase the forces between chains, but not in the form of bonds which can function as quasi cross-links in the molten ionomer.

This strong association between the ionic segments is not too surprising, since the binding energy for coulombic attraction falls off inversely with the distance of separation whereas the energy for dipoles falls off as the inverse cube of the separation. This means that during viscous deformation, more energy can be stored in the ionomer than in the corresponding un-ionised copolymer. This has extremely important practical consequences. For example, the ability of films of ionomers to be drawn down to very thin webs without 'tear-off' is due to a combination of the low elasticity with high internal viscosity which holds the segments of the chains together during flow. This behaviour is also described by saying that the ionomers have a high melt strength but low melt swell. (Melt or die swell is measured by the increase in diameter of a strand of molten polymer on emerging from an orifice compared to the diameter of the orifice.)

2.6 RELAXATION BEHAVIOUR

2.6.1 Dynamic mechanical properties

The dynamic mechanical behaviour of acid copolymers and their ionomers has been the subject of several investigations including ethylene copolymers[2, 9, 14, 25, 31, 32, 39, 53, 54, 55] and styrene copolymers.[17, 75, 76]

The effect of variation in the acid content of un-ionised copolymers of ethylene and methacrylic acid was systematically examined by Longworth and Vaughan,[14] following preliminary measurements by Rees and Vaughan.[2b] The results are shown in Fig. 33 in which the loss modulus, G'', is plotted against reciprocal temperature. In this type of plot, the strength of the relaxation is proportional to the area under a peak.[33] (Since G'' is given by the product $\Delta G'/\pi$, where Δ is the log decrement and G' the shear modulus, the rapid decrease in G' with increasing temperature has the effect of de-emphasising the height of the relaxation peak at temperatures close to the onset of rubbery behaviour.)

At the lowest temperatures, there is a large loss peak, the γ relaxation. (In the labelling of these peaks, the terminology of MacKnight et al.[9] will be used.) The molecular motion involved in this relaxation has been ascribed to motion in both the crystalline[34] and amorphous regions.[35] McKenna et al.[39] were able to resolve the γ relaxation of the ionomer and the parent copolymer into two peaks; one, γ_c, at the lowest temperature which was proportional to the degree of crystallinity; the other, γ_a, depended on the amount of amorphous material. It was found that the specific relaxation strength, i.e. the strength of the relaxation divided by the volume of amorphous material (equal to 1 − crystallinity) was independent of the degree of neutralisation and the nature of the cation. The maximum temperature of the relaxation was $-129°C$. Since the crystallinity of the copolymers described here is low, and essentially non-existent at seven mole percent (20% by weight) methacrylic acid, the relaxation can best be described by the 'crankshaft' motion of short segments of hydrocarbon chain. There is actually a gradual increase in the strength of the relaxation with increasing acid content, which is consistent with the relaxation taking place in the amorphous hydrocarbon material. The amount of this material will at first increase with decreasing crystallinity but eventually decrease with increasing acid content.

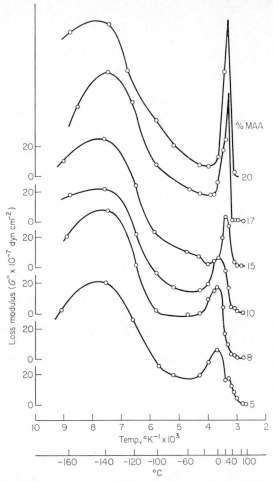

Fig. 33 Loss modulus, G″, of ethylene–methacrylic acid copolymers as a function of reciprocal temperature. Acid content % wt as indicated.[14]

At temperatures between $-60°C$ and $0°C$ the β relaxation of low density polyethylene is not found. This relaxation has been ascribed to a glass transition associated with the branch points, since it is not observed in linear polyethylene.[36] The absence of this relaxation is somewhat surprising, and reinforces the supposition that the carboxyl groups are excluded from lamellar polyethylene crystals. (There is an indication of some β relaxation in the copolymer containing 1·7 mole percent acid, but

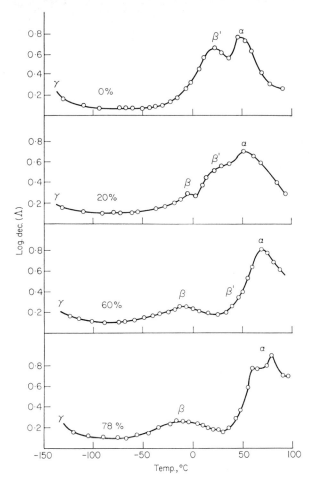

Fig. 34 Temperature dependence of the log decrement of an ethylene–methacrylic acid copolymer (4·1 mole % acid) at various degrees of neutralisation, annealed.[9]

it is overlaid by the β' relaxation which occurs near 0°C for this copolymer.)

Instead of the β relaxation, there appears a new relaxation, β', at temperatures ranging from 0°C to 50°C depending on the acid content. The magnitude and the sharpness of this relaxation increase with increasing acid content. The development of the relaxation depends on the thermal history of the sample.

The results of MacKnight *et al.*[9] are shown in Figs. 34 and 35 for a

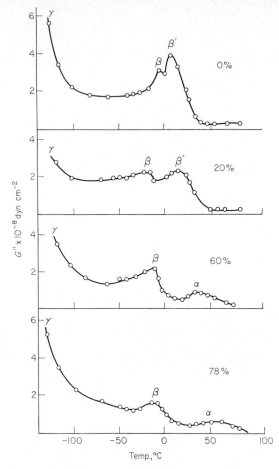

Fig. 35 Temperature dependence of the loss modulus for ethylene–methacrylic acid copolymer (4·1 mole % acid) at various degrees of neutralisation quenched.[9]

4.1 mole % copolymer. The sample quenched in dry ice from 180°C does not show the β' relaxation, which only appeared when the sample was annealed at 95°C. The samples shown in Fig. 33 had an intermediate thermal history; they were moulded at 150°C and cooled to room temperature over a period of ten minutes. The effect of annealing is to separate the β' relaxation from the α relaxation. The α relaxation in polyethylene takes place with the disordering of lamellar crystals and involves reversible viscous flow of interlamellar material. The α relaxation is not very pro-

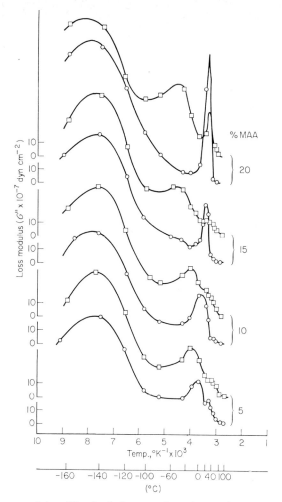

Fig. 36 Loss modulus G″ of ethylene–methacrylic acid copolymers and fully neutralised sodium ionomers versus reciprocal temperature. (○) copolymers, (□) ionomers. Acid content % wt as indicated.[14]

nounced with the samples of Fig. 33 and is only really visible for the 1·7 mole % acid sample.

The effect of complete neutralisation with sodium is shown in Fig. 36 for the same series of polymers as Fig. 33. The ionomers in this case were dry and annealed at 90°C. The β' peak of the unneutralised copolymers is drastically reduced in all cases, and there appears a peak of reduced magnitude at somewhat higher temperatures. The results of MacKnight shown in Fig. 35 indicate this to be an increase in the α relaxation. Since this α relaxation increases with decreasing crystallinity (*i.e.* from 1.7 mole % to 7.5 mole %) it is to be ascribed to viscous flow of segments not included in lamellar crystals.

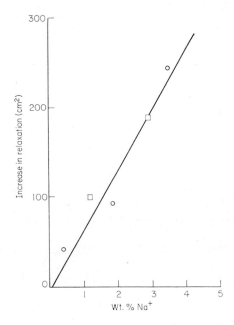

Fig. 37 Incremental differences in the strength of the β relaxation of ionomers compared to the parent ethylene–methacrylic acid copolymer. (○) 5·4 mole % acid; (□) 7·5 mole % acid.[14]

The most striking effect of neutralisation is, however, a large increase in a low temperature relaxation where the β relaxation of low density polyethylene is encountered. For the ionomers, this is also labelled a β relaxation. The increase in this relaxation is proportional to the amount of combined ions. In Fig. 37 the incremental area under this peak relative to the unneutralised copolymer is plotted against the weight of combined sodium. The effect of water on these relaxations is shown in Fig. 38. The behaviour of low density polyethylene is compared to a 5·4 mole % acid copolymer and the ionomer prepared by completely ionising with sodium.

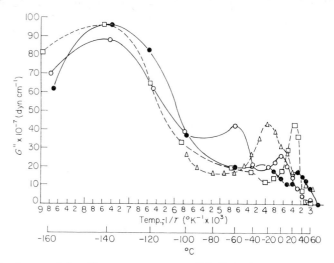

Fig. 38 Loss modulus G″ versus temperature (°K^{-1}) for low density polyethylene (△); ethylene–methacrylic acid copolymer (5·4 mole % acid) (□); fully neutralised ionomer, dry (●) and wet (○).[14]

The curves for both the dry and wet (saturated, containing 23% water) ionomers are given. The β relaxation of polyethylene can clearly be distinguished as also the β' and (barely) the α relaxations of the copolymer. The dry ionomer exhibits behaviour similar to that described above; the development of the β relaxation is seen to be in the same temperature range as the β relaxation of the low density polyethylene.

The effect of water is to shift both the α and β relaxations to lower temperatures by about 30°C and to greatly increase their strength.

MacKnight et al. note that if the ionised carboxylate groups lead to strong interchain association (as suggested by Erdi and Morawetz[38]), then it is unlikely that such a system would have a relaxation similar to the β relaxation of low density polyethylene which consists solely of branched hydrocarbon segments. They thus postulate that the relaxation is due to the motion of hydrocarbon segments containing isolated ionised residues. However, given that the β relaxation in polyethylene is associated with the motion of pure hydrocarbon segments, it is difficult to understand why the presence of highly polar residues should result in a relaxation at exactly the same temperature. Thus a more reasonable explanation of the effect of neutralisation would provide for the *formation* on neutralisation of material similar in composition to that giving rise to the β relaxation in polyethylene. Such an explanation has been offered[37] and will be described in more detail below.

The properties of ethylene–acrylic acid polymers and their ionomers have been measured by Ward and Tobolsky.[31] Relatively high levels of acid were present, up to eight mole percent, so that the samples were essentially free from crystallinity. The elastic shear moduli were measured at low frequencies from about $-60°C$ to $100°C$ and converted to values of the elastic stress-relaxation modulus. Even though these polymers were amorphous, the curves for modulus versus time could not be superposed by shifts along the log time axis; superposition could only be obtained by shifting along both the time and modulus axes. That is to say, the WLF equation could not be applied, implying that there are several distinct types of relaxation process present.

The results showed that the rate of relaxation is controlled by the degree of neutralisation, independently of the ion; the rate increasing with increasing degree of ionisation. The consequence is that the ions initially increase the modulus, but the relaxation allows the modulus to fall below that of the parent copolymer. (One would expect this effect to be more pronounced if there were appreciable crystallinity present, as for a parent copolymer containing lower amounts of copolymerised acid.)

Otocka and Kwei also have examined the behaviour of ethylene–acrylic acid copolymers[53] and ionomers.[1,54,55] A series of four copolymers was examined in which the acrylic acid varied from 0·66 to 2·78 mole percent, thus the acid contents are low compared to the studies described above, and the level of carboxyl is about the same as the frequency of side branches. The behaviour of the unneutralised copolymers was similar to the ethylene–methacrylic acid copolymers. With increasing acid content, there was an increase in the temperature of maxima in the E'' curves from $-21°C$ for polyethylene to $24°C$ for the copolymer containing 2·78 mole percent acid. This transition is the β' transition described above. The increase in T_g could be calculated from the equations of Gordon and Taylor[56] assuming that all the carboxyl groups resided in the amorphous phase. The calculated increases were less than those observed by about $20°C$. A better explanation of the data was to assume that the increase in T_g was due to the cross-linking effect of dimerised carboxyl groups. The equation of Fox and Loshaek[57] was applied and the predicted linear relation between ΔT_g (the increase in temperature of the β relaxation above that of the polyethylene) and the cross-link density was found.

The ionomers from these copolymers were prepared by neutralising completely with sodium and magnesium methoxides. The relaxation

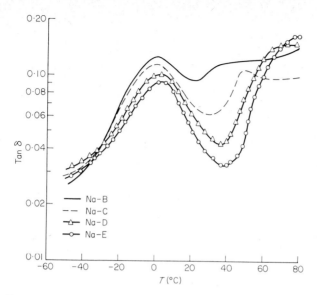

Fig. 39 *Loss tangents of sodium ionomers of ethylene–acrylic acid copolymers. Mole % acid: Na–B, 0·66; Na–C, 1·57; Na–D, 2·26; Na–E, 2·78.*[54]

behaviour is shown in Fig. 39, a plot of tan δ versus temperature for the sodium ionomers. The temperature of the relaxation maximum increases with increasing acid content, but the increase is appreciably less than for the unneutralised copolymers. The temperatures of the β transitions for the different copolymers are in the same range as described above for the ethylene–methacrylic acid copolymers.

The increase in T_β with increasing metal acrylate content is in good agreement with the increase calculated assuming the T_g of poly(sodium acrylate) to be 310°C. This temperature was obtained by extrapolation of data for the T_gs of copolymers of styrene and sodium methacrylate.[17]

The properties of ethylene–acrylic acid copolymers are essentially similar to those of ethylene–methacrylic acid copolymers, differing only by the differences due to the difference in T_g of polyacrylic acid and polymethacrylic acid and their salts. Comparing these results with the discussion above of methacrylic acid copolymers and their salts, it appears that the apparent increase in T_g of the acid copolymers (ascribed by Otocka and Kwei to cross-linking effects as well as copolymerisation), *i.e.* the transition labelled β' by MacKnight[9] is continued with increase in acid content (*see* Fig. 33).

This would argue that the molecular processes are similar in both sets of copolymers and it seems reasonable that the carboxyl groups are excluded from lamellar crystals and form an amorphous phase in which they are randomly distributed and almost completely hydrogen bonded.

In the case of the ionomers, the evidence presented earlier (Figs. 34, 35 and 36) indicates that the temperature of the β relaxation does not regularly increase with increasing acid content, and in fact remains in the region from $-20°C$ to $0°C$. In other words, the temperature of the β relaxation initially increases (from that of low density polyethylene) in copolymerised metal acrylate or methacrylate, but that with larger amounts of copolymerised metal salt, the molecular structure of the material giving rise to the relaxation remains constant and increases only in extent, thus the T_β remains constant but the magnitude of the relaxation increases.

This interpretation is consistent with the results of Phillips et al.[58a,b] for polyethylene with phosphonic acid side groups $[-(CH_2-CH_2)_n-(CH_2CHPO(OH)_2)_m-]$. In this case, T_β was independent of acid concentration and the phosphonic acid side groups were thought to be present not as strongly hydrogen bonded dimers but as loose aggregates, giving rise to a new α' relaxation at $50°C$.

The examination of the relaxation behaviour of the ionomers is of importance in any discussion of the changes in structure brought about by

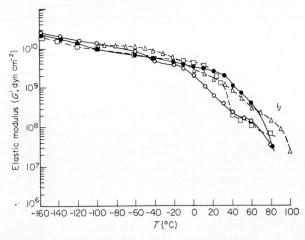

Fig. 40 Modulus G′ versus temperature of low density polyethylene ($\triangle$); ethylene–methacrylic acid copolymer ($\square$); and fully neutralised sodium ionomer dry ($\bullet$), and wet ($\circ$).[14]

the incorporation of ions into hydrocarbon polymers. However, the changes in modulus with composition and temperature are important in determining the applications of ionomers to various end uses. The moduli of the same samples as shown in Fig. 38 are given in Fig. 40. At room temperature, the dry ionomer has a slightly higher modulus, hence higher stiffness, than either low density polyethylene or the unneutralised copolymer. As the temperature is increased, above the temperature of the β' relaxation, the higher crystallinity of the polyethylene leads to its having a greater stiffness at 70°C than the other polymers. Not too surprisingly, the effect of water is to plasticise the ionomer so that its stiffness is reduced at room temperature. This means that the stiffness of mouldings made from ionomers will be somewhat sensitive to the ambient conditions of humidity and temperature.

Another feature of interest is the behaviour at moderately low temperatures, from 0°C to −60°C. The moduli of *both* the wet and dry ionomers are lower than the other two polymers in this range. A characteristic of the ionomers is their toughness under high rates of strain, greater than 66 nylon in some cases, and it can be shown[20] that this type of toughness is proportional to the average value of the modulus in this range. The explanation for this behaviour depends on the model of ionomer structure discussed below, and is further evidence supporting such a model. This behaviour is quite general for ionomers based on ethylene–methacrylic acid copolymers in the range from one to seven mole percent acid.

2.6.2 Electrical properties

Table 6 shows the changes in the dielectric constant (K) and dielectric loss factor (tan δ) with variations in amount of copolymerised methacrylic acid for free acid and fully neutralised sodium ionomers[14] (similar results were reported for an acrylic acid zinc ionomer earlier[2b]). These samples were conditioned to 50% relative humidity and the effect of water is apparent in the increased loss factor of the ionomers compared to the acid copolymers. There is also an increase in the dielectric constant with neutralisation, although for the acid copolymers there is very little change, reflecting the fact that the carboxyl–carboxyl dimer is electrically neutral and the carboxyl groups are almost entirely associated at room temperature. (For dry samples, the dielectric constants range from about 2·2 to 2·5).

A thorough examination of the dielectric relaxation of the methacrylic acid copolymers and the corresponding ionomers has been made by MacKnight and co-workers.[40,41] The copolymers contained 4·2 and 8·3 mole percent methacrylic acid and the cations included lithium, sodium and

TABLE 6
ELECTRICAL PROPERTIES OF ETHYLENE–METHACRYLIC ACID COPOLYMERS AND IONOMERS[a,b]

Mole % Acid	% Neutralisation	Temp. °C	$\tan \delta \times 10^3$	k[c]
1·7	0	23	0·9	2·54
		80	0·85	2·17
	100	23	4·15	2·78
		80	1·31	2·61
3·5	0	23	1·48	2·51
		80	0·5	2·25
	100	23	2·01	3·63
		80	12·3	4·19
5·4	0	23	1·48	2·67
		80	0·66	2·50
	100	23	1·63	3·80
		80	6·35	4·29

[a] Samples annealed at 70°C conditioned to 50% R.H. at 72°F.
[b] Sodium salts.
[c] Measured at 1000 Hz by ASTM D-150 method.

calcium. The samples were characterised by measuring the dielectric loss factor from -100°C to 100°C and identifying the various relaxation peaks, analogous with the mechanical relaxations described above.

Figure 41 shows the results for the 4·2 mole percent acid copolymer and the lithium ionomers. The acid copolymer is characterised by a β' relaxation at about 30°C. As the degree of neutralisation is increased, the β' relaxation decreases and there appear β and α relaxations. The β relaxation tends to increase in magnitude but decrease in temperature with increasing degree of neutralisation. The α relaxation is at higher temperatures and increases in magnitude and temperature with increasing degree of neutralisation. There is considerable overlap of the peaks making quantitative resolution difficult. Similar behaviour is found for the sodium and calcium salts, except that the α peak is even greater in magnitude and moves to temperatures which are above the crystalline melting point. This is shown in Fig. 42 for the sodium ionomer of the 8·3 mole % acid copolymer, which also illustrates the dependence on temperature of the β frequency.

The change in the dielectric constant with temperature is shown in Fig. 43 for the sodium and calcium salts of copolymers of high and low acid copolymers respectively. There was evidently no contribution from DC conductivity to the behaviour of any of the polymers examined.

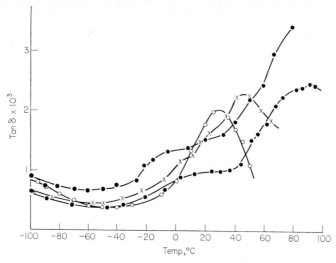

Fig. 41 Temperature dependence of the dielectric loss factor (tan δ) at 1 000 Hz for lithium ionomers of an ethylene–methacrylic acid copolymer (4·1 mole % acid). Degrees of neutralisation %: (○) zero; (×) 19; (●) 41; (◐) 72.[41]

The quantitative separation of the various peaks was made by plotting tan δ (or E'') versus frequency, the frequency of maximum loss was then plotted against reciprocal temperature to obtain the activation energies for the various relaxation processes. Examples of this procedure are given in Fig. 44 for a sodium salt of the 4.2 mole percent acid copolymer. The ranges of values of the activation energies for the various relaxation processes identified are as follows (in kcal mole^{-1}): α relaxation, 80 to 120; β′, 56 to 74; β, 7·9 to 19; γ, 6·9 to 16·0. An analysis of the various molecular processes involved in these relaxations is by analogy with the mechanical relaxation processes described above. The α relaxation is a property associated with the ionic part of the ionomer, it being absent in the unneutralised acid copolymer and increasing in magnitude and temperature with increased degree of neutralisation. Read et al.[40] suggest that the relaxation reflects the breakup of large ionic regions into smaller clusters, which is consistent with the high activation energy. (A problem worth mentioning at this point is the difficulty of describing precisely the structural differences between calcium and monovalent metal ionomers. The calcium salts show higher values of temperature and magnitude for the α relaxation, but the activation energies are similar to the sodium and lithium compounds. The calcium ionomers have measurable viscosities, for example, which means

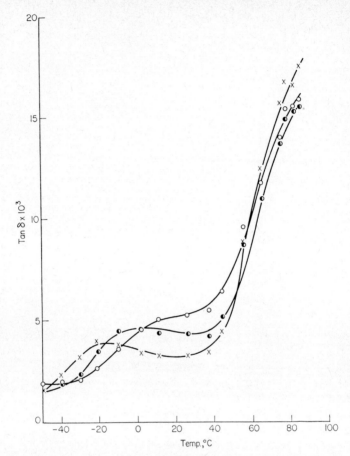

Fig. 42 Temperature dependence of the dielectric loss factor (tan δ) of a sodium ionomer (95% neutralised) of an ethylene–methacrylic acid copolymer (8·3 mole % acid). (×) 100 Hz; (◐) 1000 Hz; (○) 10 kHz.[41]

that if the calcium ions form cross-links between chains, these are dissociable.)

An interesting alternative explanation for the α relaxation is that it is due to the Maxwell–Wagner effect which is found in a dielectric consisting of droplets of a material of high dielectric constant dispersed in a medium of lower dielectric constant.[42] The change in dielectric constant with temperature is rather small, however.

The dielectric β' relaxation is similar in behaviour to the mechanical β'

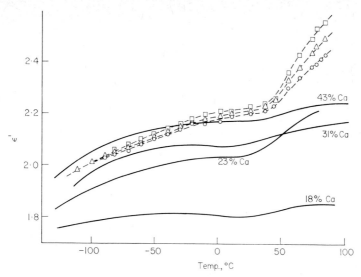

Fig. 43 Temperature dependence of the dielectric constant of sodium and calcium ionomers of ethylene–methacrylic acid copolymers. (□) 100 Hz; (△) 1000 Hz; (○) 10 kHz of 8·3 mole % acid 95% neutralised with sodium. In line form, data for calcium salts of 4·1 mole % acid copolymer (10 kHz).[41]

relaxation with respect to its strength, location and changes with composition. Since this is ascribed to the motion of amorphous hydrocarbon chains bonded by electrically neutral carboxyl dimers, the reason for the electrical activity is not immediately obvious. Calculations by Read et al.[40] show that the electrical activity can be reasonably accounted for by the presence of only about 0·4 mole percent carbonyl groups on the chains. This is quite reasonable in view of the ease with which hydrocarbon polymers can be oxidised.

The dielectric β relaxation correlates well with the mechanical relaxation in terms of dependence on degree of neutralisation and temperature location. It differs from the mechanical relaxation in having a much lower activation energy, 15 kcal mole^{-1} versus 35 kcal mole^{-1}. The mechanical relaxation has been ascribed to the motion of segments of amorphous, branched hydrocarbon chains (*see* below), which are unresponsive electrically; it is likely, therefore, that there are a few free acid or isolated salt groups attached to these segments.

As mentioned above, the mechanical γ relaxation has been divided into two transitions, γ_a and γ_c. The γ_a process is the motion of short segments

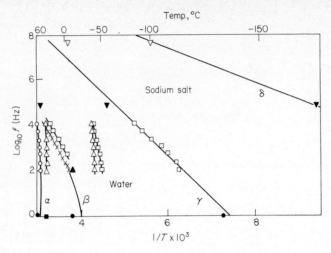

Fig. 44 Plots of frequency of maximum loss versus temperature ($°K^{-1}$) for an ionomer (sodium, 53% neutralised) of an ethylene–methacrylic acid copolymer (4·1 mole % acid). (○, ×) tan δ dielectric; (●) tan δ mechanical;[9a] (■) G″ mechanical;[9a] (▲) E″ mechanical;[9b] (▽) T_1 NMR; (▼) $T_{1\rho}$ NMR; (△) tan δ dielectric (1·09% water); (□) tan δ dielectric (2·29% water). Except where stated, points refer to dry samples. (×) refers to frequency of maximum tan δ (temp. constant). Other points denote temperatures of maximum loss or minimum T_1 or $T_{1\rho}$ at constant frequency.[40]

of hydrocarbon chains, not containing polar groups, and not included in lamellar crystals. It is this process that correlates with the γ dielectric relaxation. As with the β relaxation, the source of the polarity is not clearly understood, but may similarly be due to small numbers of carbonyl groups formed for example by thermal oxidation.

The above results were for dry ionomers. The effect of water[40] is to solvate or plasticise the metal salts; sodium ionomers may absorb up to 25% by weight water when saturated. The α relaxation is shifted by about 30°C to lower temperatures. The β relaxation, as with the mechanical β relaxation, is shifted to lower temperatures, about −40°C to −60°C. Read *et al.*[40] suggest that this is a new peak, and involves direct participation of the water molecule in the relaxation possibly within a cage of hydrocarbon surrounding solvated carboxylate groups.

2.6.3 Nuclear magnetic resonance studies

Only a few studies have been made of the nuclear magnetic resonance of ionomers. The following comments have been made by Dr. D. W. Ovenall.

The application of NMR to the study of relaxation phenomena in solid polymers has been discussed widely in the literature[71,72] and will be only briefly considered here. In the broad-line NMR technique, the absorption of radio-frequency energy by the sample is recorded, usually as the first derivative, as the magnetic field is slowly swept through resonance. In solids, the shape and the width of the resonances depend largely on magnetic dipole interactions between the nuclei. Relative motion of the nuclei tends to average out these interactions and to narrow the lines. Phase transitions which are accompanied by changes in molecular motion can be observed by taking spectra as a function of temperature.

The more recently developed pulsed NMR techniques involve the use of a short intense pulse or a series of short intense pulses of radiofrequency energy at the resonance frequency which forces the system of nuclear spins into a non-equilibrium condition. The return of the nuclear spins to an equilibrium state depends on nuclear relaxation times which are a function of the motions of the nuclei.

Read et al.[40] have reported proton magnetic relaxation studies at a radiofrequency of 30 MHz on an ethylene–methacrylic acid copolymer containing about 4·1 mole percent methacrylic acid units and its 53% ionised sodium salt. T_1, the spin-lattice relaxation time and $T_{1\rho}$, the spin-lattice relaxation time in the 'rotating frame' were measured. Connor has discussed the study of molecular motions and phase transitions in polymers via T_1 and $T_{1\rho}$ measurements in a subsequent paper.[72] T_1 and $T_{1\rho}$ are essentially rates at which the nuclear spins exchange energy with other modes of motion in the solid (referred to as 'the lattice') under certain conditions.

The occurrence of minima in the variation of T_1 with temperature is due to the effect of particular types of molecular motion on T_1. At the minimum itself, the motion is most effective in shortening T_1 and $\omega_0 \tau_c \sim 1$ where $\omega_0 = 2\pi f_0$ (f_0 is the radiofrequency in hertz) and τ_c is the correlation time for the motion. In $T_{1\rho}$ measurements, the effective frequency against which τ_c must be compared is much lower than the radiofrequency and is typically in the kilohertz region. Thus $T_{1\rho}$ minima occur at lower temperatures than the T_1 minima caused by the same relaxation processes and because of the lower correlation frequencies involved, are often more readily related to mechanical and dielectric relaxation measurements.

The temperature dependence of T_1 and $T_{1\rho}$ for the partially ionised sodium salt, measured by Read et al.[40] is shown in Fig. 45. The minimum in the T_1 curve at 0°C was identified as the γ relaxation, while an inflexion at -100°C was identified as a δ relaxation. The $T_{1\rho}$ data, which refer to a

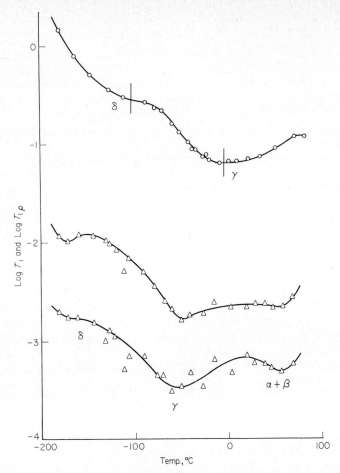

Fig. 45 Temperature dependence of T_1 and $T_{1\rho}$ for an ionomer (sodium, 59% neutralised) of an ethylene–methacrylic acid copolymer (4.1 mole % acid). (○) T_1 30 MHz; (△) $T_{1\rho}$ 25.5 MHz.[40] Upper curve, T_1; middle curve, $T_{1\rho}$ long; lower curve, $T_{1\rho}$ short.

frequency of 25.5 kHz, could be best represented by two separate relaxation times at each temperature which were tentatively assigned to nuclei in the amorphous and crystalline regions of the polymer. Three distinct minima were seen in each of the $T_{1\rho}$ curves. These were assigned to the α and β relaxations, the γ relaxation and the δ relaxation, with decreasing temperature. Similar behaviour was observed for T_1 and $T_{1\rho}$ measurements on the un-ionised copolymer. Correlation frequencies derived from the T_1 and $T_{1\rho}$ measurements on the sodium salt are plotted with data from mechanical and dielectric measurements in Fig. 44 confirming these identifications. The low temperature relaxation appears only in the NMR measurements and has an activation energy of 3·1 kcal mole^{-1}. This is believed to be due to methyl group rotation in the methacrylic acid units. Although there are relatively few of these, they provide an efficient coupling to the lattice at low temperatures and nuclear magnetisation can be conveyed to them along the polymer chains by spin-diffusion.

In order to investigate the extent of aggregation of the metal ions in an ionomer, Otocka and Davis[60] examined both the proton and lithium-7 magnetic resonances of an ethylene–lithium acrylate copolymer by the broad-line technique. In Fig. 46, the variation in line width with tempera-

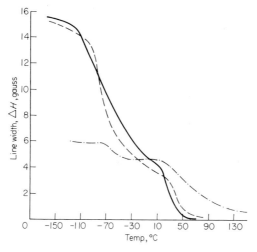

Fig. 46 Wide line magnetic resonances: ———, ^1H in ethylene–acrylic acid copolymer (4·9 mole % acid), 60 MHz; – – – ^1H lithium ionomer, 56·5 MHz; – · – · ^7Li in lithium ionomer, 16 MHz.[60]

ture is shown for the proton resonance at 60 MHz of an ethylene–acrylic acid copolymer containing 4·9 mole percent acid, for the proton resonance at 56·5 MHz of the fully neutralised lithium ionomer and for the lithium-7 resonance of the ionomer at 16 MHz. In all three cases, the NMR line width narrows from its low temperature value in two stages as the temperature is raised. The low temperature line narrowing was identified with the γ transition and the high temperature line narrowing with the β transition.

While recognising that the line narrowings observed in the lithium-7 resonances of the ionomer occur at rather higher temperatures than the corresponding line narrowings in the proton resonance, Otocka and Davis concluded that the increased motion of the lithium nuclei responsible for the decrease in line width seems to be a direct result of increased matrix mobility and they further state that, if the lithium nuclei were present in exclusive domains, the matrix mobility would not be transferred effectively. Examination of their data, Fig. 46, indicates that for both transitions the lithium-7 line narrowing occurs some 20°C above that of the proton line narrowing. This difference is by no means insignificant and it indicates that nuclei are not well-dispersed throughout the matrix as favoured by Otocka and Davis but are, to some extent, segregated. It is to be expected that, in such a model, motion in the matrix would be conveyed, albeit in a reduced form, to the segregated lithium nuclei.

The use of lithium-7 nuclei as a probe is good, although it suffers from the complication of an unknown contribution from the lithium-7 quadrupole moment. It would be of great interest to compare the relaxation times of lithium-7 nuclei and protons in a system similar to that studied by Otocka and Davis since these are frequently more informative than broad-line NMR measurements.

Pineri et al.[50] recently reported a study of butadiene–methacrylic acid copolymers neutralised by cupric ions. Use was made of the fact that pairs of cupric ions, as in cupric acetate monohydrate, give a different electron paramagnetic resonance (EPR) absorption spectrum from that of isolated cupric ions. The EPR spectrum of a copolymer neutralised with 95% Zn^{2+} and 5% Cu^{2+} was characteristic of that from isolated Cu^{2+} ions with only a slight contribution from Cu^{2+}–Cu^{2+} ion pairs. Here only one ion in twenty is a paramagnetic Cu^{2+} ion (Zn^{2+} is diamagnetic). One thus expects this behaviour since, even if all ions in the polymer were present in clusters, the probability of finding two Cu^{2+} ions together would be low. In contrast, the main fraction of the EPR spectrum of a polymer completely neutralised with Cu^{2+} was an absorption characteristic of Cu^{2+}–Cu^{2+} ion pairs, the fraction from isolated Cu^{2+} ions being much weaker. This appears to be conclusive evidence that in these ionomers a large fraction of the Cu^{2+} ions are present in pairs.

2.7 INFRA-RED SPECTROSCOPY

The infra-red spectra of ethylene–methacrylic acid copolymers and ionomers have been presented by Rees and Vaughan[2] and MacKnight et al.[52] and for ethylene–acrylic acid copolymers by Otocka and Kwei.[53]

The results at room temperature for the methacrylic acid copolymers are shown in Fig. 47. The maximum degree of neutralisation was 78% for a copolymer containing 4·1 mole percent acid. Evidence for hydrogen bonding is shown by the shoulder at 2 650 cm^{-1} (hydrogen bonded hydroxyl). There is un-ionised carbonyl at 1 700 cm^{-1} and asymmetric stretching of the carboxylate ion at 1 560 cm^{-1}. This latter band increases with increasing neutralisation.

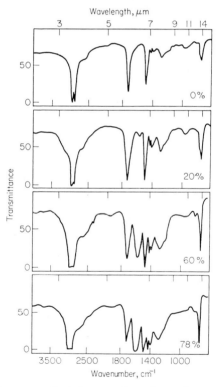

Fig. 47 Infra-red spectra of ethylene–methacrylic acid copolymer (4·1 mole % acid) and sodium ionomers. Thicknesses: 0% 1·2 × 10^{-3} cm; 20% 2·5 × 10^{-3} cm; 60% 6·1 × 10^{-3} cm; 78% 9·4 × 10^{-3} cm.[52]

The integrated absorbances were calculated from the area under the peaks. The degree of neutralisation was determined from the absorbance of the 1700 cm^{-1} band and the thickness of the film. The thickness of the film was obtained from calibration of the intensity of the band at 935 cm^{-1} with thicker films whose thickness could be measured accurately.

The temperature dependence of the spectra was used to examine the carboxyl dimerisation equilibrium. The dissociation constant, K, is defined by

$$K = \frac{[-CO_2H]^2}{[(-CO_2H)_2]}$$

The concentration of dimerised carboxyl was measured by the absorbance at 1700 cm^{-1} (hydrogen bonded carbonyl stretching vibration) and the concentration of monomeric carboxyl by the absorbance at 3540 cm^{-1} due to free hydroxyl stretching vibration. The concentration at the highest temperature was measured by the free carbonyl absorbance at 1750 cm^{-1}. A room temperature value of $9\cdot14 \times 10^5$ cm^2 mole^{-1} was compared to $12\cdot6 \times 10^5$ cm^2 mole^{-1} obtained for the model compound pivalic acid in benzene.[61] From an Arrhenius plot of log K versus reciprocal temperature, the heat of dissociation of the carboxyl dimer was found to be 11.6 kcal mole^{-1}, which agrees well with the values of $8\cdot1$ kcal mole^{-1} for the pivalic acid system[61] and 10 kcal mole^{-1} for copolymers of styrene and methacrylic acid measured above the T_g.[16]

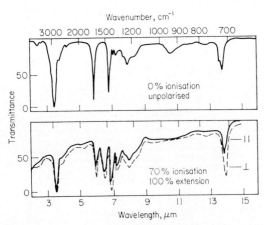

Fig. 48 *Polarised infra-red spectra of an ethylene–methacrylic acid copolymer ($4\cdot1$ mole % acid) and the 70% unneutralised sodium ionomer.*[52] —— *parallel;* ---- *perpendicular alignment.*

The results of infra-red dichroism studies are shown in Fig. 48, which compares the infra-red spectra of the un-ionised, unoriented copolymer with those of the 70% neutralised ionomer stretched 100% and aligned perpendicularly and parallel to the beam polarisation. It is apparent that appreciable polyethylene crystallinity remains, witness the behaviour of the bands at 720 cm^{-1} and 730 cm^{-1}. The 730 cm^{-1} band is characteristic of crystalline polyethylene and is polarised along the crystallographic b axis, consequently it is absent from the parallel spectrum and enhanced (as a shoulder) in the perpendicular spectrum.

There is perpendicular polarisation of the hydrogen bonded hydroxyl at 2650 cm^{-1}, indicating that the carboxyl dimers are present and are oriented perpendicularly to the main chain. Figure 49 shows the change in dichroic ratio ($A_{\parallel}/A_{\perp}$) with increasing extension. The 720 cm^{-1} band shows large perpendicular dichroism; there is some dichroism of the hydrogen bonded carbonyl (1700 cm^{-1}) but very little of the carboxylate at 1560 cm^{-1}.

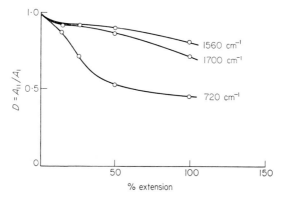

Fig. 49 Dichroic ratio versus % extension for 70% neutralised sodium ionomer of an ethylene–methacrylic acid copolymer (4·1 mole % acid).[52]

The infra-red spectra of ethylene–acrylic acid copolymers were analysed principally by means of the 940 cm^{-1} band,[53] which is assigned to the out-of-plane bending of hydrogen bonded hydroxyl. The temperature dependence of this absorption was used to calculate the extent of carboxyl dimerisation. The extinction coefficients of this band were calculated at room temperature, assuming complete association and indeed there was no systematic variation with acid content. The temperature dependence of the association showed no break at the melting point, indicating that the dimers are present in the amorphous phase.

The dichroic ratio ($A_{\parallel}/A_{\perp}$) of the 940 cm^{-1} band increased strongly with elongation for all the copolymers examined, evidence that the planes of the carboxyl dimers are perpendicular to the planes defined by the parallel main chains, as was found for the methacrylic acid copolymers.

These infra-red studies were extended by Stein et al. to examine structural features.[62] The objective was to use infra-red dichroism to characterise the behaviour on orientation of a methacrylic acid copolymer and its ionomers. A particular objective was to elucidate the structural factors giving rise to the α relaxation described earlier.[9,25] The infra-red dichroism is related to the orientation of a molecular chain by

$$f_i = C_i[(D - 1)/(D + 2)]$$

where f_i is the molecular orientation function of the ith molecular segment, defined as

$$f_i = [3 < \cos^2\theta > \text{ave} - 1]/2$$

D is the dichroism, defined as

$$D = A_{\parallel}/A_{\perp}$$

where $A_{\parallel}$ and $A_{\perp}$ are the absorbances for radiation polarised parallel and perpendicular to the stretching direction. C_i is a constant related to the angle between the stretching axis and the transition moment.

The orientation functions were calculated for various bands characteristic of the hydrocarbon segments (1470 cm^{-1}, CH$_2$ bending; 720 cm^{-1}, CH$_2$ rocking), the un-ionised acid (1700 cm^{-1}, hydrogen bonded C=O), ionised carboxyl (1560 cm^{-1}, carboxylate ion). The functions were all found to increase monotonically with degrees of orientation (stretching). The variation in orientation function with temperature at 50% elongation is shown in Fig. 50 for the acid copolymer and in Fig. 51 for the 55% neutralised sodium ionomer. For the acid copolymer, there is a steady decrease in the function above room temperature, the levelling off below room temperature indicates the onset of the β transition. For the ionomer, there is an orientation maximum at 45°C of all the bands, including those of the hydrocarbon segments. The 720 cm^{-1} band is characteristic of both crystalline and amorphous polyethylene, but the crystals show no orientation maximum by X-ray diffraction.[44] This orientation is located in the amorphous hydrocarbon regions.

The important conclusion reached by this study was that both amorphous hydrocarbon and ionised carboxyl fractions are oriented similarly at the temperature of the α transition.

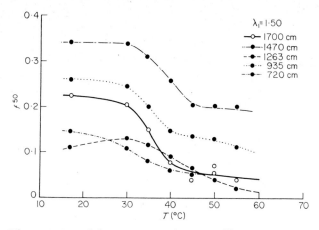

Fig. 50 The variation of the orientation function, f^{50}, with temperature for an ethylene–methacrylic acid copolymer (4·1 mole % acid).[62]

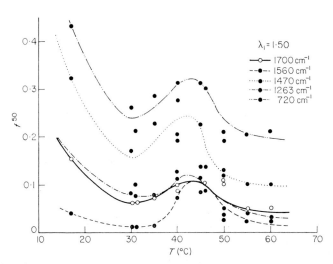

Fig. 51 The variation of the orientation function, f^{50}, with temperature for an ionomer (sodium salt, 55% neutralised) of an ethylene–methacrylic acid copolymer (4·1 mole % acid).[62]

2.8 X-RAY DIFFRACTION

X-ray diffraction measurements have been important in the analysis of the structure of ionomers. Following a preliminary report by Rees and Vaughan,[2a] an extensive examination was made by Wilson et al.;[43] subsequent measurements have tended to confirm their findings.[44,45,78]

The most characteristic features of the ionomers are shown in Fig. 52, which compares the diffraction scans of low density polyethylene, a

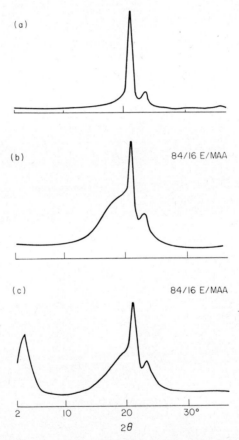

Fig. 52 X-ray diffraction scans of (a) low density polyethylene, (b) ethylene–methacrylic acid copolymer (5·8 mole % acid), (c) the 100% neutralised sodium ionomer.[43]

copolymer of ethylene with methacrylic acid (6·0 mole % acid) and the ionomer prepared by fully neutralising with sodium.

In all three cases, there is observed the polyethylene diffraction pattern, characterised by the presence of (110) and (200) peaks, superposed over a broad 'amorphous' peak. The percent polyethylene crystallinity is calculated from the ratio of the areas of the (110) plus (200) peaks to the total area. As might be expected, the crystallinity of the copolymer was considerably lower, and was found to decrease with increasing acid content; polymers containing more than 7·5 mole % methacrylic acid were amorphous. The neutralised copolymer, the ionomer, has essentially the same degree of polyethylene crystallinity as the parent copolymer. In addition, however, there is a strong peak at low diffraction angles which is not observed in the copolymer or low density polyethylene. The behaviour of this peak is particularly significant in describing the solid state structure of ionomers.

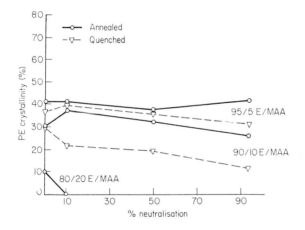

Fig. 53 Polyethylene crystallinity of ethylene–methacrylic acid copolymers and sodium ionomers. (Numbers indicate weight % acid).[43]

Before examining the behaviour of this peak, the polyethylene crystallinity is considered. In Fig. 53, the percent crystallinity is plotted against degrees of neutralisation with sodium for copolymers containing various amounts of methacrylic acid. The most significant result is that, for annealed samples, the crystallinity is virtually independent of the amount of ion. This is in spite of the fact that the appearance of the material changes from the translucence typical of low density polyethylene to essentially water-white (Fig. 1). Because of the high internal viscosity imparted by the presence of the ions, quenching is more effective in reducing the crystallinity the greater the degree of neutralisation. At the highest level of methacrylic acid, 7·5 mole percent, the presence of the ions reduces the crystallinity to zero. The clarity of the ionomers implies the absence of large spherulitic structures. Since the level of crystallinity is the same in the presence of the ions, the crystalline regions are either very small, fragmentary, lamellae or better described as fringe micelles.

Thus the crystallinity observed in these samples is entirely analogous to polyethylene homopolymer. It is reduced by quenching and increased by annealing; orientation by cold drawing has exactly the same effect as for polyethylene.

The structure of the ionomer differs from the copolymer only by the presence of the low angle peak; the behaviour of the peak was correlated with composition variables. The peak was observed with all the cations examined, including all the alkali metals from lithium through caesium, together with zinc, barium, ammonium and a quaternary ammonium ion. Thus neither the valence nor the strength of the base seems to be critical in the formation of the structure giving rise to the diffraction. The fact that the peak is observed with lithium neutralised samples indicated that the structural change is that of a rearrangement of the polymer rather than by a positioning of the cations, since lithium is too poor a scatterer of X-rays to give such an effect at the concentrations employed.

The peak persisted to temperatures of at least 300°C, the highest temperature available to the instrument, whereas the polyethylene crystallinity disappeared above about 100°C. When samples were cold drawn below 80°C, the polyethylene peaks showed the effect of orientation, but no orientation could be detected in the low angle peak.

The character of the peak was affected by the presence of water. Exposure of a dry, as-moulded sample to a 50% R.H. atmosphere enhanced the peak; saturation with water destroyed it. Lithium salts absorbed only a few percent water and in this case the peak was insensitive to humidity.

The peak occurred in the vicinity of 2θ 4·5° corresponding to a periodicity of 20 Å. The fact that there is a peak at all means that some periodicity is present; that is, there is a repeating unit consisting of a structural motif which repeats by a translation of about 20 Å. For a peak to be observed at all, there is a rule of thumb derived from optical analogues that at least five repeat units are required. This puts a lower limit of 100 Å on some linear dimension of a structural unit. Since only a single peak was observed, this implies that the perfection of the phase is low.

The magnitude and position of the peak are sensitive to the amount of acid in the parent copolymer and also to the degree of neutralisation. These effects are illustrated in Fig. 54 in which a series of copolymers containing from 1·7 mole percent to 6·4 mole percent methacrylic acid were neutralised with the same amount of base (not to the same degree of neutralisation), which neutralised the copolymer containing 1·7 mole percent acid to the extent of 90%, and the 6·6 mole percent acid to 28%. The value of 2θ

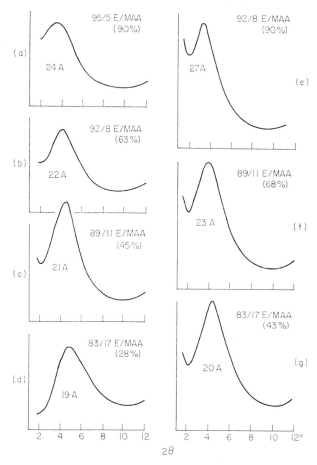

Fig. 54 Low angle X-ray diffraction peaks (polyethylene crystallinity omitted) of ionomers. Weight % methacrylic acid as indicated, (a) through (d) neutralised with same amount of sodium, (e), (f), (g) with increased amount of base.[43]

increased with increasing acid content; translated into Bragg spacing, the distance decreases from 24 Å to 19 Å along the series. However, increasing the ion content had the reverse effect. Samples (b), (c) and (d) were further neutralised to form a set in which the degree of neutralisation of the sample containing 2·7 mole percent acid was increased from 63% to 90%. In each case, the spacing increased. The magnitude of the peak depends primarily on the amount of ion. Thus samples (a), (b) and (d) have the same intensity (sample (c) is unusual) as do samples (e), (f) and (g), but with greater intensity.

The conclusion drawn from these observations was that there is present in this type of ionomer a phase containing the ionic fraction, which is distinct from a crystalline polyethylene phase. This description is elaborated below, but it was noted that a similar low angle peak was observed in dry sodium salts of polyacrylic and polymethacrylic acids (*see also* Ref. 45), except that in these cases the diffraction maximum was at 2θ 7·5° corresponding to a repeat distance of 12 Å.

The polyethylene crystallinity of an acid copolymer and its sodium ionomer was determined by Kajiyama et al.[44] The crystallinity was measured by X-ray diffraction and also by differential scanning calorimetry (DSC). As with the earlier observations, they found the crystallinity of the acid copolymer (4·1 mole % methacrylic acid) and its ionomer (78% neutralised with sodium) to be almost the same, about 50% at room temperature, and to decrease at the same rate with increasing temperature. Curiously enough, the level of crystallinity was about that of a low density polyethylene, even though there are shorter average sequences of methylene groups in the copolymer and ionomer, resulting in a lower melting point of 100°C versus 115°C for low density polyethylene.

Similarly, they found that the acid copolymer could be quenched to a crystallinity of about 25%, which is not possible with polyethylene. This ability, of course, reflects the higher glass temperature T_β' of the copolymer.

The wide angle scattering of ionomers made from caesium salts of an ethylene–acrylic acid copolymer was examined by Roe.[46] The procedure was to compare the differences in radial distribution functions of the copolymer and the ionomer as a function of angle (or spacing) to see if there was any periodicity in the occurrence of Cs ions. As shown in Fig. 55 differences were found at spacings less than 8 Å, caused by the order imposed on Cs–O and Cs–Cs pairs. No differences were observed at spacings greater than 8 Å and up to about 20 Å. The smallest angle of observation was such that spacings of 50 Å and upwards would not have been observed. The study was intended to provide evidence for or against

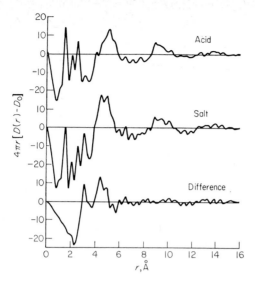

Fig. 55 Radial distribution functions for an ethylene–acrylic acid copolymer (4·6 mole %) and its ionomer and their difference vs. r, distance of atomic separation.[46]

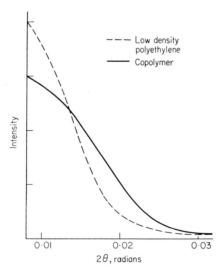

Fig. 56 Angular dependence of low angle X-ray scattering pattern from low density polyethylene and an ethylene–methacrylic acid copolymer (4·1 mole % acid).[47]

the formation of clusters of Cs ions in regions greater than about 100 Å in diameter and actually failed to resolve this question. A spacing of Cs ions of, say, 10–20 Å would only have been noticed had the Cs ions been arranged on a lattice rather than randomly arranged about an average spacing in this range.

Delf and MacKnight[47] measured the X-ray diffraction of a caesium ionomer of an ethylene–methacrylic acid copolymer down to 2θ 0·01°. The copolymer and low density polyethylene exhibited a similar steady increase in intensity with decreasing angle (Fig. 56). By contrast, the ionomer had a strong peak at 2θ 0·02°, corresponding to a length of 83 Å, (Fig. 57). This was interpreted as being due to the presence of large aggregates containing caesium ions. The peak was insensitive to annealing, which is known to affect the nature and extent of polyethylene crystals and spherulites. The corresponding lithium ionomer did not show this peak, which is to be expected, since lithium scatters X-rays only weakly.

MacKnight et al.[78] extended the low angle X-ray diffraction measurements to a series of the caesium ionomers of ethylene–methacrylic and –acrylic acid copolymers. The radial distribution functions were also calculated in the region of the ionomer peak not examined by Roe. In

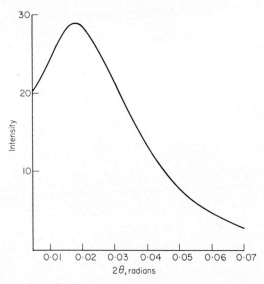

Fig. 57 Angular dependence of low angle X-ray scattering pattern from ionomer (59% neutralised Cs salt of 4·1 mole % ethylene–methacrylic acid copolymer).[47]

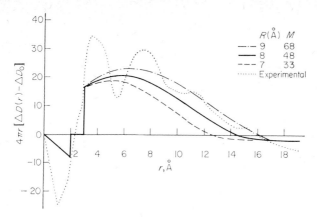

Fig. 58 Differences between the radial distribution functions of an ethylene–methacrylic acid copolymer (3·8 mole % acid) and its Cs ionomer, compared to the theoretical predictions of a hard sphere model. R is the radius of the sphere and M the number of Cs atoms per cluster.[78]

Fig. 58, the experimental differences in the functions between the ionomer and the copolymer are compared to the theoretical predictions of a hard sphere model of randomly arranged clusters for various values of the radius of the sphere and numbers of metal ions per sphere. It is evident that the hard sphere model does not account for the experimental function in the range 3–12 Å, and there must be present some internal structure. Furthermore, the curve becomes negative above 16 Å indicating an upper limit of 16 Å to a cluster. (These clusters are similar to the 'multiplets' postulated by Eisenberg[74] discussed in detail below.)

An extensive study of X-ray scattering of various ionomers has been made by Marx et al.[45] Two types of composition were examined, ionomers made from ethylene–methacrylic acid copolymers (synthesised similarly to those of Wilson et al.[43]) and butadiene–methacrylic acid copolymers. The cation was sodium. Variations in acid content (up to 7 mole % acid) and degree of neutralisation were examined together with the effect of plasticisers including water, methanol and carboxylic acids (formic, acetic and free methacrylic acids). X-ray scattering at wide and low angles was determined and also melting behaviour by differential scanning calorimetry. Care was taken to condition and anneal samples and to locate as precisely as possible the changes in intensity and angle of the 'low angle' ionomer peak.

The results were interpreted according to the equation

$$d_{\text{Bragg}} = C(V'f^{-1})^{1/3}$$

d_{Bragg} is the spacing corresponding to the measured Bragg angle, C is a constant of order unity, V' is the volume per carboxyl group (calculated from the composition) and f^{-1} is defined as the number of carboxyl groups per scattering site. The volume per scattering site was calculated from V' by picking integral values of f^{-1} such that a plot of d_{Bragg} versus V yielded a line of slope $\frac{1}{3}$ on a log–log plot. The plot is shown in Fig. 59. As with the earlier work,[43] the spacing decreased with increased acid content, down to a value of 11·7 Å for polymethacrylic acid. The data for both types of ionomer could be combined in this fashion.

The values of f^{-1}, the number of carboxyl groups per scattering site, increased from two for copolymers containing two mole percent acid to three for compositions between three and five mole percent and four for compositions from five to seven percent. The number was not dependent on the degree of neutralisation. A value of seven carboxyls per site was found for polymethacrylic acid two-thirds neutralised with sodium. At low angles, the behaviour observed by Delf and MacKnight was not found, as shown in Fig. 60. Although there is some scattering by the ionomer, the peak is smaller than the unneutralised copolymer and very much weaker than low density polyethylene. The conclusion was that the scattering was caused by polyethylene lamellae and that the influence of the ions was to reduce the size and extent of these lamellae, and it was suggested that the results of Delf and MacKnight were symptomatic of the presence of water. However, Delf and MacKnight made efforts to use a dry sample, and noted that the peak was quite pronounced for the caesium ionomer but could not be seen for the lithium salt, although all ionomers exhibited similar wide angle peaks.

Marx *et al.* explain the behaviour of the ionomers in terms of a so-called 'aggregate' model of the structure. The justification for this proposal will be discussed below in comparison with other models of the structure. Their evidence for this model comes from (a) the calculations described above of the carboxyl groups per scattering site, based partly on reinterpretation of earlier data;[43] (b) the absence of an ionomer peak at low angles (0·01 radians) corresponding to spacings of the order of 100 Å, as shown in Fig. 60; and (c) the absence of large grains in electron micrographs of films of ionomers from methacrylic acid–butadiene copolymers.[48] However, other studies indicate that a low angle peak is fairly general; for example, such a peak has been found with ionomers prepared from ethylene–methacrylic acid copolymers,[47] styrene–methacrylic copolymers[49] and

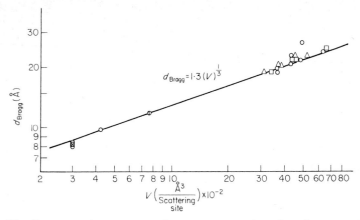

Fig. 59 Bragg spacing versus volume per scattering site. Ionomers from ethylene–methacrylic acid copolymers (○, △); butadiene–methacrylic acid copolymers (□); polymethacrylic acid (⊙); sodium acetate (⊖), one-third neutralised.[45]

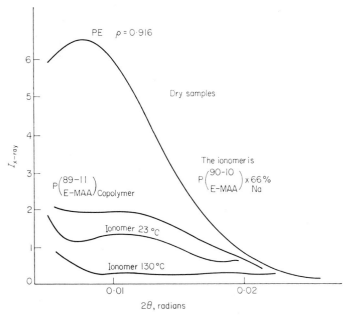

Fig. 60 Small angle X-ray scattering of low density polyethylene (PE), an ethylene–methacrylic acid copolymer (3·5 mole % acid) and the 60% sodium neutralised ionomer.[45]

carboxy terminated polybutadienes;[50] these peaks have been interpreted as evidence for ionic clusters of about 100 Å size. That is to say, that the existence of a secondary low angle peak outside the main peak (*i.e.* at a wide angle) is indicative of regularity in the spacing between the scattering centres.

2.9 OPTICAL PROPERTIES OF IONOMERS

The drastic reduction in haze with neutralisation of ethylene–methacrylic acid copolymers has been noted earlier. Since these ionomers retain a substantial degree of polyethylene crystallinity, the simplest explanation is that the ions nucleate the formation of crystallites and also that the high internal viscosity slows down the growth of these crystallites into spherulites; with lengthy annealing, spherulites do grow.[63] Similar behaviour has been noted by Wissbrun[80] for ionomers based on polyoxymethylene copolymers. When the amount of ionic comonomer exceeded 5 mole %, there was a sudden drop in haze from 100% to 20%, whereas the crystallinity (X-ray) declined linearly over this range only from 80% to 60%.

A more detailed examination of the optical properties of ethylene–methacrylic acid copolymers and ionomers has been made by Stein and coworkers.[32,63,64] The principal objective of these studies was to understand the structural features of ionomers giving rise to the α transition, which is distinguished from the α relaxation in polyethylene, since it increases with acid and ionic content and decreasing crystallinity. The effect of this transition on the mechanical and dielectric properties has been reviewed earlier.

Kajiyama *et al.*[32] measured the static and dynamic birefringence of ionomers derived from an ethylene–methacrylic acid copolymer containing 4·1 mole % acid. The static strain optical coefficient, K_s, was calculated from the slope of the (linear) plot of birefringence versus elongation. The dynamic strain optical coefficient, K^*, was separated into real (K') and imaginary (K'') parts.

The birefringence was essentially independent of time, and did not change appreciably when a stretched specimen was held for several hours even at a temperature above the α relaxation temperature.

The variation in K_s with temperature is shown in Fig. 61 for various ionomers and the free acid copolymer. (For a copolymer containing 8 mole % acid, the peak in K_s was much greater and also at a slightly higher temperature.) For the acid copolymer, K_s decreases monotonically with

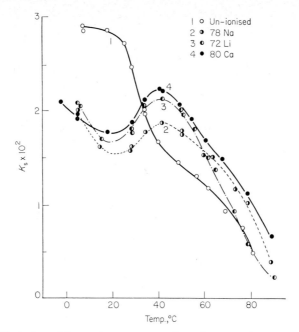

Fig. 61 Variation of the static strain-optical coefficient with temperature for an ethylene–methacrylic acid copolymer (4·1 mole % acid) and ionomers, metals and degree of neutralisation as shown.[32]

temperature whereas all the ionomers exhibit a maximum at 42°C. The birefringence and mechanical relaxation of the copolymer and calcium ionomer are compared in Fig. 62. The temperature of the dynamic mechanical peak is somewhat higher than the dynamic birefringence peak, which is in turn higher than the static birefringence peak. This latter difference reflects the fact that the dynamic birefringence occurs over a shorter time scale than the static test and involves only reversible deformation, whereas K_s includes irreversible deformation.

These results can be explained as being due to two processes. As the transition temperature is approached, there is an increase in orientation of both the hydrocarbon and ion containing segments,[62] corresponding to a softening of the latter. It is considered that the acid groups function as quasi cross-links, either by hydrogen bonding or by electrostatic attraction. Since the strain optical coefficient is proportional to the number of these cross-links, there is a general decrease in K_s with increasing temperature. However, the carboxylate cross-links are 'frozen in' at lower temperatures

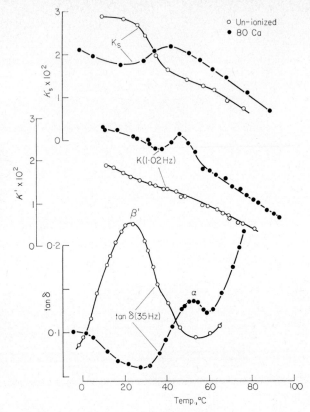

Fig. 62 A comparison of the variation with temperature of the static strain-optical coefficient, the dynamic strain optical coefficient and the mechanical loss tangent for an ethylene–methacrylic acid copolymer (4·1 mole % acid) and the calcium ionomer (80% neutralised).[32]

and do not contribute to the birefringence until the α temperature is reached.

The variations in K' could be treated by temperature frequency superposition. However, both vertical and horizontal shift factors were needed, the results being shown in Fig. 63. A plot of the log of the horizontal shift factor versus reciprocal temperature could be fitted by a single straight line for the acid copolymer, but two lines were necessary for the ionomer; similar behaviour was already noted for the mechanical results. This implies the existence of two distinct processes for the ionomer.

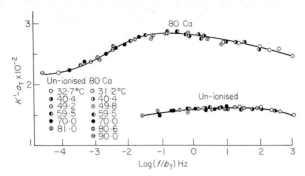

Fig. 63 Master curve of the variation of the reduced dynamic optical coefficient with reduced frequency for an ethylene–methacrylic acid copolymer (4·1 mole % acid) and the ionomer, 80% neutralised with calcium, reference temperature 30°C.[32]

Static[63] and dynamic[64] light scattering studies have been made by Prud'homme and Stein. The technique was to illuminate a film of the plastic with an He–Ne laser, vertically polarised, and record the scattered light polarised horizontally, H_v, and vertically, V_v. From the shape and intensity of the pattern of scattered light the shape and size of the scattering particles could be calculated.[65]

The pattern for the ethylene–methacrylic acid copolymer containing 4·1 mole % acid indicated spherulitic structures. From this and earlier measurements[44] it was concluded that lamellar crystals of the copolymer form spherulites similar to those of low density polyethylene, but that because of the acid groups the rate of crystallisation is slower, implying that the acid groups are excluded from the crystals on to the surfaces of the lamellae. By contrast, the air-cooled ionomer (55% neutralised with sodium) gave a pattern characteristic of anisotropic rods, which is indicative of disordered or incompletely formed spherulites. When the ionomer was annealed at 91°C, spherulites were formed even after 0·5 h, but more definitely after 18 h. Photometric scans of the intensity of the scattered light versus scattering angle are shown in Figs. 64 and 65 for the acid copolymer and the quenched and annealed ionomer. The spherulites of the acid copolymer had radii of 2·3 μm, whereas those formed by annealing the ionomer were appreciably larger, 4·1 μm.

Further information about the spherulitic structure was obtained by dynamic light scattering measurements,[64] in which a sinusoidal (2·9 Hz) strain (1%) is imposed on an elongated (5%) sample. The direction of elongation was parallel to the polarisation of the incident light and the

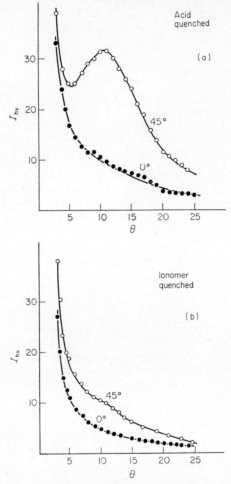

Fig. 64 Photometric scans of the scattering intensity at $\psi = 0°$ and $45°$ for samples of (a) ethylene–methacrylic acid copolymer (4·1 mole % acid) and (b) sodium ionomer 55% neutralised.[63]

horizontally polarised scattered light recorded (H_v mode). To develop the spherulitic structure, the sample of ionomer was slow-cooled overnight from 125°C. Values of the in-phase and out-of-phase components of the scattered light were obtained at temperatures up to 70°C. The values of the in-phase (real) components are given in Fig. 66 for the acid and the ionomer (sodium). The behaviour of the ionomer is drastically different, in fact

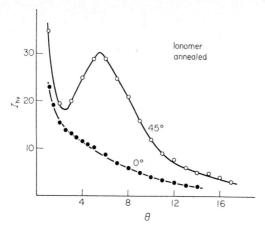

Fig. 65 Photometric scans of the intensity of the scattered light at $\psi_1 = 0°$ and $45°$ for an annealed sodium ionomer (ethylene N methacrylic acid copolymer 4·1 mole % acid 55% neutralised).[63]

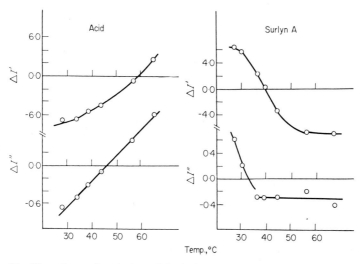

Fig. 66 Experimental variation of the $\Delta I'$ and $\Delta I''$ components of the dynamic light scattering as a function of temperature at $\theta = \theta_{max}$ (angle of maximum scattering) azimuthal angle $\mu = 45°$, for an ethylene–methacrylic acid copolymer (4·1 mole % acid) and the ionomer 55% neutralised with sodium.[64]

almost the reverse, from the copolymer. This behaviour was explained in terms of the Van Aarsten–Stein theory[65b] of the deformation of spherulites which considers two distinct types of process, tilting of crystallites or lamellae towards the optic axis and twisting, predominantly in the equatorial region.

For the acid copolymer, twisting predominates at low temperatures but as the temperature increases the inter-lamellar forces decrease and tilting becomes more important. For the ionomer, the tilting process at low temperatures implies that the deformation of the lamellae is controlled by high modulus ionic material. As the α relaxation temperature is approached this ionic material becomes rubbery, and the strain is distributed throughout the material and twisting processes can occur.

2.10 THERMAL PROPERTIES OF IONOMERS

The crystallisation of ethylene–acid copolymers and ionomers as determined by calorimetry has been examined by several authors, beginning with Rees and Vaughan.[2a] They showed that the melting point of both the parent copolymer and the ionomer was lowered to about 100°C from 115°C, characteristic of low density polyethylene. A premelting endotherm was also identified qualitatively, whose magnitude and temperature depended on both composition and thermal history.

MacKnight et al.[9,44] compared calorimetry data with degree of ionisation for a copolymer containing 4·1 mole percent methacrylic acid. Figure 67 shows typical heating and cooling curves and Table 7 shows the results; degrees of ionisation were determined by infra-red. The melting point, T_m, is virtually independent of degree of neutralisation, as is T_1, the peak temperature of the endotherm (except for the 78% ionised sample). However, the values of T_2, the maximum of the exotherm, show considerable supercooling, which increases with degree of ionisation from 18°C (zero ionisation) to over 40°C for 78% ionisation. It was suggested that this increase is caused by the increase in melt viscosity on ionisation, which is in line with the enormous increase in 'internal viscosity' noted earlier (Section 2.5.3). The degree of crystallinity was quite small compared to low density polyethylene.

Similar findings were reported by Otocka and Kwei[53,54] who studied variations in acid content of ethylene–acrylic copolymers. A comparable low density polyethylene had short chain branching of 3·5 CH_3/100 C and a melting point (defined as in Fig. 67) of 108°C. The degree of crystallinity

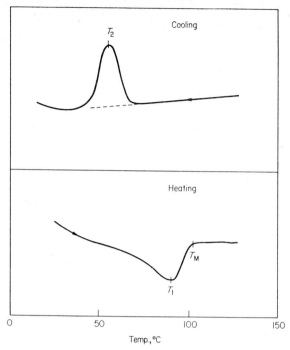

Fig. 67 Schematic representation of DSC data showing the melting endotherm and crystallisation exotherm of an ionomer.[9]

TABLE 7

SUMMARY OF IR AND DSC DATA[a]

% Ionisation	T_m (°C)	T_1 (°C)	T_2 (°C)	ΔH_f[b] (cal g^{-1})	wt% crystallinity[c]
0	100	91	73	9.9	14.9
20	101	91	70	10.7	16.2
60	100	91	54	8.2	12.5
78	99	86	42	4.6	6.9

[a] Cooling and heating rates in the differential-scanning calorimeter were each 10°C/min.
[b] ΔH_f = heat of fusion.
[c] Based on a heat of fusion of 66 cal. g^{-1} for the 100% crystalline polymer.

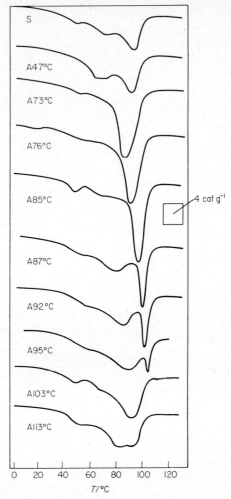

Fig. 68 Effect of annealing (A) on the ionomer DSC curves. Following slow cooling (S) from the melt, samples were annealed at the temperatures indicated. Ethylene–methacrylic acid copolymer (5·7 mole % acid) 37% neutralised with sodium.[51] □ = area for calibration.

when annealed was 40% and the supercooling, 18°C. As the acid content was increased to 2·8 mole percent (four samples in all) the melting point decreased to 95°C and the supercooling was constant at 18°C and the crystallinity decreased to 16%. For the ionomers (100% neutralised) the changes were more drastic. At the highest level of acid, the melting point was 91°C, the supercooling was 42°C and the crystallinity only 5%.

The results were interpreted by means of Flory's equation

$$\frac{1}{T_m} - \frac{1}{T_m^0} = \frac{-R}{\Delta H_u} \ln N$$

where T_m is the melting point of the copolymer, T_m^0 that of the homopolymer, N the fraction of crystallisable units and ΔH_u the heat of fusion of homopolymer crystals. Linear plots of $1/T_m$ versus log N were obtained and a value of 750 cal/mol. obtained for the heat of fusion of —CH_2— was obtained. However, only the carboxyl and salt groups were included in N, and three different intercepts (for T_m^0) were obtained for the acid, sodium

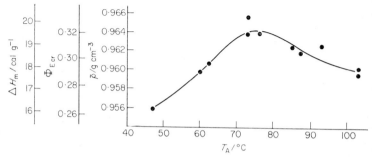

Fig. 69 Melting enthalpy (ΔH_m), ionomer volume fraction crystallinity ($\Phi_{E,Cr}$) and calculated density (ρ) as a function of annealing temperature (T_A). For an ionomer, ethylene–methacrylic acid (5·7 mole % acid) 37% neutralised with sodium.[51]

and magnesium ionomers. The tentative explanation for this was a change in either the fold surface free energy of the lamellar crystal or the lamellar thickness.

Marx and Cooper[51] made a detailed examination of the effect of annealing on an ionomer prepared from a 6 mole percent ethylene–methacrylic acid copolymer, 37% neutralised with sodium. The results are shown in Fig. 68; the effect of annealing is to cause the melting endotherm to appear at a slightly higher temperature than the annealing temperature. The degree of crystallinity was calculated from the peak area, which is proportional to the melting enthalpy. The values are shown in Fig. 69, which also gives derived values of the volume fraction crystallinity and density. The data show a maximum similar to high density polyethylene.[67] When the calculated density curve is compared to the experimental densities, the latter form a plateau lower than the calculated values. This is interpreted as there being voids present, caused by lamellar thickening straining the amorphous regions.

Wissbrun[80] examined the crystallisation behaviour of polyoxymethylene ionomers; the results were qualitatively similar to the ethylene based materials. Interestingly, the peak melting temperatures for a variety of ionic and non-ionic copolymers fell on the same curve, decreasing from 175°C to 150°C at 8·5 mole percent comonomer. However, the freezing temperatures for the ionomers were about 10°C lower than the corresponding non-ionic copolymers.

2.11 THE STRUCTURE OF IONOMERS

The foregoing sections have described various measurements of the physical structure of ionomers derived from ethylene–acid copolymers. There will now be considered interpretations which have been made to account for the results of these measurements.

The discussion starts with the structure of low density polyethylene. Numerous investigations[68] have established that linear, or high density, polyethylene consists of lamellar crystals approximately 100 Å in thickness and several microns in extent containing folded chains of polyethylene molecules with the chain axis perpendicular to the plane of the crystal. The lamellae are arranged in regularly twisted ribbons which are oriented radially from a core to form crystalline spherulites. These spherulites can vary from 10 μ to 1000 μ in diameter depending on the method of crystallisation and the degree of nucleation. The amorphous fraction is material consisting of defects of various kinds: folded sections of chains, chain ends, and sections of chains connecting lamellae; that is, all of the material could in principle be incorporated into lamellae, as it is in lamellae crystallised from solution. In the case of low density polyethylene, there is a substantial fraction of the material in the branch points of long and short chains. These occur with a frequency of 1 to 10 per thousand carbon atoms for the long chain branch points and 1 to 10 per 100 carbons for the short chains. The effect of these branch points is to increase the T_g of the amorphous, non-crystalline material so that there is a transition around −20°C to −40°C, the β transition.

There is considerable discussion as to whether these branch points are located in or out of the lamellar crystals. The evidence favours their exclusion. Thus it is to be expected that the side groups of a vinyl comonomer will also be excluded, particularly when these groups are polar and capable of hydrogen bonding, as are carboxyl groups.

The observed behaviour of the ethylene copolymers that have been discussed here is consistent with this viewpoint. The crystallinity as shown by X-ray diffraction is quite well-defined even at high levels of comonomer content albeit the amorphous fraction is increased considerably (compare Figs. 52a and b). The electron micrographs show clearly evidence for lamellar structure up to six mole percent copolymerised acid. Evidence from mechanical relaxations[9,14,53] shows that the β relaxation of low density polyethylene is shifted to higher temperatures, from $-20°C$ to about $40°C$ with increasing acid content. To a greater or lesser extent, similar behaviour is observed for all ethylene copolymers (*e.g.* methyl methacrylate and vinyl acetate copolymers, respectively).

It is apparent that the structural changes taking place on the ionisation of the acid groups are much more complicated than for other copolymers. The features which cannot be explained by comparison with the behaviour of other copolymers may be summarised as follows:

(1) Electron microscopy shows disruption of spherulitic structure even at low (2 mole percent) levels of ionised acid (although extensive annealing can develop spherulitic structure to a limited extent[63]). This results in high transparency in thick sections (Fig. 1).

(2) X-ray diffraction shows the presence in the ionomers of a characteristic low angle peak in addition to the polyethylene diffraction peaks. This peak corresponds to a spacing of about 20 Å; is dependent on the nature of the ion; is stable to temperatures well above the polyethylene melting point ($300°C$); is only slightly dependent on the degree of ionisation (rather more so on the total acid content); is not affected by orientation.

(3) With copolymerisation, the β (amorphous) relaxation of polyethylene is reduced and there appears a β' relaxation which increases in maximum loss temperature with increasing comonomer content. On ionisation, the β relaxation reappears at approximately the same temperature as in polyethylene, and the β' relaxation decreases with increased degree of ionisation. There is also an α relaxation which is at a higher temperature than for the un-ionised copolymer.

(4) Orientation experiments[32,44,64] show that the free carboxyl groups orient as expected in a copolymer, but the ionised groups behave anomalously.

(5) The effect of changing the ion from mono- to divalent is relatively minor. In the ionomers in the solid state, both ions are effective as 'vulcan-

ising' agents, yet the explanation of why a monovalent ion does provide cross-linking in the solid state is not obvious.

Furthermore, the modulus of the ionomers at low temperatures (0°C to −40°C) is less than that of either the parent copolymer or low density polyethylene itself, whereas one would have expected it to be greater.[6a,14]

The behaviour of solutions of simple electrolytes in solvents of low dielectric constant may be noted briefly.[69,81] The potential energy of an ion pair is given by the Coulomb formula, $e^2/2aD$, (e is the electronic charge, a, the separation and D, the dielectric constant). For aqueous solutions at 25°C this is about equal to kT, the thermal energy, and so an ion pair is not very stable. When D falls to <10, the dissociation energy per ion pair is around 3 kcal mole^{-1} and considerable association is found. The conductance of simple electrolytes in solvents such as benzene thus decreases rapidly with increasing concentration as electrically neutral ion pairs are formed and then rises erratically as triplets are formed, levels off with quadrupole formation and then rises again. The formation of quadrupoles occurs when the concentration is not too great, about 0·01 molar.[70]

Consequently, it is to be expected that in an ionomer, where there are ions embedded in a macromolecular solvent of low dielectric constant, there will be association to form clusters of ions. The idea that association to ionic clusters occurs in ionomers was apparently first proposed by Bonotto (noted in Ref. 30), following a suggestion by Rees and Wilson. However, the simple idea of an ionic cluster leads to the concept of a structural unit which is a sort of labile cross-link (labile because the ionomers are truly thermoplastic). The material thus cross-linked is the amorphous hydrocarbon excluded from lamellae. It seems likely that such material would behave similarly to a simple copolymer of ethylene with the salt of a multifunctional acid. The results described above imply a more complicated structure.

A model which accounted for the experimental evidence available at that time was proposed by Longworth and Vaughan.[14,37] This model is represented schematically in Fig. 70. In (a) is a sector of a spherulite of an ethylene–methacrylic acid copolymer, with lamellar crystals aligned radially. The methacrylic acid groups are shown as being excluded from the lamellae. (The area of the blocks of folded chains is approximately the degree of crystallinity.) The matrix of hydrocarbon plus methacrylic acid forms a more or less uniform copolymer of T_g which will increase with acid content to temperatures above the β transition temperature of low density polyethylene.

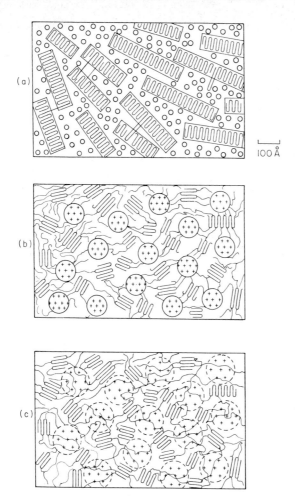

Fig. 70 Schematic representation of the structure of ethylene copolymers (a) folded chains of polyethylene segments interspersed with carboxyl groups; (b) the dry ionomer; (c) the wet ionomer.[37]

When the acid groups are ionised, a profound structural reorganisation takes place; there are formed clusters of ions into regions about 100 Å in diameter or greater (*b*). The inter-ionic forces leading to the formation of these clusters are powerful enough to prevent the formation of lamellae and the crystallites of polyethylene resemble fringed micelles. The crystalline and ionic regions are connected by a matrix of hydrocarbon. The effect of water, (c), is to plasticise the ionic segments, thus allowing hydrocarbon to be incorporated into crystalline material and also lowers the T_g of the ionic material.

This model accounts for the observations listed above, in particular there is qualitative agreement with the electron microscopic observations; the low angle X-ray diffraction peak; and the dynamic mechanical behaviour, exemplified by the β transition which is thus a process involving the rubbery hydrocarbon segments formed between glassy ionic domains on the one hand, and crystalline lamellae or micelles on the other.

The identification of the ionic regions with the low angle X-ray diffraction peak requires further comment. The peak occurs at an angle 2θ 4·5°, corresponding to a separation of about 20 Å. The existence of this peak implies a periodicity of some sort; optical analogues indicate that at least five repeat units are necessary for any periodicity at all to be observed; this sets the lower limit of the ionic regions at 100 Å. Since only one peak has been observed, the perfection is low. The fact that the peak is observed for lithium ionomers as well as for larger ions implies that the structural change wrought by the ions is of the macromolecules themselves (including the pendant carboxylate groups). Lithium is much too poor a scatterer to give this effect at the concentrations employed. The X-ray diffraction results alone, only give a lower limit to the extent of the ionic domains.

Alkali metal ions have a tendency for six-fold co-ordination, *i.e.* three carboxylate groups can be arranged in an octahedron around the metal ion. It is noteworthy that many of the physical properties which distinguish the ionomers from low density polyethylene or the acid copolymer are substantially achieved at 33% neutralisation. Thus it is proposed that these co-ordinated ions have a tendency to cluster together and that this tendency imposes an order on the interconnecting hydrocarbon segments. It is the ordered chains which give rise to the scattering and not an ordered arrangement of the ions.[43]

In summary, this model provides for the existence of three structural features rather than two: crystalline polyethylene, amorphous (branched) polyethylene and ionic copolymer.

The scheme shown in Fig. 70b represents the ionic regions as being isolated from the crystalline material by hydrocarbon segments. However, since it is quite likely that there is a considerable concentration of methacrylic acid groups close to the surface of the crystals, from which they are excluded, a physically more realistic model is the modification proposed by Stein *et al.*[44,62,63] in which the crystalline, amorphous and ionic regions are interlaced, so that orientation of the material results in extension of segments within each region to an extent depending on the relative moduli. The regions are connected in parallel rather than in series.

This model has been used with considerable success to explain the results described above of measurements of dielectric and mechanical relaxation,[9,37,40] X-ray diffraction,[43,44,47] infra-red dichroism and optical birefringence[32,62,64] and electron microscopic examinations.[15,59] However, contrasting evidence from mechanical relaxation,[54] X-ray diffraction,[46] nuclear magnetic resonance,[60] and electron microscopy[48] has been used to argue that the ionised carboxylate groups are uniformly dispersed throughout the non-crystalline regions. This model is thus a two-phase concept of the structure of the ionomers.

Otocka and Kwei[53,54] observed that the temperature of maximum loss of the low temperature transition increased from the value characteristic of low density polyethylene ($-21°C$, T_β) with increasing amount of copolymerised acrylic acid (T_β'). The increase was accounted for as partly due to an increase in the glass temperature of the amorphous regions, and partly due to a cross-linking effect of the carboxyl dimers. The range of compositions was rather narrow, 0·7 to 3·0 mole percent, and the increase in T_β, ΔT_β, was from 16°C to 45°C. The calculated increases in T_g due to copolymerisation alone were from 4°C (0·7 mole percent acid) to 16°C (3·0 mole percent acid).

In the case of the ionomers prepared from these copolymers, the values of ΔT_β were relatively much smaller ranging from 7°C to 19°C for the sodium salts. However, the T_g of poly(sodium acrylate) was estimated (by extrapolation of the T_g's of the styrene acrylate ionomers) to be 230°C, which is much greater than the T_g of polyacrylic acid (106°C). These values are only slightly larger than the values for the acid copolymer calculated for copolymerisation effect. Although the agreement between the calculated and measured ΔT_β's is good, it is clear that for these materials the calculation is fairly insensitive to the value of T_g of the comonomer, *i.e.* a difference of over 100°C in the T_g's of polyacrylic acid and poly(sodium acrylate) results in a difference of only 3°C in the calculated increment at the three mole percent level.

Furthermore, the data of MacKnight *et al.*[9] and Longworth and Vaughan[14] show that the T_β of the ionomers does not increase continuously with acid content and in fact remains close to T_β of low density polyethylene even at 7·0 mole percent acid. The position of T_β is indeed affected by annealing and thermal history and particularly by water content. Thus it seems reasonable that the T_β values observed by Otocka and Kwei, which are consistent with other data, can be explained on the basis of formation of segregation of ionic material from amorphous hydrocarbon.

Roe's X-ray diffraction study of an ionomer from an ethylene–acrylic acid copolymer does not examine the size range that is of interest, that is clusters or ionic regions 100 Å or greater in extent. The electron microscopic observations of Marx et al.[48] were of butadiene–methacrylic acid copolymers, which are not directly comparable to ethylene–methacrylic acid copolymers, particularly because of the absence of crystallinity in the former. The photomicrographs revealed domain structures, which were about 20 Å in extent, in both the free acid and ionised acid copolymers, but any larger structures were not observed. Because of the differences in the nature of the hydrocarbon backbone and the absence of crystallinity, these results cannot be considered as contrary evidence to the electron microscopy results from the examination of ethylene copolymers. These results do show that copolymers of highly polar monomers with hydrocarbons develop supramolecular structures, however.

The X-ray diffraction results of Marx et al.[48] include data from butadiene–methacrylic acid copolymers as well as extending and reinterpreting the results of Wilson et al.[43] The interpretation is to calculate from the composition and the Bragg spacing the number of carboxyl groups per cluster or aggregate. A similar calculation has been made by Binsbergen and Kroon.[73] The conclusion drawn from these calculations is that such ionic aggregates are uniformly distributed throughout the non-crystalline region. This conclusion does not differ significantly from the earlier discussion of Wilson et al.[43] who interpreted the single low peak as indicating a rather imperfect periodicity of about 20 Å. They mentioned that a complex of three carboxyl groups per metal ion would provide a suitable scattering site. Their results only indicated the minimum number of these sites to be about five to give rise to the scattering. They did not imply the existence of lattice planes adjacent to the ions. The point at issue is whether these scattering centres occurring at an average distance of 20 Å apart (depending on composition and sample preparation) entirely occupy the material not actually in lamellar crystals or whether there exists homogeneous hydrocarbon in addition.

This question does not, for this system, appear to be resolvable by X-ray diffraction alone, and the evidence from mechanical relaxation and optical studies is more to the point.

A theoretical treatment of the problem of ion association and clustering in polymers has been made by Eisenberg,[74] as already discussed briefly in Chapter 1. Assuming that the ions exist as ion pairs, he calculated the maximum size of ionic 'multiplets' from geometrical considerations. The result, for an ethylene–sodium methacrylate copolymer, was that up to eight pairs ($-COO^-Na^+$) could be accommodated in an 'ionic drop' before the surface of this drop was completely coated with non-ionic hydrocarbon.

Making various reasonable assumptions, the electrostatic energy released on these multiplets aggregating to form clusters was balanced against the net elastic energy necessary to stretch and contract the hydrocarbon chains between ions. The temperature at which these energies balanced, T_c, was identified with the temperature of the α transition in ionomers, about 50°C.[9] If the multiplets contain n_0 ion pairs and are separated by a distance R_0, then on clustering there are n ion pairs per cluster separated by a distance R. n is given by the equation

$$n = \rho \frac{N}{M_c} \left[\frac{4l^2}{3kT_c} \frac{\bar{h}^2}{\bar{h}_0^2} \frac{M_c}{M_0} \frac{k'}{K} \frac{1}{4\pi t_0} \frac{e^2}{r} + 2 \left(\frac{n_0 M_c}{\rho N} \right)^{2/3} \right]^{3/2}$$

where ρ is the density; N Avogadro's number; M_c the molecular weight between carboxyl groups; M_0 the molecular weight per repeat unit; l the length of a C—C bond, $\bar{h}^2$ the mean square end-to-end distance of a free chain and $\bar{h}_0^2$ the mean square end-to-end distance of a freely jointed chain, K the dielectric constant; k' will depend on the particular cluster geometry chosen. Values of n and R were calculated for various cluster geometries.

A cluster was visualised as a central ionic multiplet surrounded by a non-ionic hydrocarbon skin around which are arranged various ion pairs and the multiplets in such a way as to minimise the electrostatic energy. Quantitative calculations of such a model are difficult and three simpler arrangements were considered.

(1) An arrangement of ion pairs (acting as dipoles) on a cubic lattice. k' was found to be 0·020, n was then calculated to be 160 ion pairs and the cluster separation, R, 55 Å.

(2) The octet-pair model, in which below T_c four ion pairs are arranged at the corners of a cube to form octets which are connected along the body diagonals by single ion pairs to form clusters. Above T_c the octets are separated from the tie pairs. The result was $k' = 0.063$ whence $n = 800$ and $R = 95$ Å.

(3) In this case, the multiplets have an excess of one or more charges. Because of the greater distance and sizes, the aggregation into clusters is much weaker than for ion pair formation. The calculation was made for octets of approximately four ion pairs with an excess of one charge of

either sign. k' was approximately equal to 0·008 whence $n = 82$ and $R = 44$ Å.

These calculations are instructive as they show that various models will give rise to supramolecular structures containing ionic multiplets which are about 100 Å in diameter. This dimension corresponds to the size found by Delf and MacKnight from low angle X-ray scattering.[47] However, the persistence of the low angle peak[43] to temperatures of 300°C in the case of ethylene based ionomers, which is over 200°C above T_c is not accounted for. No change in the intensity or spacing of this peak was observed with increase in temperature through the α transition temperature for any given composition. This suggests a model resembling the earlier proposal[43] in which the ionic drops or multiplets, of diameter 6 Å or 8 Å are arranged in a paracrystalline array at an average distance of about 20 Å, which would allow space for two hydrocarbon chains between the hydrocarbon coating of the droplet. Such a suggestion has recently been made by MacKnight et al.[78] from low angle X-ray diffraction studies.

Eisenberg considered qualitatively the effect of reducing the ionic concentration by decreasing the spacing, i.e. by lowering the amount of copolymerised acid. When the spacing is less than a certain amount, the elastic retractive forces will reduce the size of the multiplets down to single ion pairs. Their ability to approach each other to form clusters is hampered by entanglements between chains. The onset of this condition is when M_c, the distance between ionic groups is about equal to the critical molecular weight for the development of entanglements (i.e. the chain length above which the 3·4 power viscosity law holds[4]).

The equations predict that $n \propto M_\text{c}^{1/2}$, $T_\text{c}^{-3/2}$ and $R \propto M_\text{c}^{1/2}$, $T_\text{c}^{-1/2}$. Because of the dependence on the square root of M_c, the effects of change in composition will not, therefore, be very marked above the critical composition. (Note that in the electron micrographs presented above, it appears that the critical limit is around 0·5 to 1·0 mole % acid for the ethylene ionomers.)

2.12 WATER ABSORPTION AND PLASTICISATION

From the above discussion of the clustering of ionic multiplets it is to be expected that water and other hydroxylic compounds will affect the properties of ionomers. Wilson et al.[43] noted that a small amount of water sharpened and intensified the ionomer peak in X-ray diffraction at

2θ 4·5°. Saturation with water hydrated the ions (presumably the multiplets) to the extent that the peak was eliminated.

This effect was studied in greater detail by Marx et al.[48] For sodium ionomers of methacrylic acid copolymers of both ethylene and butadiene, the results are similar, the intensity of the low angle peak is reduced with increasing amount of water and there is increased scattering at smaller angles.

In addition to altering the character of the low angle peak, water lowers the temperature of both the α and β relaxations (Fig. 38). The lowering of the α transition temperature can be interpreted straightforwardly as a

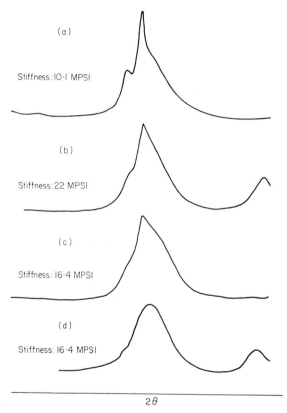

Fig. 71 X-ray diffraction scans shortly after quenching. (a) Ethylene–methacrylic acid copolymer (3·1 mole % acid); (b) the fully ionised sodium ionomer; (c) ethylene–methacrylic acid copolymer (5·8 mole % acid); (d) the fully ionised sodium ionomer.[79] (1 mpsi = 1000 psi.)

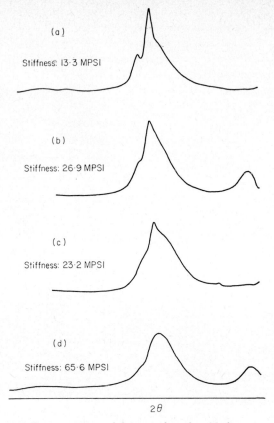

Fig. 72 X-ray diffraction scans of dry samples after 21 days at room temperature. Samples as in Fig. 71.[29] (1 mpsi = 1000 psi.)

plasticising of the ionic multiplets. The effect on the β' transition is more complicated, since this relaxation is thought to be a glass–rubber transition of primarily hydrocarbon segments between ionic clusters and crystalline polyethylene. However, any un-ionised carboxyl groups and probably some isolated carboxylate groups that are still present will hydrogen bond to or become solvated by the water and the relaxation also increases, and this is because some of the segments tied to ionic clusters are freed and take part in the β relaxation.

The modulus of the entire ionomer will be the result of contributions from the crystalline, amorphous hydrocarbon and clustered ionic regions. Some understanding of how these can change with thermal and environ-

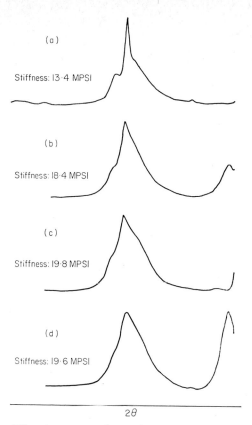

Fig. 73 X-ray diffraction scans of samples conditioned 24 hr. at 80% R.H. Samples as in Fig. 71.[79] (1 mpsi = 1000 psi.)

mental history is of practical importance. Experiments by Longworth and Wilson[79] examined the effect of variations in the conditioning on two copolymers of ethylene containing 3·1 mole percent and 5·0 mole percent methacrylic acid and their fully neutralised sodium ionomers. The flex modulus was determined as the stiffness, according to *ASTM* D747 (Ref. 11, p. 255) which because of the larger strain gives values about 30% lower than the flex modulus measured at small strains, and is measured with bars 2·5 in. × 0·5 in. × 0·075 in.

The bars were moulded at 180°C and quenched into iced water, keeping the water away from the surface of the bars. As soon as possible after moulding, the stiffnesses were measured and X-ray diffraction scans were

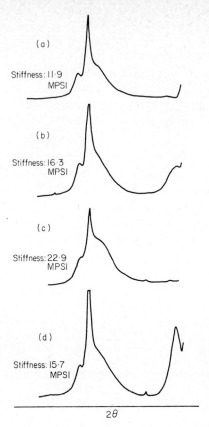

Fig. 74 X-ray diffraction scans of conditioned (80% R.H.) annealed (70°C/24 hr) sample. Samples as in Fig. 71.[79] (1 mpsi = 1000 psi.)

made. The results are shown in Fig. 71. The stiffnesses are all quite low and the polyethylene crystallinity of both the ionomers is markedly lower than the corresponding copolymer. Both of the ionomers exhibit the low angle peak.

The bars were then kept in a desiccator for twenty-one days, the stiffnesses and X-ray diffraction being measured at intervals. The final results are shown in Fig. 72, although most of the changes occurred during the first day. The changes in the stiffnesses are rather small, except for the ionomer sample (d), and there is no observable increase in the polyethylene crystallinity of the ionomers. The bars were then suspended in an atmosphere of 80% relative humidity (above saturated ammonium sulphate) for

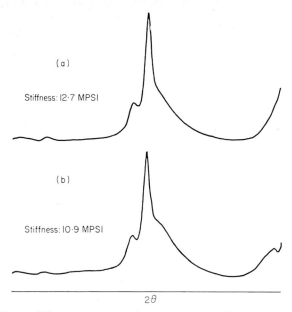

Fig. 75 X-ray diffraction scans of water saturated ionomers (annealed); ethylene–methacrylic acid copolymers fully neutralised with sodium. (a) 3·1 mole % acid; (b) 5·8 mole % acid.[79] (1 mpsi = 1000 psi.)

twenty-four hours, Fig. 73. While there is little or no change in the parent copolymers, for the ionomers there is a decrease in the stiffness, which is quite marked for the 6·0 mole percent copolymer, and a decided increase in the low angle peak. For the 6·0 mole percent copolymer, the polyethylene crystallinity shows signs of development. After 18 days at this condition, there were no significant changes. The bars were sealed in aluminium foil to prevent loss of moisture and then annealed overnight at 70°C. The results in Fig. 74 show that the polyethylene crystallinity has increased for all samples, but that the stiffness has not changed very much, being even lower on the whole. There were no changes in the low angle peak and so the bars were immersed in water for seven days at room temperature. The results for the ionomers are shown in Fig. 75; now the low angle peak has virtually disappeared and the stiffness is close to that of an elastomer.

Finally, the bars were dried in vacuum (0·1 mm) at 73°C for seven days, Fig. 76. The low angle peak has decreased in intensity and comparison with Fig. 74 shows that the scattering at still lower angles has decreased, as noted by Marx et al.[45] The stiffnesses are much higher than previously.

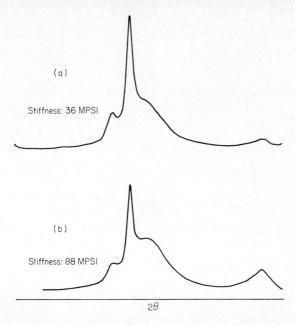

Fig. 76 X-ray diffraction scan of dry, annealed ionomers; ethylene–methacrylic acid copolymers fully neutralised with sodium. (a) 3·1 mole % acid; (b) 5·8 mole % acid.[79] (1 mpsi = 1000 psi.)

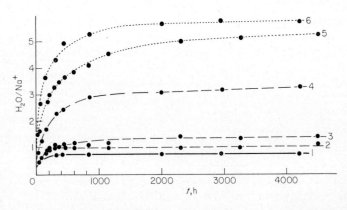

Fig. 77 Number of water molecules per sodium ion absorbed by polymer vs. time for styrene–methacrylic acid copolymers neutralised with sodium. Mole % methacrylic acid: (1) 3·7; (2) 1·9; (3) 5·5; (4) 7·9; (5) 9·7; (6) 9·1.[76]

The range of these values (36 000 psi to 88 000 psi) is much greater than is typical of low density polyethylene (22 000 psi to 33 000 psi for the density range 0·918 g./cm^3 to 0·923 g./cm^3). Thus the modulus of an ionomer is quite dependent on its exposure to heat and water, and it is the effect of these on the ionic clusters which determines the mechanical properties; the polyethylene crystallinity is of much less importance. These changes reflect the shift in the α transition temperature from about 50°C to below room temperature with increased water content. At saturation, the 6·0 mole percent ionomer absorbed 29% water by weight, corresponding to a molar ratio H_2O/—CO_2Na of 10·6.

The water absorption behaviour of ionomers made from styrene–methacrylic acid copolymers was determined by Eisenberg and Navratil.[76] Their results are shown in Fig. 77. For this system, time–temperature superposition does not apply to compositions above 6·0 mole percent acid, and this indicated the onset of clustering. The water absorption curves show that at lowest levels of acid, one mole of water per sodium ion is absorbed in about 300 hours, whereas a continuously increasing water uptake is found for the clustered samples containing above 6·0 mole percent acid.

Plasticisers other than water have not been investigated extensively. Marx et al.[48] studied the effect on the low angle X-ray diffraction peak of absorption by methanol, formic and acetic acids. The effect was qualitatively similar to water for both ethylene– and butadiene–methacrylic acid copolymers.

2.13 CONCLUSION

The commercial introduction of ionomers has stimulated a rather large number of structural studies which have been reviewed here. The combination of the extremes of ionic pairs and crystallisable hydrocarbon leads to unique structures. Undoubtedly, these and similar materials will be of increasing importance in the future.

Eisenberg,[77] in an interesting general discussion of ionomers, has suggested several areas in which further structural studies would be of value. These include: the existence and nature of clusters and multiplets, the kinetics of viscous flow, and the exact nature of the α transition. In addition, the function of water is not well understood and the effect of other polar plasticisers in so far as the physical and mechanical properties are changed has not been examined systematically. The usefulness of ionomers other than ethylene based materials has not been demonstrated.

ACKNOWLEDGMENTS

It is a pleasure to acknowledge the help of the following colleagues: Dr. E. T. Pieski for numerous discussions of the behaviour of polyolefins, Drs. F. C. Wilson and H. A. Davis for collaboration in the study of ionomers, Dr. D. J. Vaughan for stimulating our interest in the structure of ionomers, and Miss Marie J. de Brabander for patient and careful preparation of the manuscript. Thanks are also due to John Wiley & Sons, Inc., and to the American Chemical Society for permission to reproduce a number of diagrams.

REFERENCES

1. Otocka, E. P. (1971). *Macromolecular Science—Rev. Macromol. Chem.*, **C5** (2), 275.
2. (a) Rees, R. W. and Vaughan, D. J. (April, 1965). *ACS Polymer Preprints*, **6**, 287.
   (b) Rees, R. W. and Vaughan, D. J. *ibid.*, p. 296.
   (c) Rees, R. W. (September 1964). *Modern Plastics* **42**, 98.
3. US Patent 3,264,272 to E. I. du Pont de Nemours & Co., Inc.
4. Berry, G. C. and Fox, T. G. (1968). *Advances in Polymer Science*, **5**, 261–357.
5. Watkins, J. M., Spangler, R. D. and McKannan, E. C. (1956). *J. Applied Phys.*, **27**, 685.
6. (a) Bonotto, S. and Purcell, C. L. (March 1965). *Modern Plastics*, **42**, 135.
   (b) Bonotto, S. and Bonner, E. F. (1968). *ACS Polymer Preprints*, **9**, 537.
   (c) Bonotto, S. and Bonner, E. F. (1968). *Macromolecules*, **1**, 510.
7. Brandrup, J. and Immergut, E. H. (1966). *Polymer Handbook*, Interscience, New York, vol. 2, p. 341.
8. Harwood, H. J. and Ritchey, W. M. (1964). *Polymer Letters*, **2**, 601.
9. (a) MacKnight, W. J., McKenna, L. W. and Read, B. E. (1967). *J. Applied Phys.*, **38**, 4208.
   (b) MacKnight, W. J., Kajiyama, T. and McKenna, L. W. (1968). *Polym. Eng. Sci.*, **8**, 267.
10. Pieski, E. T. private communication.
11. See *ASTM* D1601, ASTM Annual Book of Standards, Part 27 (A.S.T.M., Philadelphia).
12. Canadian Patent 655,298 to E. I. du Pont de Nemours & Co., Inc.
13. Sakamoto, K., MacKnight, W. J. and Porter, R. S. (1970). *J. Polym. Sci.*, **A2**, 277.
14. Longworth, R. and Vaughan, D. J. (1968). *ACS Polymer Preprints*, **9**, 525.
15. Davis, H. A., Longworth, R. and Vaughan, D. J. *ibid.*, 515.
16. Longworth, R. and Morawetz, H. (1958). *J. Polym. Sci.*, **29**, 307.
17. Fitzgerald, W. E. and Nielsen, L. E. (1962). *Proc. Roy. Soc.*, **A282**, 137.
18. Blyler, L. L. and Haas, T. W. (1969). *J. Appl. Polym. Sci.*, **13**, 2721.
19. Pieski, E. T. unpublished results.
20. Longworth, R. and Vaughan, D. J. unpublished results.
21. Trifan, D. S. and Terenzi, J. F. (1958). *J. Polym. Sci.*, **28**, 443.
22. Ferry, J. D. (1961). *The Viscoelastic Properties of Polymers*, 2nd edn, Wiley, New York, Chapter 11.

23. Bueche, F. (1962). *The Physical Properties of Polymers*, Interscience, New York, p. 104.
24. Eisenberg, A., Farb, H. and Cool, L. G. (1966). *J. Polym. Sci.*, **A2**, 855.
25. MacKnight, W. J., Kajiyama, T. and McKenna, L. W. (1968). *Polym. Eng. Sci.* **8**, 267.
26. Dunleavy, M. T. and Longworth, R., unpublished results.
27. Cox, W. P. and Merz, E. H. (1958). *J. Polym. Sci.*, **28**, 619.
28. Strella, S. (1962). *J. Polym. Sci.*, **60**, S-9.
29. Schwarzl, F. and Staverman, A. J. (1952). *Physica*, **18**, 791; Ref. 22, Ch. 4.
30. Alfrey, T. Jr., (1945). *The Mechanical Behaviour of High Polymers*, Interscience, New York, p. 166.
31. Ward, T. C. and Tobolsky, A. V. (1967). *J. Appl. Polym. Sci.*, **11**, 2403.
32. Kajiyama, T., Stein, R. S. and MacKnight, W. J. (1970). *J. Appl. Phys.*, **41**, 4361.
33. Read, B. E. and Williams, G. (1961). *Trans. Farad. Soc.* **57**, 1979.
34. Sinnott, K. M. (1966). *J. Polym. Sci.*, **C14**, 141.
35. Schatzki, T. F. (1966). *J. Polym. Sci.*, **C14**, 139.
36. Nielsen, L. E. (1960). *J. Polym. Sci.*, **52**, 357.
37. Longworth, R. and Vaughan, D. J. (1968). *Nature*, **218**, 85.
38. Erdi, N. Z. and Morawetz, H. (1964). *J. Colloid Sci.*, **19**, 708.
39. McKenna, L. W., Kajiyama, T. and MacKnight, W. J. (1969). *Macromolecules*, **2**, 58.
40. Read, B. E., Carter, E. A., Connor, T. M., and MacKnight, W. J. (1969). *British Polymer J.*, **1**, 123.
41. Phillips, P. J. and MacKnight, W. J. (1970). *J. Polym. Sci.*, **A2**, 727.
42. Smyth, C. P. (1955). *Dielectric Behaviour and Structure*, McGraw-Hill, New York, p. 73.
43. Wilson, F. C., Longworth, R. and Vaughan, D. J. (1968). *Polymer Preprints*, **9**, 505.
44. Kajiyama, T., Oda, T., Stein, R. S. and MacKnight, W. J. (1971). *Macromolecules*, **4**, 198.
45. Marx, C. L., Caulfield, D. F. and Cooper, S. L.
    (a) (1973). *Macromolecules*, **6**, 344.
    (b) (1973). *Polymer Preprints*, **14**, No. 2, 890.
46. Roe, R.-J. (September 1971). *Polymer Preprints*, **12**, 730.
47. Delf, B. W. and MacKnight, W. J. (1969). *Macromolecules*, **2**, 309.
48. Marx, C. L., Koutsky, J. A. and Cooper, S. L. (1971). *Polymer Letters*, **9**, 167.
49. Eisenberg, A. (August 1973). *Polymer Preprints*, **14**, 871.
50. Pineri, M., Meyer, C., Levelut, A. M. and Lambert, M. (1974). *J. Polym. Sci., Polymer Physics*, **12**, 115.
51. Marx, C. L. and Cooper, S. L. (1973). *Makromol. Chem.*, **168**, 339.
52. MacKnight, W. J., McKenna, L. M., Read, B. E. and Stein, R. S. (1968). *J. Phys. Chem.*, **72**, 1122.
53. Otocka, E. P. and Kwei, T. K. (1968). *Macromolecules*, **1**, 244.
54. Otocka, E. P. and Kwei, T. K. (1968). *Macromolecules*, **1**, 401.
55. Otocka, E. P. and Kwei, T. K. (1968). *Polymer Preprints*, **9**, 583.
56. Gordon, M. and Taylor, J. S. (1952). *J. Appl. Chem.*, **2**, 493.
57. Fox, T. G. and Loshaek, S. (1955). *J. Polym. Sci.*, **15**, 371.
58. Phillips, P. J., Emerson, F. A. and MacKnight, W. J.
    (a) (1970). *Macromolecules*, **3**, 767.
    (b) *ibid.* p. 771.
59. Phillips, P. J. (1972). *Polymer Letters*, **10**, 443.
60. Otocka, E. P. and Davis, D. D. (1969). *Macromolecules*, **2**, 437.
61. Chang, L. S.-Y. (1955). Ph.D. Thesis, Polytechnic Institute of Brooklyn, N.Y.

62. Uemura, Y., Stein, R. S. and MacKnight, W. J. (1971). *Macromolecules*, **4**, 490.
63. Prud'homme, R. E. and Stein, R. S. (1971). *Macromolecules*, **4**, 668.
64. Prud'homme, R. E. and Stein, R. S. (1973). *J. Polym. Sci., Polymer Physics*, **11**, 1347.
65. (a) Stein, R. S. (1964). In *Newer Methods of Polymer Characterisation*, Chapter 4. ed. Ke, B., Interscience, New York.
    (b) van Aarsten, J. J. and Stein, R. S. (1971). *J. Polymer Sci.*, Part A-2, **9**, 295.
66. Otocka, E. P., Kwei, T. K. and Salovey, R. (1969). *Makromoleculare Chem.*, **129**, 144.
67. Harland, W. G., Khadr, M. M. and Peters, R. H. (1972). *Polymer*, **13**, 13.
68. Geil, P. H. (1963). *Polymer Single Crystals*, Interscience, New York.
69. Fuoss, M. and Accascina, F. (1959). *Electrolytic Conductance*, Chapter 18, Interscience, New York.
70. Fuoss, R. M. and Kraus, C. A. (1933). *JACS*, **55**, 3614.
71. Powles, D. G. (1960). *Polymer*, **1**, 219.
72. Connor, T. M. (1969). *Brit. Polym. J.*, **1**, 116.
73. Binsbergen, F. L. and Kroon, G. F. (1973). *Macromolecules*, **6**, 145.
74. Eisenberg, A. (1970). *Macromolecules*, **3**, 147.
75. Eisenberg, A. and Navratil, M. (1972). *J. Polym. Sci.*, **B10** (7) 537.
76. Eisenberg, A. and Navratil, M. (1973). *Macromolecules*, **6**, 605.
77. Eisenberg, A. (August 1973). *ACS Polymer Preprints*, **14**, 871.
78. MacKnight, W. J., Taggart, W. P. and Stein, R. S. *ibid.*, 880.
79. Longworth, R. and Wilson, F. C. unpublished results.
80. Wissbrun, K. F. (1968). *Makromol. Chem.*, **118**, 211.
81. Pettit, L. D. and Bruckenstein, S. (1966). *JACS*, **88**, 4783.
82. Ogura, K., Sabue, H. and Nakamura, S. (1973). *J. Polym. Sci., Polymer Physics*, **11**, 2079.
83. Phillips, P. J. and MacKnight, W. J. (1970). *Polymer Letters*, **8**, 87.
84. Ide, F. and Hasegawa, A. (1968). *Chem. High Polymers (Japan)*, **25**, 825.

CHAPTER 3

CARBOXYLATED ELASTOMERS

D. K. JENKINS AND E. W. DUCK

3.1 INTRODUCTION

Elastomeric polymers in general are characterised by weak inter-chain forces and lack of symmetry or order within the molecular chain. Incorporation of unsaturated carboxylic monomers into the chain increases the inter- and intra-chain forces, resulting in increased polymer tensile strength, but simultaneously decreases the extension and recovery properties. If sufficient carboxyl groups are present in a given chain length, all rubbery properties are lost and the polymer resembles a fibre forming material.

Commercially, latices are the most important form of carboxylated elastomer. The carboxyl functional group confers important advantages such as the ability to use sulphurless curing systems, *e.g.* a metal oxide, cross-linking with other functional polymers and high adhesive strength. The increased polarity of the polymer increases its compatibility with, and affinity for, polar surfaces encountered in fibres and inorganic fillers.

No information is available on the consumption of carboxylated rubber but on the following page a list is given of the companies offering this material commercially. Assuming a production level of 10 000 tons per annum for each manufacturer, this leads to an estimated world-wide output of approximately 0·5 million tons per annum.

3.2 HISTORICAL

The first patent for a carboxylic elastomer was issued to I.G. Farbenindustrie in 1933,[1] but although much research was carried out and several

COMPANIES PRODUCING CARBOXYLATED RUBBERS
(emulsion type) (1972)

Major type	Company	Location of plant	No. of types
Nitrile (dry rubber)	Firestone	USA	1
	Goodrich	USA	6
	Standard Brands	USA	6
Nitrile (latex)	Goodrich	USA	9
	Bayer	W. Germany	5
	BP Chemicals	UK	4
	Doverstrand	UK	8
	Standard Brands	USA	9
	Synthomer	W. Germany	15
	Synthetic Latex	S. Africa	3
	Plastimer	France	2
	Montecatini	Italy	2
	Japanese Geon	Japan	1
	Goodyear (France)	France	2
	Goodyear	USA	2
	AKU-Goodrich	Netherlands	1
	Takeda	Japan	3
	Uniroyal	USA	1
	USSR	USSR	1
S.B.R. (latex)	Uniroyal	USA	9
	Uniroyal	UK	3
	Goodrich	USA	2
	Japanese Geon	Japan	3
	BP Chemicals	UK	2
	Synthomer	W. Germany	12
	Doverstrand	UK	17
	AKU-Goodrich	Netherlands	2
	ISR	UK	2
	Takeda	Japan	4
	Standard Brands	USA	5
	Resistol	Mexico	4
	Sumitomo Naugatuck	Japan	1
	ANIC	Italy	3
	Synthetic latex	S. Africa	9
	Plastimer	France	2
	JSR	Japan	1
	W.R. Grace	USA	4
	Polymer Corp.	Canada	5
	Polymer Corp.	France	3
	Firestone	France	2
	Firestone	USA	3
	USSR	USSR	1

patents issued in the ensuing years, it was not until 1949 that the first commercial product was introduced. This was a butadiene–styrene–acrylic acid terpolymer latex marketed by B. F. Goodrich under the trade name Hycar 1571. A good account of the early work is given in a review by H. P. Brown.[2] During this period, incorporation of carboxylic groups was only used as a means of altering the polarity of a polymer and hence its properties. Thus Farmer and Bacon[3] prepared materials ranging from polar elastomers through fibrous products to brittle resins. A major advance was made in the early 1950s when Brown[4] in the USA and later Dolgoplosk[5] in the USSR used the carboxyl as a functional group to achieve sulphurless cures with metal oxide systems and cross-linking with other functional polymers. By 1960 the most widely used products were those based on diene rubbers including butadiene, butadiene–acrylonitrile or butadiene–styrene combined with acrylic or methacrylic acid and those based on acrylic elastomers. At this time Cuneen[6] carried out the first efficient carboxylation of natural rubber by grafting thioglycollic esters on to latex using hydroperoxide initiators and forming the acid by hydrolysis. Previous attempts by this route had given non-elastomeric products[7] or low grafting efficiencies.[8] Over the last ten years, several theories have been advanced to explain carboxylate cross-linking and the number of industrial applications has increased rapidly. Very few new types of carboxylated elastomer have been evolved however. It has proved difficult to graft carboxyl containing reagents on to the new EPR and EPDM (ethylene–propylene–diene terpolymer) rubbers in hydrocarbon solution, although there have been reports in the patent literature of radical grafting of maleic anhydride on to the raw rubber during milling.

EPDM rubbers containing ethylidene norbornene have been metallated[9] in solution at tertiary hydrogen centres of the termonomer group. Treatment of the resulting alkali metal salt with CO_2 generated the carboxylate salt of the rubber. An alternative method involves emulsification of the base rubber prior to grafting with polyacrylic acid.[10] The most interesting future prospect may stem from the work of Gaylord[11] where it appears possible to terminate a living styrene–maleic anhydride copolymer by reaction with an EP polymer backbone during a milling/extrusion operation.

The most novel carboxylated rubbers to appear in recent years have been the fluorinated carboxy-nitroso-rubbers developed by Thiokol Co.[12] for use under extreme conditions in aerospace applications.

3.3 PREPARATION

The majority of commercially available materials are prepared by emulsion polymerisation. Compared to conventional polymerisation, of dienes for example, free radical copolymerisation with unsaturated free acids presents some special problems. In an alkaline medium, the acids become converted to salts which do not polymerise and are not soluble in the oil phase. It is therefore necessary to work with an acid type recipe using emulsifiers such as alkyl or alkyl aryl sulphonates or dodecylamine hydrochloride and to shortstop and coagulate with substances which do not react with and destroy the acid groups. A typical recipe would be; monomers (100 parts), sodium alkyl aryl polyether sulphate (1 phm), potassium persulphate (0·3 phm) and water (188 phm). Tertiary dodecyl mercaptan is used as modifier to regulate molecular weight if required. HCl, or methanolic HCl is used as coagulant.

The most common polymers employ a combination of monomers selected from butadiene, styrene, acrylonitrile, and acrylic esters together with the saturated acids.

The amount of carboxyl function and its distribution can be controlled by choice of comonomer acid, choice of monomers and monomer charge ratios. The conversion level to which the polymerisation is taken and whether a mixture of carboxy monomers is used also influences the final product. The unsaturated acid is partitioned between the oil and water phases. The distribution in the polymer depends on its solubility in the oil phase, the amount charged and the effect of the other components of the recipe. Acrylic acid is more soluble in the aqueous phase than the oil phase and only half the amount charged enters the copolymer. Therefore in order to obtain the required level in the polymer it may be necessary to use a higher charge ratio, together with incremental butadiene addition or restricted conversion. Methacrylic acid is about five times more soluble in the oil phase than acrylic acid at 10 wt % charge and thus incorporates in the copolymer very readily. To get a uniform acid distribution, a mixture of acrylic and methacrylic acids in a 2:1 ratio may be employed. Sorbic acid, unlike acrylic or methacrylic acid, has only a limited solubility in the system and hence the initial polymer composition is limited to around 5%, although the insoluble acid acts as a reservoir, supplying extra acid as the soluble portion is polymerised. Similar incorporation of acid for low and high conversions in a butadiene–acrylonitrile–methacrylic acid system suggested a uniform distribution of acid. As the ratio of butadiene to acrylonitrile increased, however the distribution became less uniform and more like that of butadiene alone. The comparable reactivities of acrylates and acrylic acids provides ready preparation of the carboxylated acrylates, and incremental addition of monomers can lead to a very uniform carboxyl distribution.

In order to impart specific properties to the copolymers, it is not unusual to find the unsaturated dibasic acids such as itaconic acid or fumaric acid being used as the carboxy monomer constituent. Monobasic acids tend to give better physical properties in the final polymer but are poorer in heat ageing characteristics, and incorporation of dibasic acids improves the heat ageing properties.

In the case of elastomers used as latices it is unlikely that the level of total functional monomers will exceed 5 or 6 wt % in the final product. Flexibility and stiffness are most commonly adjusted by means of the ratio of butadiene to styrene or acrylonitrile. Obviously for latices which must be transported in the liquid form a high conversion and solids concentration is desirable. By control of the emulsion polymerisation and particle size, latices are prepared to 50% solids concentration.

Carboxylated elastomers are not only produced by emulsion polymerisation. Unsaturated acids can be grafted on to preformed polymers by two methods. These are solution grafting or grafting on to milled solid rubber. Both methods had been used on natural rubber to give the first carboxylated elastomers, but it was not until 1950 that carboxylation of synthetic rubbers was successfully carried out.

In solution grafting, a polymer is treated with peroxide initiator and the reactive acid entity. This may be a thiol acid or unsaturated acid. Thus, for example, a 6% solution of polybutadiene in benzene can be treated with benzoyl peroxide and thioglycollic acid[13] at 50°C for 24 hours to give 85% incorporation of the acid. In the particular case of thioglycollic acid[14] the system can be simplified to blowing air through a mixture of the thiol acid and rubber in toluene, containing some piperidine. This technique produces the peroxidic constituent *in situ* and efficiencies of up to 90% reaction of the thiol acid have been achieved.

Grafting methods on the mill depend on the same chemistry. β-mercaptopropionic acid was grafted on to polybutadiene on a two roll mill by adding benzoyl peroxide and the acid to the banded rubber, and milling for 30 minutes at 38°C. Grafting on to a saturated polymer is more difficult than grafting on to an unsaturated polymer. With saturated polymers, the sites which must be attacked are the tertiary carbon atoms in the chain, for it is here where hydrogens may be detached and free radicals formed. Ethylene propylene rubber in chlorobenzene solution has been grafted with maleic anhydride in the presence of chlorine and UV light.[15] In this case, the product contained 21·6% chlorine and acid groups equivalent to a saponification number of 69. Polyacrylic acid chains[10] have been grafted on to ethylene propylene copolymer in an emulsion polymerisation by adding acrylic acid and benzoyl peroxide to the emulsified copolymer.

When scrap vulcanised rubber was treated under specific conditions in an extruder,[16] a replasticised, oxide-vulcanisable carboxylated rubber was obtained. Further variations on the grafting method have recently been described[11] in which a live styrene–maleic anhydride alternating copolymer can be injected into an extruder containing ethylene propylene rubber to terminate on the rubber and form a carboxylated terpolymer on hydrolysis.

A few specialised carboxylates can be prepared via solution metallation reactions, followed by treatment with CO_2 and acidification of the metal salt which is formed. Polybutadiene has been reacted at temperatures up to about 50°C with a 1:1 ratio of lithium butyl to tetramethylethylenediamine. The result was a polymer metallated in positions allylic to the double bonds[17] which could be reacted with CO_2 to form the lithium salt of the carboxylated elastomer. At 50°C a considerable amount of chain scission accompanied the metallation, but this decreased with decreasing reaction temperature. Ethylene propylene rubbers containing the termonomers ethylidene norbornene, endomethylene hexahydronaphthalene or 1,4-hexadiene proved difficult to metallate. The lithium butyl:TMEDA reagent only metallated to about 5%. It was found that better results were obtained if a 1:1 complex of lithium butyl and potassium tertiary butoxide was used,[9] giving polymer acid numbers up to about 20.

The final preparative method which may be mentioned is based on hydrolysis of a preformed polymer to provide carboxyl groups. This partly applies to the copolymers of maleic anhydride previously mentioned, where hydrolysis would be required if the free carboxyl groups are not generated by the preparation itself. Butadiene can be copolymerised with a variety of comonomers such as acrylate esters, methacrylamide and acrylonitrile, which can be hydrolysed to the acid by heating with alkaline solutions. Ethylene and propylene have been solution polymerised with zinc and magnesium salts of long chain unsaturated acids[18] and the resultant polymer hydrolysed with methanolic hydrochloric acid. The preparation of a polyisobutylene by copolymerisation of acryloyl chloride with isobutylene[19] and subsequent hydrolysis of the acid chloride has also been described. Controlled hydrolysis of acrylate esters has been used to form carboxylated acrylates although these are readily obtainable by direct polymerisation. The advantages of using hydrolytic latex preparations are first that the polymers may be obtained by emulsion polymerisation in alkaline media using common anionic soaps; secondly the different monomers may have a more favourable reactivity ratio in the non-acid form and thirdly, there may be a different distribution of carboxyl groups in the polymer.

3.4 PROPERTIES

3.4.1 Unvulcanised elastomer

Modified rubbers containing less than 0·1 equivalents phr of carboxyl have very similar physical properties to the base polymer. As the carboxyl level is increased, the rubber becomes tougher, less elastic and more thermoplastic as shown by the increase in tensile strength (see Fig. 1), increased 300% modulus and decrease in elongation at break. In butadiene–methacrylic acid copolymers for example, all elastic properties are lost if incorporation of the acid reaches 40–45 wt % of the total polymer. Practical

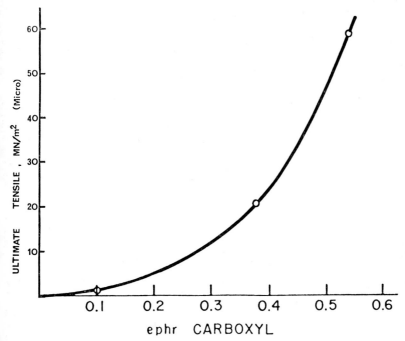

Fig. 1 Ultimate tensile strength of raw copolymers of butadiene with methacrylic acid. (Reproduced from Ref. 20 with permission.)

advantages of carboxyl incorporation are increased hardness, crumb strength and resistance to hydrocarbon solvents, better milling properties, easier film formation and higher temperature limit of elasticity. The main disadvantages are increased tendency to swell or react in aqueous ammoniacal or alkaline solutions and increased oxidative attack leading to degradation and cross-linking. The acid from which the carboxyl is derived has little effect, and similar physical properties are obtained from butadiene copolymers of acrylic, methacrylic or sorbic acids, although it has been claimed that the ratio of modulus to elongation at break can be altered by using acids giving different pendant chain lengths.

The carboxyl groups provide sites for curing with added reagents but in the absence of the latter, under severe conditions, self cure via anhydride formation is possible. Thus a butadiene polymer containing 0·11 ephr (equivalents per hundred parts rubber) carboxyl had a tensile strength of 24·1 MN/m^2, compared to less than 1·4 MN/m^2 for an equivalent non-carboxylated polymer, when both were heated with carbon black to 260°C for two hours.[20]

Natural rubber grafted with thiol acid groups[21] shows a very much reduced crystallisation rate, leading to longer storage life, partly due to restricted chain motion by the carboxyl groups, but also affected to some extent by double-bond cis-trans isomerisation occurring concurrently with the grafting. This reduced rate of crystallisation also improves the low temperature properties of natural rubber to give products, the so-called 'Arctic' or anti-crystallising rubbers, which remain flexible under conditions of extreme cold.

3.4.2 Vulcanisation with metal oxides

Carboxylic elastomers can be rapidly vulcanised by heating in the presence of divalent metal oxides to give rubbers with a very high gum strength compared to corresponding sulphur vulcanised non-carboxylic elastomers. However, the addition of carbon black or whiting does not cause reinforcement but reduces tensile strength. This contrasts with the effect on normal sulphur vulcanisates of carbon black where the reinforcing effect is considerable. Although zinc and magnesium oxides impart the best physical properties, many metal oxides, hydroxides and salts of organic acids weaker than acetic, including fatty acid salts, have given some degree of curing. Some of the more novel compounds which have been described are dibutyl tin oxide, calcium silicate, sodium phosphoaluminate, aluminium isopropoxide and basic zinc maleate. The suitability of and rate of vulcanisation with any particular compound depend on its valence, ionic character, compatibility with the carboxylated rubber and ease with which any undesirable reaction products can be removed from the rubber. A comparison of some divalent metal oxides and hydroxides has been given by Dolgoplosk,[22] and some results are shown in Table 1. No account is taken of the varying excess oxide levels on a stoichiometric basis. This excess of oxide, least with PbO and greatest with MgO, should lead to a slight increase in modulus and tensile, by acting as a filler.

As noted previously, the raw carboxylated rubbers have a high tensile strength and this is enhanced even by vulcanisation with monovalent metal ions. Thus a butadiene–methacrylic acid copolymer containing 0·12 ephr COOH with less than 0·7 MN/m² tensile and 1600% elongation, had a tensile of 11·7 MN/m² and e.a.b. of 900% after treatment with aqueous NaOH followed by heating,[20] and a tensile of 41·4 and e.a.b. of 400% after zinc oxide vulcanisation.

Like the raw polymer, the tensile strength of the cured polymer increases with increase in % COOH in the chain. For a butadiene–acrylic acid copolymer cured with excess zinc oxide, the increase is linear.

In theory, addition of metal oxide or salt equivalent to the amount of carboxyl present should result in complete cure and maximum effect on physical properties. In practice about twice the calculated amount of oxide is required to give optimum vulcanisation judged by mechanical properties, as shown in Fig. 2. Phthalic anhydride is included in this recipe to prevent precure.

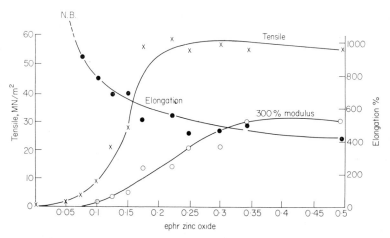

Fig. 2 Effect of ZnO level on butadiene–acrylonitrile–methacrylic acid polymer. 55/33/10 copolymer of 73% conversion, 0·099 ephr COOH. (Reproduced from Ref. 2 with permission.)

TABLE 1
MECHANICAL PROPERTIES OF GUM COMPOUNDS OF SKS-30-1 [a] VULCANISED WITH DIVALENT METAL OXIDES (10 PHR).
(Reproduced from Ref. 5 with permission)

Property	MgO	ZnO	CaO	PbO	CdO	$Mg(OH)_2$	$Zn(OH)_2$	$Ca(OH)_2$	$Ba(OH)_2$
Vulc. time, min	20	10	100	30	120	20	10	80	60
300% Modulus, MN/m^2	4·4	1·8	2·2	3·0	2·3	2·9	2·9	5·5	3·7
Tensile, MN/m^2	38·9	15·7	13·2	12·8	19·0	22·0	24·1	39·4	24·8
Relative elongation %	850	800	760	740	890	835	660	770	675
Residual elongation %	22	10	22	14	23	15	2	28	18

[a] Butadiene–styrene–methacrylic acid with 1·5 wt % of COOH.

Such a result suggests that only a proportion of the oxide is used in effective chemical cross-linking at the equivalent oxide to —COOH ratio, leaving some free COOH groups, some basic —COO M—OH unlinked metal bonds and some intra-molecular bonds. Brown has shown that the amount of zinc chemically bonded to the elastomer is equivalent to the number of carboxyl groups in a polymer vulcanised with zinc palmitate, and this is confirmed by the absence of a zinc oxide X-ray pattern in an oxide vulcanised butadiene sorbic acid copolymer. In a network swelling study, however, Poddubnyi found only 15% of the available carboxyl groups in the vulcanisation network but no details of the polymer system used are available. Similarly, experiments to determine the number of cross-links per cm^3 of vulcanised rubber by the equilibrium gel method[23] indicated a much lower cross-link density for the carboxyl rubber, compared to a sulphur vulcanisate of a non-carboxyl rubber, where the two rubbers had the same average segment molecular weight between carboxyl groups and between two sulphur cross-links respectively.

A lack of stable cross-links in the metal oxide vulcanisates is also indicated by stress-relaxation experiments (*see* Fig. 3).

In latices, the effectiveness of zinc oxide cure is dependent on the pH of the latex, extent of neutralisation and cure time. Optimum cure occurs at pH 10–11 and excess oxide or overcure can result in reversion. These effects have been associated with increasing uncoiling of the polymer chains caused by repulsion between neighbouring ionised salt groups. At complete neutralisation, the chains are at their greatest extension with the maximum number of carboxyl sites exposed for reaction. Correspondingly, chain entanglement is also at a maximum and a latex dried in this state has optimum properties by virtue both of chemical cross-linking and chain entanglement. The presence of excess oxide tends to reduce effective ionic charge density through ion pair formation, allowing the chains to recoil.

One shortcoming of the simple oxide vulcanisation system is the tendency to 'Mooney scorch' or precure. Under certain conditions, compounded carboxylic rubbers containing zinc oxide can partially cure in 48 hours at room temperature. It has been found that certain organic acids and anhydrides, particularly phthalic and maleic anhydride, prevent precure (*see* p. 186). Three reasons have been advanced for their effectiveness. First, the added compound can exchange with metal–acid links present in the vulcanisate, leading to redistribution and a more evenly distributed cross-link network. Secondly, reaction with any partially neutralised —COO$^-$ M$^+$—OH groups can occur, either forming additional cross-links, or

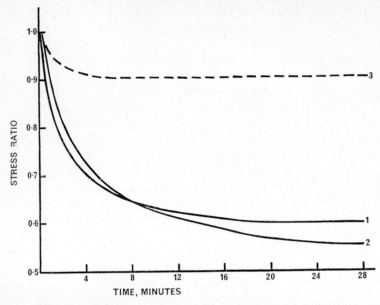

Fig. 3 Stress relaxation of metal oxide vulcanisates of carboxylic rubber SKS-30-1. 1, with MgO (10 parts by wt); 2, with CaO (10 parts by wt); 3, sulphur vulcanisate of SKS-30-A. (N.B. SKS-30-1 is 70:30 SBR with 1·5% COOH; SKS 30A is 70:30 wt ratio SBR.) (Reproduced from Ref. 23 with permission.)

simply completing neutralisation. Finally any free metal oxide left in the polymer, due to poor mixing or poor compatibility, would be neutralised.

Other disadvantages of oxide vulcanised carboxylated rubbers are related to poor compression set and high stress-relaxation, giving rise to low flex resistance, very low hysteresis and fluidity at higher temperatures or repeated deformation. In carboxylated rubbers treated with monovalent salts, some kind of weak cross-linkage is suggested by the increase in gum tensile at room temperature, but if the temperature is raised to 70–100°C such a structure is completely destroyed and the tensile falls to zero. Vulcanisation with divalent metal salts gives a network containing definite chemically bonded cross-links and some secondary bonding. Such vulcanisates still show appreciable reversible fluidity at temperatures of 100°C or higher. These properties can be used in the design of thermoplastic elastomeric materials.

Metal oxide vulcanisates can be readily reclaimed by treatment on the mill with an amount of organic acid in excess of the theoretical require-

TABLE 2

CHANGE IN MECHANICAL PROPERTIES OF CARBOXYLIC RUBBERS[a]
(30 PARTS ADDED CHANNEL BLACK) WITH TEMPERATURE
(*Reproduced from Ref. 5 with permission*)

Temp. of experiment °C	Modulus at 300% MN/m^2	Tensile strength MN/m^2	Elongation %	
			relative	residual
20	11·6	34·0	735	22
60	11·0	23·6	680	24
80	6·9	17·3	685	28
100	4·6	12·8	670	25

[a] Probably SKS-30-1 carboxylated SBR.

ments. Stearic acid has been used successfully although, for economy, di- and polybasic acids may be preferred. The resultant plastic product can be revulcanised to a good tensile strength by the addition of more oxide.

3.4.3 Mixed vulcanisation

The most useful property of carboxylated elastomers cured with metal oxide systems is the high tensile strength obtained in the absence of reinforcing fillers such as carbon black. Thus a standard butadiene or SBR rubber has a gum strength of 1·9–5·8 MN/m^2 using a simple sulphur cure recipe whereas a similar base rubber containing 1·5% copolymerised methacrylic acid can be cured to give a gum tensile of around 29·0 MN/m^2, with magnesium oxide.

Many recipes used for curing non-carboxylated rubbers contain sulphur, an organic accelerator and zinc oxide activator. When these are used to cure carboxylated rubbers, features of both sulphur cure and metal oxide cure are evident. With short cure times, the effect of metal oxide cure predominates giving high tensile strength and modulus but poor compression set and stress-relaxation properties. With increasing cure times, sulphur type cross-linking becomes increasingly evident giving rise to lower tensiles but better compression set. Thus a butadiene–cinnamic acid copolymer containing 15% cinnamic acid was cured at 144°C in a tread type recipe containing EPC black (40 phr), zinc oxide (5 phr), sulphur (2 phr) and MBTS (benzthiazyl disulphide) (1·75 phr) to give tensiles of 39·7 at 25 minutes decreasing to 25·4 MN/m^2 at 150 minutes, with a corresponding drop in compression set from 3% to 1%.

For many applications either a standard sulphur cure recipe, containing little or no metal oxide, gives satisfactory properties or an optimum recipe can be formulated from a mixture of oxide and sulphur systems. For instance, the tendency of a simple metal oxide vulcanisate to lose strength (65% at 100°) and flow at higher temperatures can be counteracted by superimposing a light sulphur cure. Some work has been carried out on polyisoprene[24] containing 0·15% carboxyl groups. In the uncured state, green strengths were higher than in a comparable natural rubber compound but tack remained at the same level as that of the untreated polyisoprene. The main features of vulcanisates containing 40 pphr HAF black, cured with CBS (n-cyclohexyl benzothiazole-2-sulphenamide) and sulphur, were slightly greater heat build-up than for natural or polyisoprene rubber and a bond strength to steel greater than that of polyisoprene alone and approaching that of natural rubber. In mixtures of polyisoprene containing 30% of the carboxylated polymer, cohesive strength increased nine-fold and bond strength to steel increased twenty-fold, using the same vulcanisation recipe.

An initial problem encountered with zinc oxide containing sulphur cure recipes was precure via the oxide–carboxyl reaction. This gave rise to poor mould flow of the compounded rubber, viscosity changes on storage at room temperature and 'Mooney scorch'. One method of countering this is to add the metal oxide component as late as possible in the mixing cycle, particularly where long mixing periods or stocks with especially high heat build-up are involved. An alternative method is to use a 'retarder'. Strong monobasic acids reduce scorch by reacting with the oxide and eliminating its effect, to give the equivalent of a straight sulphur cure. However, anhydrides of polybasic acids, particularly maleic, act as true retarders which, although increasing cure time, do not affect the physical properties of the vulcanisate. Magnesium stearate has been proposed as an improved scorch retarder which has no effect on cure time and improves the properties of the vulcanisate.

Many cross-linking reactions can be carried out via the carboxyl groups by using covalent linking agents. Polyamines, polyalcohols, epoxy resins and carbodiimides are among the more important reagents and can be varied through the number of active groups or by inter-group chain length. Reactions with amines and polyepoxides have been used with solid carboxylated rubber, but imines, carbodiimides and polyisocyanates have been widely used for films, coatings, impregnation and cements because of their extremely rapid cure. Intra-molecular linkages are kept at a minimum by keeping the number of carboxyl groups per chain in the region of around one carboxyl per 30 to 100 carbon atoms. In all cases, the carboxylated polymer cross-linked with a covalent bond is at best equivalent to the corresponding sulphur or peroxide vulcanisate and possesses none of the advantages of the salt type vulcanisate. Covalent links are advantageous, however, where rapid cure is required at lower temperatures than those used for sulphur and peroxide. Only those covalent cross-links which have been used in combination with metal oxide vulcanisation will be considered in this review.

The simplest covalent link is formed by peroxide vulcanisation to join chains directly through attack at tertiary hydrogen sites. Carboxyl groups in the chain have little or no effect on the peroxide cure, but mixed peroxide–metal oxide vulcanisation results in a compromise of properties. This is illustrated for a nitrile rubber in Table 3.

In general, the combination of peroxide plus metal oxide gives a better blend of properties than that of sulphur plus metal oxide. For ethylene propylene rubbers, vulcanisation with peroxide alone gives vulcanisates with somewhat unsatisfactory physical properties. Addition of small amounts of co-agents such a maleic anhydride, maleic acid or fumaric acid, increases the degree of cross-linking as measured by increased modulus and lower swelling ratio. Further marked improvements are obtained by incorporating polyvalent metal oxides. It has been suggested that this is the result of salt forming reactions of the unsaturated acid components grafted (about 1–1·5% by weight) on to different copolymer chains during vulcanisation. This view is supported by the fact that no improvement over straight peroxide cure is obtained if saturated acids are used *i.e.* no grafting and no effect from metal oxide addition is observed if esterified unsaturated acids are used.

Carboxylic elastomers can be cross-linked with suitable di- and polyamines, such as hexamethylenediamine, or their salts. Treatment at 80°C gives a product with high tensile and poor compression set analogous to a metal oxide vulcanisate. This is due to formation of the tertiary amine salt structure —COO$^-$ H$_3^+$N—R. With increasing heat history and temperatures of about 120°C lower tensiles and better compression set values are obtained as covalent amide cross-linkages are formed, —CONH—R. Increasing amide formation renders the polymer equivalent to the sulphur cured analogue. A direct conversion to amide may be obtained by starting with the acid chloride instead of the free acid. Although mixed metal oxide–amine vulcanisates have been described in the literature, it is doubtful whether any real advantages accrue, but a better balance of stress strain properties has been claimed for a metal oxide–polyamine vulcanised tyre tread recipe. Polyols and polyepoxy compounds have been used for covalent

TABLE 3

CROSS-LINKING OF CARBOXYLIC NITRILE ELASTOMER WITH PEROXIDE AND SALT FORMING SYSTEMS WITHOUT FILLER

(Reproduced from Ref. 20 with permission)

Recipe	Salt cross-link				Salt/peroxide cross-link			
		100% Mod. MN/m²	Tensile MN/m²	Elong. %		100% mod. MN/m²	Tensile MN/m²	Elong. %
Polymer	100				100			
stearic acid	1				1			
zinc oxide	9.3				9.3			
dicumyl peroxide	—				5			
Cure at 154°C								
15 min		1.9	35.1	520		3.27	13.9	280
30 min		2.0	39.6	560		3.5	10.8	250
45 min		2.2	30.5	490		5.6	8.9	150
Comp. set (ASTM-B) (45 min at 154°C):								
70 hr 100°C, %			95				7	
Hardness (Duro A)			70				76	
Tear resistance (ASTM-D264 Die C) (30 min at 100°C)								
With grain (kg/m)			3754				1964	
Properties measured at 121°C (30 min at 100°C):								
Ult. tens., MN/m²			1.3				3.0	
Ult. elong., %			1000+				180	
Tear-die C (kg/m)			715				179	

TABLE 4

EFFECT OF ZINC AND CADMIUM OXIDES IN EPOXY VULCANISATION OF 85–15 BUTADIENE–METHACRYLIC ACID COPOLYMER
(Reproduced from Ref. 25 with permission)

Basic Recipe. Polymer, 100 parts; HAF black, 40; Epon 828, 4; stearic acid, 2; N-phenyl-2-naphthylamine, 2; variable metal oxide. Cure, 60 minutes at 150°C

Test temp. °C		Parts of metal oxide/100 of polymer						
		0	2ZnO	5ZnO	8ZnO	6CdO	8CdO	11CdO
23	200% modulus MN/m²	4·8	6·0	15·5	19·3	18·9	24·6	23·2
	Tensile MN/m²	21·7	22·7	34·2	32·7	29·3	29·3	30·3
	Elongation %	500	520	340	280	280	250	260
150	200% modulus MN/m²	4·9	1·8	2·1	3·8	4·1	5·2	5·8
	Tensile MN/m²	4·9	4·9	6·2	8·3	9·2	8·9	12·1
	Elongation	200	390	430	400	320	280	320
204	200% modulus MN/m²	—	—	2·3	2·7	—	—	—
	Tensile MN/m²	1·3	2·1	5·5	6·4	2·9	4·8	7·5
	Elongation	80	180	380	380	130	190	260

cross-linking of carboxylic elastomers via esterification reactions, although the vulcanisates are only equivalent or inferior to analogous sulphur vulcanisates. For the epoxy systems it may be necessary to add to some amine as a catalyst. Combinations of epoxy and metal oxide cures have resulted[25] in polymers with improved high temperature properties up to 260°C, but no data was reported for this polymer cured with oxide alone (see Table 4).

One disadvantage of such systems is an increased tendency to scorch during compounding. This can be avoided by using a compound which only hydrolyses during the vulcanisation step. One such system which gave good results was a butadiene–methylmethacrylate copolymer vulcanised with barium hydroxide octahydrate. Both ionic and covalent cross-links contribute to the final properties. With no epoxide, the compounded mixture is thermoplastic and with no metal oxide, the high temperature tensile is lost.

3.5 NATURE OF THE CROSS-LINK IN OXIDE VULCANISATION

The earliest descriptions of bonding in oxide-vulcanised carboxylic elastomers involved inter- and intra-molecular formation of neutral metal salts to give strong ionic bonds. Secondary association through partially reacted basic metal salts, free carboxyl groups, hydrogen bonding and Van der Waals forces was also postulated. Dolgoplosk,[22] amplifying an earlier proposal by Kargin, suggested that optimum physical properties in the vulcanisate required a reversible equilibrium between stable neutral salts $-COO^- Zn^{++} OOC^-$ and labile basic salts $-COO^- Zn^+ -OH$. Fluidity with increased temperature was ascribed to increasing 'solution' of the relatively incompatible metal salt bonds in the hydrocarbon matrix. Cooper,[26] following Tobolsky's work on polymer network stress-relaxation, put forward the related idea that the high strength of the oxide vulcanisates was due to exchange reactions between neighbouring groups which prevented the formation of local stresses and hence premature failure. At any given time, sufficient of the chains remained strongly linked to maintain the network structure during rearrangement.

Gelation in the ionic systems is determined by the ability of organic solvents to separate ionic aggregates. This is borne out by the fact that the polymers containing the more polar magnesium linkages are less easily solubilised than those containing zinc.

Halpin and Bueche[27] also related the high strength of the oxide vulcanisates to a relaxation process but postulated a viscoelastic, rather than a chemical mechanism. The arguments against chemical relaxation were first that breaking a strained cross-link and reforming an unstrained link should eventually lead to complete relaxation and therefore a high permanent set, which is not observed experimentally, and secondly there was no difference in apparent activation energy for relaxation between comparable metal oxide and sulphur vulcanisates. In support of a physical relaxation process, both Zakharov[23] and Halpin obtained results, for stress-relaxation and creep, for oxide vulcanised carboxylates, which were identical to those obtained for lightly sulphur-vulcanised conventional elastomers. It was concluded that high tensile strength was a general feature of low cross-link density and not a unique feature of the ionic bonding.

Viscoelastic studies on non-rubbery polymers containing carboxyl groups particularly those prepared from ethylene and an unsaturated organic acid, have indicated that mechanical behaviour and strength are directly related to the formation of ionic clusters in an essentially amorphous matrix. In the light of these results, Tobolsky[28] suggested that the properties of carboxylated elastomers could best be explained on a similar basis. The energetically unfavourable situation of isolated ionic sites scattered throughout a non-polar medium could be relieved by aggregation of the ions to form hard clusters. The resulting two-phase system, with the ionic cluster bonded directly to the elastomer should give a structure similar in character to a normal rubber containing reinforcing fillers. The situation is analogous to the reinforcing effect of styrene domains in a butadiene matrix in ABA thermoelastomers, even to the extent of fluid flow at elevated temperatures. The increased tensiles of the raw carboxylated compared to non-carboxylated elastomers suggest that association of un-ionised carboxyl groups also takes place. Much of Tobolsky's evidence for ion clusters is related to modulus versus temperature plots and stress-relaxation experiments.

The pattern of the modulus–temperature plot for the zinc salt of a butadiene–acrylonitrile–methacrylic acid polymer (B), compared to that of the untreated carboxylated polymer (A), shows a transition region extending over a much wider temperature range (Fig. 4). The untreated un-ionised elastomer has a steep curve indicating only a low level of acid group interaction. This behaviour of the two materials is similar to that shown by the free acid and corresponding zinc salt of the polymers represented by curves C and D.

The viscoelastic behaviour of the postulated ionic clusters in sulphur-cured and metal-oxide cured carboxylic elastomers is shown in Fig. 5. The

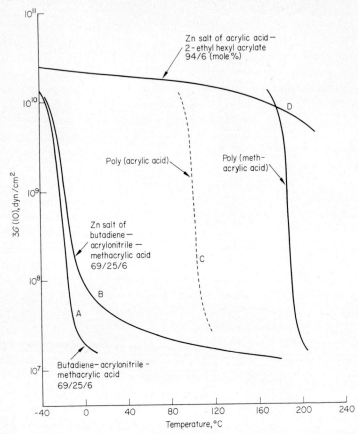

Fig. 4 Modulus–temperature curves of poly(acrylic acid), a butadiene–acrylonitrile–methacrylic acid terpolymer (69:25:6), and their zinc salts. (Taken from Ref. 28 with permission.)

oxide-cured elastomer has a characteristically high modulus at ambient temperature which slowly decreases with increasing temperature.

The high stress-relaxation rate of the oxide-cured polymer, corresponding to relatively labile cross-links or movement of whole ionic clusters, has been compared with the sulphur-cured polymer containing stable covalent cross-links (Fig. 6). The curve for mixed metal oxide/sulphur cure shows the previously mentioned compromise between the two types.

At the present time, there is little evidence to show conclusively whether the postulated clusters in metallo-carboxylated rubbers are simple pairs of

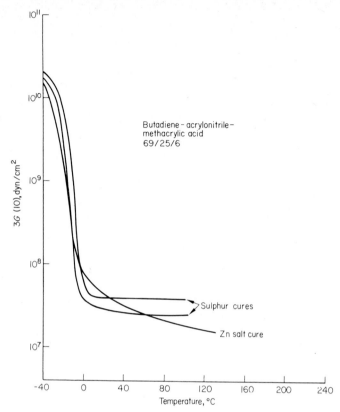

Fig. 5   Modulus–temperature curves of the butadiene–acrylonitrile–methacrylic acid terpolymer with sulphur cures and a zinc salt cure. (Taken from Ref. 28 with permission.)

quartets or whether more complex groupings are involved. In the field of ionomers, Otocka[29] favours the former type while MacKnight and Longworth propose that low angle X-ray scattering data is consistent with a complex 100 Å cluster. Eisenberg[30] has suggested that either situation may apply at the correct ion concentration, and that there is a limiting concentration of carboxyl groups below which cluster formation is energetically unfavourable. His theoretical calculations indicate a maximum of eight closely packed ion pairs, or multiplets, at low concentration but aggregation of such units at higher concentration, via weak dipolar interactions, to give clusters. It is suggested that in these, the multiplets are

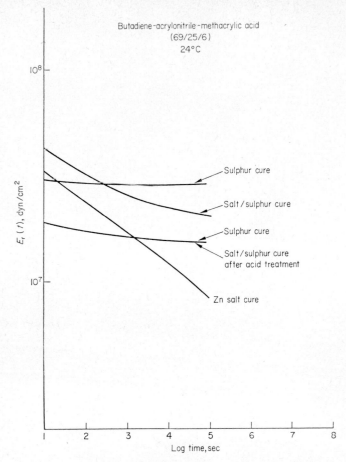

Fig. 6 Stress-relaxation curves of the butadiene–acrylonitrile–methacrylic acid terpolymer with sulphur cures, a zinc salt cure and a combination of zinc salt–sulphur cure. (Taken from Ref. 28 with permission.)

separated by at least the thickness of one hydrocarbon chain to give a structure in which 30% of the cluster is ionic.

3.6 USES OF CARBOXYLATED RUBBER

3.6.1 Solid and solution rubber

The high growth rate of carboxylated rubber sales has stemmed mainly from the use of these elastomers in the form of a latex. The advantages of latex over the solid elastomer will be discussed in the next section. Use of solid rubbers has not been exploited commercially to any great extent although preliminary work has shown that carboxylated material can advantageously replace ordinary elastomers in applications where non-sulphur curing, good adhesion or specific physical properties are required.

Tyres

Some work has been carried out [31] on the use of carboxylated SBR in tyre tread formulations. Dolgoplosk [22] had originally shown the very high resistance to cut growth and tear under static and dynamic conditions and suggested that this was a result of orientation of the polymer chains upon stretching. By X-ray diffraction, this orientation effect in carboxylated rubbers has been shown to increase both with carboxyl and metal oxide content. Under test, bending and stretch flexing of a carboxylated SBR did not cause cut or puncture growth after 360 000 deformation cycles, while an analogous non-carboxylate was destroyed in 130 000–140 000 cycles. The high abrasion resistance of carboxylated rubber tyre treads is related to the above factors. Tyre abrasion is a combination of mechanical wear, determined by tensile and tear strength, and thermally induced oxidative breakdown at the point of surface contact coupled with a fatigue process from stretching over irregularities in the abrasive surface. Stress orientation strengthening should both improve the tear resistance and reduce fatigue. A tyre tread formulation [32] consisting of carboxylated SBR (100 pts), HAF black (45 pts), oil (10 pts), MgO (2 pts) as main vulcanising agent, TMTD (tetramethyl thiuram disulphide) (1 pt) and ZnO (1 pt) was compared with an oil extended SBR and a natural rubber treadstock. The carboxylated SBR tread gave the best combination of high modulus and break elongation, showed exceptional crack growth resistance with no cracking along the tread groove and gave longer life and better abrasion resistance in service. Some initial trouble was experienced with tread joint separation, but this was overcome by buffing the tread layer to remove surface scorch and using an adhesive based on carboxylated SBR.

Adhesives

Although prepared in nearly all cases by emulsion polymerisation, carboxylic elastomers may be used for adhesives in the form of latices, low molecular weight solids or solutions in organic solvents. Latices are used for sticking porous or easily wetted materials such as paper, textiles, wood, etc. while solutions are preferred for bonding to smooth hydrophobic surfaces, particularly metal and glass. Occasionally the carboxylic polymer may be prepared directly in solution for applications where the presence of soap or emulsifier would be harmful. Improved adhesion with carboxylic, compared to non-carboxylic, elastomers results from interaction of the COOH groups with the substrate surface, increased penetration of polar surfaces and higher inherent gum strength. In many cases the carboxylic elastomers need no further pigments, resins or fillers to enhance their adhesive properties. Although one adhesive may bond a wide variety of materials, optimum bond strength between specific surfaces often requires a unique formulation. In addition to the modifications previously otbainable through changes in polymer type, copolymer type and distribution, and molecular weight distribution, the inclusion of the carboxyl group provides a further set of variables dependent on the nature of the acid, carboxyl content, distribution and extent of neutralisation.

The unsaturated rubbers have proved useful in adhesives for bonding rubber to metal. The polar COOH groups interact with the metal surface while the unsaturation sites can be covulcanised with the rubber substrate. When prevulcanised rubber substrates are used, a high degree of adhesion is simply obtained by reaction between the carboxyl groups of the adhesive and metal oxide fillers in the rubber. In early work reported by Frank,[33] simple copolymers of dienes and acrylic or methacrylic acids were made up to a brushing consistency as a 3% solution in cyclohexanone or dioxan. Maximum adhesion of a rubber stock vulcanised against steel required about 20% acid in the copolymer, as shown in Fig. 7.

It was suggested that some of the bond strength was due to hydrogen bonding between the carboxyl group and a surface oxide layer on the metal. Good rubber to metal adhesion has also been obtained with cements from blends of carboxylated butadiene–acrylonitrile copolymers and phenolic resins dissolved in MEK. These gave better adhesion than the analogous non-carboxylates, with cohesive failure in the rubber base. The non-carboxylic rubber showed failure at the metal surface. Similar results were obtained[24] in press-cured bonding of methylstyrene rubber to fabric glued with a sandwiched layer of butadiene–methacrylic acid copolymer latex containing triethanolamine as curing agent. As the carboxyl content increased or as the degree of neutralisation with amine increased, the bond strength increased to a maximum and then decreased. Concurrently the separation changed from cohesive failure, to mixed cohesive/adhesive failure and then to increasing proportion of adhesive failure. This behaviour agrees with the theory that with a low concentration of salt cross-links, molecules readily diffuse into the carboxylated rubber phase to give adhesive forces greater than the cohesive force derived from the cross-linking. At higher cross-link densities, the ionic bonding gives increased

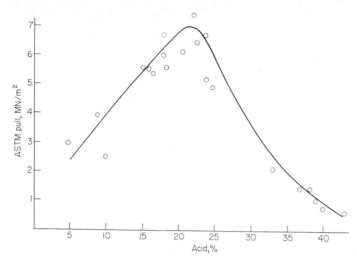

Fig. 7 Effect of acid content on adhesion (butadiene–methacrylic acid polymer). (Taken from Ref. 33 with permission.)

cohesion but decreased compatibility, and hence lower degree of diffusion, and reduces the adhesive force. With fabric glued to steel with carboxylated latex containing zinc oxide, overall bond strengths were much lower, presumably due to lack of diffusion from the film into the metal surface.

For metal to metal adhesion, there is no necessity for a curing cement, which widens the range of usable rubbers. Equivalent carboxylated and non-carboxylated butadiene–acrylonitrile copolymers gave very similar bond strengths for steel to steel addition, with the advantage of a faster bonding rate for the carboxylate system.[34] However, for a series of equivalent polyethylacrylate cements, the bond strength increased steadily from 0·70 MN/m^2 to 6·3 MN/m^2 with increase in acrylic acid content of the acrylate from 4 to 12 parts.

Most carboxylic elastomers adhere well to glass and may be used for glass laminating or bonding glass to other materials. The carboxylated acrylates have outstanding advantages in optical applications because they have refractive indices similar to those of glass and, in contrast to the diene based cements, the absence of unsaturation prevents degradation and colour formation on ageing.

A further use of the solution based adhesives is in the bonding of plastic films or paper to metal foil. The requirement is for a permanently tacky film with a high cohesive bond strength between the laminate surfaces. Non-carboxylated acrylic elastomers are weak adhesives for most pairs of metal and plastic surfaces but are markedly improved by introducing some proportion of carboxyl groups, while diene based polymers tend to be non-adhesive until carboxyl groups are incorporated. The carboxylated acrylate elastomers have been used to bond aluminium, tin and other foils to polystyrene, cellophane, cellulose acetate, polyvinyl chloride and polyethylene. Variations of carboxyl content, acrylate ester group and molecular weight have been used to optimise bond strength between specific surfaces. The application of permanently tacky adhesive to one side of a plastic strip is the basis of the pressure sensitive sticky tape. The adhesive should be stretchy, elastic and have good adhesion to a wide variety of surfaces including the backing tape but most important it should have sufficient cohesion to prevent separation of the elastomer layer on to the reverse side of the film when coiled into a roll. Tapes prepared from simple polyacrylate esters were inadequate for severe applications such as fastening heavy objects to vertical surfaces, sealing thick paper bags or attaching stiff labels to curved surfaces. These limitations were overcome by using copolymers of acrylates and unsaturated acids deposited from solution on to cellophane or cellulose acetate backings. It has been shown[35] that tack of a metallo-carboxylate based pressure sensitive adhesive decreases with increasing ionic character, as measured by the infra-red intensity of the ionised carboxyl group at 6·0 to 6·5μ. In peel tests, adhesive tapes which peeled cleanly from a steel surface at room temperature, failed cohesively at elevated temperatures. The temperature for the onset of cohesive failure increased with increasing concentration of ionic bonds.

Shrink proofing

Shrinkage and felting in knitted or woven woollen articles can be prevented by treatment with carboxylated diene elastomers[36] used in latex or solution form. A latex application would fit best into present manufacturing methods, but treatment with a solution of the elastomer gives equivalent results at a lower pick-up and consequently has less effect on the 'handle' of the material. Overall shrinkage is due to (a) shrinkage relaxation to relieve stress set up during processing or weaving, (b) felting shrinkage resulting from natural elasticity and the scaly nature of the fibre surface. Mechanical agitation during washing aids both of these mechanisms via a one way compaction of the fibres. Since wool is essentially a network of amino acids linked by disulphide groups, there are several types of site which could be chemically modified either to smooth out the surface scale structure or 'weld' crossing fibres to prevent movement. Carboxylated elastomers appear to react with the amino groups of the wool to give satisfactory shrink proofing at a pick-up level of 2–4% using a 5% carboxylated polymer. An elastomer prepared from polybutadiene solution grafted with thioglycollic acid gave better results than one derived from a simple organic acid and both types were improved by partial conversion of the free acid groups to the acid chloride or anhydride.

Miscellaneous
A few miscellaneous uses for solution carboxylates have been claimed but do not appear to have found a large market. Low pressure polyethylene mixed with 30% butadiene–styrene–methacrylic acid terpolymer and 5% magnesium oxide gave a film[37] with improved strength, elasticity and durability, derived from oxide cross-linking of the carboxylate. The polar carboxyl groups bonded strongly enough to abrasive oxides to give usable rubber bonded grinding wheels.[38] Finally, in the footwear industry, carboxylated rubbers have been used for both uppers and soles. Artificial leather uppers[39] were prepared from carboxylated SBR, pore-forming diisocyanates, cotton filler, and stearic acid to increase the pore size. Shoe soles were obtained using SBR containing 0·8% methacrylic acid cross-linked with an equivalent amount of epoxy resin and filled with leather fibres. In a more conventional recipe based on Hycar 1072,[40] a carboxylated diene-nitrile rubber, shoe soles had better abrasion and tear resistance and higher tensile strength than a high quality non-carboxylated SBR sole, although rebound and flex were slightly inferior.

Application of the thermoplastic-elastomeric properties of reversibly cured carboxylated rubbers still remains to be exploited fully. The main problem is to design a polymer composition which has the necessary flow characteristics for precision moulding at temperatures around 150°C but with negligible permanent set or creep at the ambient working temperatures.

3.6.2 Latex rubber

Where possible, latices are used in preference to solutions of rubber in organic solvents. The main advantages of a latex over a solution are (a) very low viscosity at high concentration, *e.g.* at 50% concentration a latex will be less than 1000 cps compared to a solution which may be over 30 000 cps, and (b) a cheap, non-flammable medium (water) is present. Most carboxylated elastomers are prepared initially as latices, therefore it is more economic to use them as such, rather than to coagulate and redissolve with consequent recovery of solvent.

In addition, many industrial applications require the deposition of a thin film of polymer on to a suitable substrate and this, and any required compounding, is most easily accomplished with a latex. At 30–50% solids, the latex contains dispersed particles which pack into irreversible contact with one another on drying to 50–75%. In final drying of a rubbery latex, the particles coalesce to form a tough coherent film. The latices are also more easily applied to hydrophilic surfaces to give better saturation and penetration than solutions. There are several advantages specific to carboxylated latices. Mechanical stability is improved, partly due to hydrophilicity of the carboxyl group, partly to anionic emulsifiers used, and the low temperature stability is improved. The COOH group provides a site for sulphurless cures to give colourless, odourless products with the possibility of metal oxide cure at room temperature, although in most cases temperatures are high enough to allow water removal. In many cases the metal compound required for vulcanisation would be derived from the various fillers normally incorporated as dispersions into the compounded latex, particularly in such applications as carpet backing. The low viscosity and good wetting properties of the carboxylated latices allow high levels of filler to be added as a dry powder, but the filler is not essential for high strength. In many applications, carboxylated latices based on dienes are in direct competition with those derived from polyacrylates. The better colour stability and ageing properties of the acrylates are balanced against the lower price of the diene materials.

Non-woven materials
Carboxylated latices have proved extremely good binders for papers and non-woven fabrics.[41] Non-wovens are prepared from a web of interlocking fibres, most popularly of cotton, rayon, wool, polyesters or polyamides. In contrast to knitted and woven articles, where interlocking fibres give strength and allow slippage to provide good drape and flexibility, the non-wovens must be treated with a binder in sufficient quantity to give mechanical strength without affecting the feel of the material. Successful binders have been based on thermoplastic powders or fibres, which bind on heating, or on deposited elastomers from latex. The latex may be applied to a wet or dry laid web by saturation, by spraying the web surface or by printing on the latex from embossed rollers. The non-wovens are potentially cheap because mixed length fibres or waste fibre materials may be used. The requirements for an ideal binder are that the binder strength should be

greater than the fibres, it should only be located at fibre cross-over points, have high adhesion, enhance elastic recovery and should have good crease and light resistance and be unaffected by wet or dry cleaning. Although latices of non-carboxylic elastomers fulfil many of these requirements, carboxylated elastomers, being polar, increase the polymer/fibre bond strength while improving resilience and liveliness or 'bounce'. Their main function, however, has been to dramatically increase resistance to washing and dry cleaning processes when cured. Carboxylic nitrile latices with around 5% carboxyl groups can be cured with ZnO alone but for best cleaning resistance, cross-linking with a resin or methylol acrylamide is necessary. Because the web formation, type of fibre and type of binder can all be varied, the non-wovens can be tailored to suit specific purposes including garment interlinings, wadding, industrial filters, cleaning rags, sanitary articles, packaging materials, book jackets, and disposables such as hospital sheets, industrial overalls. In most cases, a polymer with a high non-rubbery content is preferred, *i.e.* in carboxylated SBR the styrene content is greater than 50%. The main carboxylated latex types used are SBR, butadiene–acrylonitrile, polyvinylacetates and polyacrylates. Diene polymers give the best drape and handle, the more polar polymers adhere better to the more polar fibres and the acrylates give the best cleaning resistance and colour fastness. A typical formulation[42] for a non-woven binder is 40% latex with high nitrile to butadiene ratio and 5% carboxyl modification (100 pts dry weight), zinc oxide (9 pts), melamine formaldehyde condensate (5 pts), ammonium sulphate (0·5 pts) and polyethylene oxide condensate (1 pt). The emulsion is diluted to 10–20% solids and after application is dried at 85°C and cured for 5 minutes at 150°C.

Paper
Manufacture of paper is closely related to non-wovens, since the initial stage is the laying of a cellulose fibre web, and again the advent of carboxylated latices has made a great impact.[43,44] The latices are added by three methods, namely beater addition where latex is slurried with the wet cellulose fibres during formation of the sheet, saturation of the paper with 30% dry weight of latex, and coating the paper with about 5% dry weight of latex. Prior to the use of latex, the traditional coating materials were starch, casein and soya protein and early non-carboxylated latices were formulated to replace part of these materials, but starch in particular tended to cause latex agglomeration with consequent loss of pick and rub resistance. Agglomeration by starch and also by mechanical shear or reaction with multivalent metal ions is prevented, without the use of stabilisers,

by using carboxylated elastomers. Unmodified SBR latex is thixotropic and the resultant increased flow at high shear reduces the rate of high-speed knife and roller application. The carboxylic polymers on the other hand, have Newtonian flow leading to faster application rates. In addition, the low viscosity of the carboxylated material allows acceptance of higher solids levels leading to improved gloss and smoothness. By using carboxylated SBR latices with a higher level of styrene, 65% compared to the more usual 50%, a higher gloss can be obtained by supercalendering than with the softer polymers. Addition of a few percent of hydroxyethyl acrylate improves the cross-linking with starch in the surface coating.

Drying temperatures do not generate much thermal cross-linking, but further reaction is possible via the fillers and reactive cellulose groups to give a compact elastic structure with improved water resistance, adhesion and tensile strength. Non-carboxylated latices give poor penetration, particularly of thick papers such as the Kraft type, and although this can be overcome by using beater addition, the resulting paper quality is lower. Carboxylated latices give much increased penetration because of their lower viscosity and improved wetting characteristics. Thus Hycar 1571, a carboxylated nitrile latex,[41] penetrated a thick paper uniformly in 5 seconds whereas Hycar 1561, the corresponding non-carboxylated, did not strike through in 300 seconds. Although SBR latices are used in the cheaper papers, acrylic latices are preferred for top grade paper and offset printing because of their excellent adhesion, colour stability, flexibility and pick resistance (surface disturbance by type face). A soft flexible acrylate becomes, in the carboxylate form, easily cross-linked and shows no after-cure tack.

Foam and carpeting
Another area in which carboxylated elastomers, in this case mainly SBR, are taking the place of non-carboxylates is the manufacture of foam. Neither the Dunlop nor the Talalay process[45] is directly applicable to carboxylated latex prepared at low pH because the sodium silicofluoride gelling agent is ineffective, but in 1965, a process was introduced[46] for providing thin layers of carboxylated foam which could be supported on a variety of substrates. The carboxylated latex is mixed with melamine formaldehyde, to give improved cure and rigidity, and a frothing agent. The mixture is whipped up with air and spread on to backing material to give layers up to 12mm thick. Thicker layers do not dry and cure uniformly, due to the good insulation of the surface layers. Such a system gives consistent batch-to-batch properties because of its basic simplicity, with the

ingredients all mixed in before foaming commences. The carboxylated elastomer foam has the unique properties of uniform fine cell structure, velvety hand, whiteness, minimum strike through and absence of copper staining. It is suitable for foam coating on cellulose, wool, natural and synthetic fibres and in some cases a second layer of fabric can be laid on top of the wet foam. This type of foam has been used in clothing interlinings, bathing suits, shoe insoles, carpets, cushion coatings and car lining. It has a major outlet in carpeting of various kinds.

Originally, woven carpets of the Axminster or Wilton type were back sized with starch and initially some of the starch was replaced by latex to improve non-slip properties and handle. Replacement of ordinary SBR latex by carboxylated latex gave improvements in tuft lock and anti-fray properties. A typical formula for such back sizing would be carboxylated latex (100 pts dry weight), filler (300 pts) and thickener (1 pt). In the simplest method, this is spread on the carpet back and dried by radiant heat or forced air circulation.

When tufted carpets were introduced, carboxylated latex was found to give the highest tuft anchorage values compatible with other desirable features such as good 'quick grab', firm handle, low thickener requirements and rapid drying. In more expensive tufted carpets a secondary backing of jute can be applied to the wet initial latex coating, usually by passing the carpet, latex and jute sandwich through nip rollers to improve adhesion and penetration. An alternative improvement was to apply a layer of carboxylated latex foam to the back of tufted carpets both to anchor the tufts and provide a soft, pliable, built-in underlay.[47] Carboxylated latex is also used to give the chemical bonding in needlefelt carpeting. A web of needle-punched felt made from nylon or polypropylene fibres is strong enough to withstand treatment by spraying, or saturation, with the latex binder at levels of 10–40% by weight of the untreated fibre. The final carpet has extremely good scuff and wear resistance coupled with low dirt pick-up. In all cases, the carboxylated latex gives the best adhesion to the fibres, tolerates the greatest amount of filler and gives the highest rate of application. It acts as an adhesive and penetrant to lock the tufts, as an adhesive for secondary backing and as a flexible backing which contributes to the 'handle' of the carpet. Where necessary, a foam layer on hessian can be used as a separate underlay.

Adhesives
Carboxylated latices are used as adhesives in a number of applications for sticking and laminating the more polar and more porous surfaces such as

paper, wood and leather, having the ability to wet or combine with water-wetted surfaces. Since the bonding is largely mechanical, the type of latex is relatively unimportant. Typical uses are in milk cartons, straws, gummed paper etc. Russian workers carried out detailed investigations into the use of carboxylated SBR for improved adhesion of tyre cord to rubber in tyres. They found that carboxylated SBR plus a resorcinol resin gave better adhesion and cohesion than either SBR latex alone or SBR latex with resorcinol. At the present time, however, the majority of tyre cord dips are based on polyvinyl pyridine latices. Some carboxylated latex is used for laminating textiles, and in the main this is acrylic elastomer.[48] Tricot linings have been laminated to tweed to give a smooth inner surface and some very delicate fabrics have been strengthened by lamination. Some problems have been encountered with differential shrinkage and dry cleaning stability.

Miscellaneous uses

Scrap leather has been reconstituted into a usable form by treatment with carboxylated latex.[41] The leather is ground up in the dry state, to prevent formation of an unfilterable pulp, then ground in water. Since the produce is not very fibrous, either cotton flock is added or the processed leather is deposited on the mesh backing. To a 5% leather pulp solution, the latex is added and coagulated by dropping the pH to about 4·5 with alum. Sheets of the product are dried at 55°C and finished by calendering, moulding or embossing. In another process, the surface of normal leather is improved with respect to abrasion, and chemical resistance, flexibility, ageing and gas/liquid permeability by treating with a carboxylated elastomer in a penetrating solvent such as diacetone alcohol or isopropanol.

Although emulsion paints are based mainly on PVA, some carboxylated latex is used, usually acrylic or high styrene diene. Such paints are easily applied, dry rapidly, can be water-washed and are stable to pigments. Latices with high carboxyl contents are also used as thickeners and film formers. Recently it has been suggested that they may be of growing importance in electrocoating paints. Carboxylated latices have also been used to make textile printing pastes where the carboxyl groups give added adhesion to the textile fibres. A typical formulation includes the elastomer, a dye, a lubricant and a reactive dispersant, such as a half ester of a maleic anhydride copolymer, which becomes insoluble when the paste cures.

In some specialised applications, carboxylated latex can be used for dipping formation of a thin complex shape by building up successive layers of rubber over a former. These materials are formed as a gumstock and therefore must have strength without reinforcing filler, a property conferred by the carboxyl group. Of particular interest are oil and chemical resistant gloves made from carboxylated nitrile rubber by a coacervant dipping process and cured via a sulphur/zinc oxide system. The requirement for oil resistance is met by the nitrile base rubber and high tensile and tear strength is conferred by the carboxyl groups. In addition, choice of the correct acid may allow a high tensile strength coupled with a low modulus at intermediate elongation.

3.7 SUMMARY

The increasing commercial use of carboxylated elastomers, particularly carboxylated latices, illustrates their importance as speciality rubbers. Their specific properties should further increase the range of applications. In general, the carboxyl groups increase the strength of the base polymer at the expense of elasticity, elevate the temperature range of elasticity, enhance film forming properties and resistance to hydrocarbon solvent, improve hardness and abrasion resistance, and cause improved adhesion to many surfaces. Chemical reactions with the carboxyl groups lead to cross-linking via polyvalent metal ions or difunctional covalent molecules. In particular, cross-linking with metal ions produces a sulphur-free vulcanisate which is equal or superior to comparable reinforced sulphur vulcanisates in stress–strain properties. Decrease in strength at higher temperatures and poor set characteristics of the metal oxide vulcanisates can be overcome by superimposing a light sulphur cure. The rapid cure rate in oxide vulcanisation may be retarded by addition of various organic compounds in order to prevent precure.

The general properties looked for in carboxylated latices are:

(a) Good total solids/viscosity relationship, *e.g.* 200 to 400 cps at 50% solids content.
(b) Low odour.
(c) Fast cure (must be good in 3 minutes) and the associated properties of adhesion, 'quick grab', rate of drying and washability/water tolerance.
(d) High filler tolerance.
(e) Dry loading ability with clay, whiting etc.
(f) Compatibility with other carboxylated polymers.
(g) High thickener efficiency at high and low filler levels.

(h) Ageing resistance to heat, light and gases.
(i) High cold bond strength and wet bond strength.
(j) Improved mechanical stability.

At the present time the biggest potential for carboxylated latex is in the textile and paper fields.

Over the last decade, theories about the nature of the structure of the salts formed from carboxylic elastomers have gradually changed. The most recent ideas concerning ionic clusters situated in a hydrocarbon matrix and the similarity between the carboxylates and segmented ABA elastomers, suggest that a future use may be in the preparation of thermoplastic elastomers.

REFERENCES

1. French Patent 701,102.
2. Brown, H. P. (1957). *Rubber Chem. Tech.*, **35** (5), 1347.
3. Bacon, R. G. R., Farmer, E. H., Morrison-Jones, C.R. and Errington, K. D., Rubber Tech. Conf., London, Preprint No. 56, p. 256 (May 1968).
4. Brown, H. P. and Duke, N. G. (1954). *Rubber World*, **130**, 784.
5. Dolgoplosk, B. A. *et al.* (1959). *Rubber Chem. Tech.*, **32**, 321.
6. Cuneen, J. I., Moore, C. G. and Shephard, B. R. (1960). *J. Appl. Polym. Sci.*, **3** (7), 11.
7. Holmberg, B. (1932). *Chem. Berichte*, **65**, 1349.
8. Serniuk, G. E., Banes, F. W. and Swaney, M. W. (1948). *JACS*, **70**, 1804.
9. Amass, A. J., Duck, E. W., Hawkins, J. R. and Locke, J. M. (1972). *European Polym. J.*, **8**, 781.
10. Crugnola, A., Pegoraro, M. and Severini, F. (1969). *J. Polym. Sci.*, Part C, **16**, 4547.
11. Gaylord, N. G., Takahashi, A., Kikuchi, S. and Guzzi, R. A. (1972). *Polymer Letters*, **10**, 95.
12. Levine, N. B. (1969). *Rubber Age*, **101** (5), 45.
13. US Patent 2,662,874.
14. British Patent 1,030,196.
15. British Patent 1,148,855.
16. Green, J. and Sverdrup, E. F. (1957). *Rubber Chem. Tech.*, **30**, 689.
17. Minoura, Y., Shiina, K. and Harada, H. (1968). *J. Polym. Sci.*, **A1**(6), 559.
18. British Patent 980,988.
19. US Patent 2,671,074.
20. Brown, H. P. (1963). *Rubber Chem. Tech.*, **36**, 931.
21. Cuneen, J. I. and Shipley, F. W. (1959). *J. Polym. Sci.*, **36**, 77.
22. Dolgoplosk, B. A. *et al.* (1959). *Rubber Chem. Tech.*, **32**, 328.
23. Zakharov, N. D. (1963). *Rubber Chem. Tech.*, **36**, 568.
24. Morrell, S. H. and Moseley, R. J. (1973). *Rubber J.*, **155**(1), 37.
25. Hayes, R. A. (1960). *J. Chem. Eng. Data*, **5**(1), 63.
26. Cooper, W. (1958). *J. Polym. Sci.*, **28**, 195.
27. Halpin, J. C. and Bueche, F. (1965). *J. Polym. Sci.*, **A3**, 3935.

28. Tobolsky, A. V., Lyons, P. F. and Hata, N. (1968). *Macromolecules*, **1**(6), 515.
29. Otocka, E. P. and Eirich, F. R. (1968). *J. Polym. Sci.*, **6**, 921 and 933.
30. Eisenberg, A. (1970). *Macromolecules*, **3**(2), 147.
31. Brodskii, G. I., Sakhnovskii, N. L., Reznikovskii, M. M. and Evstratov, V. F. (1960). *Soviet Rubber Techn.*, **19** (8), 22.
32. Buiko, G. N., Sakhnovskii, N. L., Evstratov, V. F., Smirnova, L. A., Levitina, G. A. and Katkov, V. I. (1961). *Soviet Rubber Techn.*, **20**(3), 8.
33. Frank, C. E., Kraus, G. and Haefner, A. J. (1952). *Ind. Eng. Chem.*, **44**(7), 1600.
34. Brown, H. P. and Anderson, J. F. (1962). *Handbook of Adhesives*, Ed. I. Skeist, Chap. 19, p. 255, Reinhold, New York.
35. Satas, D. and Mihalik, R. (1968). *J. Appl. Polym. Sci.*, **12**, 2371.
36. British Patent 1,064,660. British Patent 1,067,903.
37. Denisenko, I. S., Mishustin, I. V., Semenova, A. I. and Rezvaya, G. (1971). *Nauch.-Issled. Tr. Vses. Nauch.-Issled. Inst. Plenoch. Mater. Iskusstv. Kozhi*, **21**, 62.
38. USSR Patent 204,180.
39. Ostrovskii, V. I., Khromova, N. S. and Pavlov, S. A. (1970). *Kozh-Obuv. Prom.*, **12**(1), 54.
40. Hycar Service Bulletin H-21 June 1956.
41. Hycar Latex Manual, B.F. Goodrich Chemical Co. 1956.
42. Blackley, D. C. (1966). *High Polymer Latices*, Vol. 2, p. 672. Applied Science, London.
43. Wilson, J. H. (1964). *Rubber Plastics Age*, **45**(10), 1195.
44. Law, C. (1968). *Rubber World*, **158**(6), 79.
45. Calvert, K. O. and Newnham, J. L. M. (1966). *Rubber Age*, **98**(2), 73.
46. Zimmerman, R. L., Hibbard, B. B. and Bailey, H. R. (1966). *Rubber Age*, **98**(5), 68.
47. *Symposium*, 'Application of Latices to Carpets' (1966). *Rubber Journal*, **148**(12), 67.
48. Kantner, G. C. (1968). *Resin Review*, **18**(2), 3.

CHAPTER 4

RIGID, HIGHLY CARBOXYLATED, IONIC POLYMERS

A. D. WILSON AND S. CRISP

This chapter covers in general those materials that are formed by the interaction of cations with highly carboxylated polyacids: that is polycarboxylic acids like poly(acrylic acid) and the poly(alkenyl carboxylic) acids in general. These polyacids are characterised by a mole ratio of COOH:C of about 1:3. In poly(acrylic acid) this ratio is exactly 1:3 while in poly(itaconic acid) it becomes 2:5. These ionic polymers are diverse in character. They may be formed in the solid state as with Nielsen's ionic polymer salts, or in aqueous solution as is the case with the dental ionic polymer cements and certain soil stabilisers. Fabrication can be at moderately high temperatures (Nielsen's salts), by acid–base reaction at room and oral temperatures between preformed polyacids and cation releasing bodies (ionic polymer dental cements) or by polymerisation of a metal acrylate *in situ* (soil stabilisation).

4.1 IONIC POLYMER CEMENTS

The ionic polymer cements are the product of the hardening reaction that occurs when aqueous solutions of highly carboxylated acid polyelectrolytes, in particular poly(acrylic acid) are brought into intimate contact with certain simple or complex metal oxides. These oxides act as sources of cations which serve to ionically cross-link the polyanionic chains to form a hard gel matrix. These materials are of interest in being intermediate between a filled organic polymer and an inorganic cement. They are novel as polymers in that the matrix is a hydrocolloid and can be correctly described as an ionic polymer gel. These cements, at present, find applications only in dentistry but analogies may be drawn with certain biomaterials particularly dental plaque. They also have certain formal similarities with the alginate impression materials, where aqueous solutions of the polysaccharide, alginic acid, act as the acid polyelectrolyte and cations are supplied by calcium or lead salts (the use of the latter is declining, and is to be deprecated because of the toxic nature of lead).

The ionic polymer cements are also related to other dental cements, which although not ionic polymers, have the same type of setting reactions. In this reaction simple or complex oxides react with a liquid proton donor to form a salt-like gel matrix. The most common metal oxide is zinc oxide and the proton donating liquid may commonly be an aqueous solution of phosphoric acid,† eugenol or 2-ethoxybenzoic acid. The binding matrices of these cements are, chemically, pure salt-like structures bound solely by ionic bonds. The ionic polymer cements, however, are unique in having a linear covalent backbone, ionic bonds serving only to cross-link covalent linear chains.

The possibility of developing new cements based on the inorganic salts of polyelectrolytes was first suggested by McLean.[1] At present two forms of ionic polymer cement have been developed. In both cases poly(acrylic acid), in aqueous solution, is used as the cement forming liquid:

(a) The zinc polycarboxylate cement described by Smith[2,3] where zinc oxide is used as a source of cations.
(b) The glass ionic polymer or aluminosilicate polyacrylate cement of Wilson and Kent,[4-6] where a vitreous fluoroaluminate silicate acts as a source of cations.

4.1.1 Ion binding and molecular configuration

The ionic polymer dental cements harden as the result of the neutralisation of poly(acrylic acid) and salt formation with multivalent cations, leading to gelation. These reactions need to be considered in the light of fundamental behaviour of polyacids in solution. There are two aspects. The binding of ions to the chains and configurational changes caused by neutralisation of the polyacid.

The molecular configuration adopted by poly(acrylic acid) and similar poly(alkenylcarboxylic acids), which are largely un-ionised in aqueous solution, is one where the molecular coils are arranged in a compact spherical form.[7-9] However, when these polyacids are partially neutralised by a

† The phosphoric acid referred to in this chapter is orthophosphoric acid.

metal base, carboxylate groups become highly ionised, causing the molecular coil to expand considerably because of electrostatic repulsion, an effect reflected in the rise of viscosity of solutions.[10] These configurational changes play a role in the setting mechanism, but there are other factors to consider.

Solutions containing cations and poly(acrylate) and similar poly(alkenyl carboxylate) ions have interesting properties. The polyanionic chain exerts an electrostatic attraction on the small cations, the counterions, which thus are constrained to remain in the vicinity of the polyanion. This phenomenon is known as ion binding and there is a considerable body of experimental evidence for it. For example, during electrophoresis, some of the counter cations migrate with the polyanion,[11] a classical observation well verified.[12] Bjerrum in 1926,[13] first pointed out that ions of opposite charge could under certain circumstances and in media of low dielectric constant, associate to form ion pairs. The treatment of ion association in dilute solutions has since been refined by Fuoss.[14] This phenomenon is more marked, when one of the ions carries a double charge, especially if these are separated by a short chain ('bolaform ions') when association can occur even in aqueous solutions. This phenomenon is exaggerated further in the case of polyions, where the binding of some of the counterions can be extremely tight as has been shown in counterion diffusion using radioactive tagging.[15,16] The extent of ion binding for univalent cations depends on the average density of the ionisable groups[17] and their strength[18,19] but not their distribution.[17] Multivalent cations are bound more firmly and extensively than are univalent ions.[20]

The nature of ion binding has been the subject of much speculation and can result from either general or localised electrostatic fields, giving rise to either general or localised ion-pair binding. The binding of hydrogen ions and sodium ions to poly(acrylic acid) is apparently of the general type.[18] By contrast multivalent cations tend to be bound to specific sites on the polyanion chain.[20] Specific site binding is enhanced by chelate formation for which there is some evidence in the case of cupric and polyacrylate ions.[21-25] More relevant to the field of ionic polymer dental cements is the evidence that zinc and magnesium form complexes[22] involving two carboxylate ions.

However, Jacobson[26] considers that the behaviour of a counterion strongly bound to a poly(acrylate) chain need not be explained in terms of specific binding and chelation; configurational changes in the polymer chains providing an alternative explanation. There are also associated changes in the structure of water consequent on ion binding.[27] Multivalent ions such as Ba^{2+} and La^{3+} tend to displace water molecules oriented around individual functional groups causing partial dehydration, an effect not observed for Na^+ ions. The extent of the displacement of water is dependent on the degree of ion binding which is related to the ionic potential of the counterion.

Ion binding gives rise to phase changes, gelation and precipitation of salts of polyelectrolytes. The ionic polymer dental cements are based on this phenomenon. Wall and Drennan[28] were amongst the first to study this topic, and considered that the gelation of poly(acrylic acid) by divalent cations was to be attributed principally to ion association with the consequent formation of salt bridges. However, they recognised that there were factors other than simple coulombic attraction. Naturally occurring polycarboxylic acids, alginic and pectic acids, form precipitates with Ca^{2+} ion which are said to be due to cross-linking.[29,30] Michaeli[30] considers that whatever the binding mechanism may be, divalent cations cause the polyanionic moelcule to contract and that the precipitated gel consists of spherical polyelectrolyte molecules in which are embedded cations. This contrasts to the model of Jacobson,[26] where expansion of the macromolecule is attributed to progressive ionisation of the polyelectrolyte chain. Neutralisation of poly(acrylic acid) with a base containing a weakly bound cation would result in the expansion of the macromolecular coil. However, it must be admitted that if ion binding was tight enough or if complex formation took place, then such an event would tend to negate the unwinding of the molecular coil. Quite clearly, complex configurational changes can be envisaged, which will be dependent on the nature of the ion binding and have yet to be definitely established.

Ikegami and Imai[20] investigated the effects of hydration of the polyion and specific ion binding on precipitation, and were able to distinguish between different multivalent ions. They distinguished two types of binding (a) a salt bridge formation (COO—M—OOC) with considerable dehydration of the polyanion which was characteristic of Ca^{2+} and Ba^{2+} ions and (b) a pendant half salt (COO—M$^{+\,-}$OOC) where dehydration of the polyanion was small and characteristic of the Mg^{2+} ion. It was found from conductance measurements that carboxylate groups desolvate alkali metal ions in a range of solvents.[19] This phenomenon was not related to the strength of the ion binding but depended on the nature of the cation, Mg^{2+} ions being more hydrophilic than the Ca^{2+} or Ba^{2+} ions. With hydrophilic ions, the tendency is for the dehydrated salt bridge to be broken by hydration and this affects the mode of precipitation. These authors ascribed precipitation to the dehydration of the polyion by salt formation, and show that expansion of the polymer chain becomes small near the precipitation point. This work tends to resolve the conflicting models of Jacobson and Michaeli. Polyanion expansion would appear to occur as more divalent cations become bound to the chain, especially if they are hydrophilic. However, with the formation of definite tightly bound salt bridges near the precipitation point, the negative charge on the polyanion is neutralised locally, inhibiting further expansion of the macromolecule.

4.1.2 Zinc polycarboxylate cement

Zinc polycarboxylate cement, the first ionic polymer cement, was developed by Smith[2] because of the defects of existing dental luting agents, in particular the inability to form true adhesion bonds to enamel and dentine. It is the product of the hardening reaction that occurs when finely powdered zinc oxide is mixed with aqueous solutions of poly(acrylic acid). Poly(acrylic acid) may be replaced by similar poly(alkenylcarboxylic acids).[31] Smith's cement can be regarded as one of a family of dental cements based on the ability of zinc oxide to react with a wide range of proton-donating liquids to form hard salt-like gels which act as cement-bonding matrices. The generic chemical reaction may be represented thus:

$$\underset{\text{proton acceptor}}{ZnO} + \underset{\text{proton donor}}{H_2A} = \underset{\text{salt-like gel}}{ZnA} + \underset{\text{water}}{H_2O}$$

Liquid cement formers include aqueous solutions of phosphoric acid,[32] eugenol,[32] 2-ethoxybenzoic acid[33] and other liquid chelating agents.[34] However in this case the binding gels are purely of a salt-like character, because they are bound solely by bonds which are highly ionic in character. Smith's cement differs qualitatively from other zinc oxide cements in that the matrix is bound by a covalent chain, as well as by ionic linkages in the macromolecule and thus it has the character of an ionic polymer as well as a salt. Although somewhat weaker than the zinc phosphate cement, it has the advantage of not severely irritating tissues and is stronger than the bland zinc oxide chelate cements. It has the outstanding advantage over these other cements in showing true adhesion to tooth material and base metal, and thus represents a considerable advance in the technology of dental cements.

The thickness or the consistency of the cement is controlled by the powder:liquid ratio employed in the mix, and by the viscosity of the liquid; which in turn is dependent on the concentration and molecular weight of the polyacid. Although some commercial products on the market employ a universal liquid, in which case consistency is entirely controlled by the powder:liquid ratio, other manufacturers prepare liquids of different viscosities for different purposes. One manufacturer achieves this by varying the molecular weight of the polyacid while keeping its concentration constant, while another fixes the molecular weight and varies the liquid concentration.

The zinc polycarboxylate cement is at present used exclusively as a dental material, finding no application outside dentistry. The manipulative characteristics of the system can be adjusted to suit various applications by changing the formulation of the liquids in respect of molecular weight and concentration. This cement is used as a cavity liner or base, to protect dental pulp against thermal, mechanical and chemical insult and for the cementation of stainless steel orthodontic appliances. For these applications, the cement is mixed to a thick consistency. The cement is also used for the cementation of inlays, crowns and bridges when a thinner consistency is used. In these luting operations, it is desirable that the cement flows freely and that its film thickness is kept to a minimum. A crown or inlay is made to fit the tooth accurately, and the luting cement is required to make up minor imperfections of fit and to aid retention mainly by flowing and setting in these imperfections thus ensuring mechanical interlocking. The film thickness of a cement is a function not only of the particle size of the powder but also of the consistency of the cement mix.

Materials

All experimental and commercial examples come as a two component pack consisting of a powder and a liquid. However, a number of variations are possible leading to several different types. They may be designed for mixing by hand or using a mechanical vibrator. The nature of the two components can differ, for although all cements of the type are based on the hardening reaction between a metal oxide and a polycarboxylic acid in aqueous solution, there are several ways of attaining the objective.

(1) By supplying metal oxide powder for mixing with a strong aqueous solution of polycarboxylic acid. Because of the high viscosity of the liquid these materials are not suitable for mechanical mixing, in general.

(2) By supplying an intimately blended powder containing both metal oxide and polycarboxylic acid for mixing with water or a neutral buffered solution. This system is preferred for mechanical mixing.

Other variations are possible. Although in nearly all examples poly-(acrylic acid) is the polyacid employed, an example is known where an alternative polyacid is used.

Bertenshaw and Combe[35-37] have examined commercial examples available and their data is summarised in Table 1. Using this information these materials may be classified as follows:

(i) *Type I*. The basic type is one where cements are prepared, by hand mixing, from a powder consisting chiefly of zinc oxide and a strong aqueous solution of poly(acrylic acid). Analysis of five examples (A, B, C, D, E) on the market shows that the powders consist of zinc oxide as the major component with 4·73–9·19% of magnesium oxide. The poly(acrylic acid) liquids have a concentration range between 32·4–42·9% and a mean molecular weight range between 23 000–50 000. Two manufacturers market two types of material, one for lining cavities (*i.e.* thick consistency paste)

TABLE 1

COMPOSITION OF COMMERCIAL ZINC POLYCARBOXYLATE CEMENTS

	Liquid PAA %	M_w of polyacid	Powder composition			
			ZnO	MgO	Al_2O_3	PAA
TYPE I (*for hand mixing*)						
A[a]	42·9	45 000	87·0	9·2	—	—
A[b]	42·6	23 000	87·0	9·2	—	—
B	39·1	48 000	96·7	4·7	—	—
B[b]	32·4	49 000	96·7	4·7	—	—
C	41·3	35 000	89·5	9·0	—	—
D	41·5	50 000	96·8	5·0	—	—
E	41·5	35 000	85·2	9·0	—	—
TYPE II (*for hand mixing*)						
F	31·8[c]	n.d.	87·9	10·1	—	—
TYPE III (*for mechanical mixing*)						
G	28·7	n.d.	52·1	2·8	43·7	—
TYPE IV (*for mechanical mixing*)						
H	—	22 000	77·1	7·6	—	15·2
J	—	24 000	83·4	0·4	—	16·0
K	—	31 000	80·8	0·2	—	18·2

[a] For lining.
[b] For luting.
[c] Copolymer.

and the other for luting (*i.e.* thin consistency). One reduces the viscosity of the liquid (concentration 42%) by reducing the M_w from 45 000 to 23 000; the other manufacturer varies the concentration of polyacid in the liquid from 39·1% to 32·4%, keeping the M_w at 48–49 000.

(*ii*) *Type II*. Similar to Type I but the poly(acrylic acid) is replaced by a copolymer of acrylic and itaconic acids. Only one example is known (F), the concentration of the solution is 31·8% and the zinc oxide powder contains 10·06% of magnesium oxide.

(*iii*) *Type III*. Similar to Type I but formulated for mechanical mixing. To facilitate this operation, the concentration of the poly(acrylic acid) solution is reduced to 28·6%, in the only example of this type at present available (G). The powder too is unusual, containing 52·1% zinc oxide, 2·8% magnesium oxide and 43·7% aluminium oxide, the latter added presumably as a reinforcing filler.

(*iv*) *Type IV*. These materials are specifically formulated for mechanical mixing. The poly(acrylic acid) is added as a solid to the metal oxide powder and the liquid is essentially water buffered by a few percent of a sodium dihydrogen phosphate. This low viscosity liquid is more suitable for mechanical mixing than the viscous poly(acrylic acid) solutions. Three examples of this type have been analysed (H, J, K) and the composition of the powders is as follows: 77·1–83·4% zinc oxide, 0·20–7·65% magnesium oxide, 15·2–18·2% polyacrylic acid.

The metal oxide powder employed in the zinc polycarboxylate cements has to be prepared under closely controlled conditions of temperature, so that when mixed with the chosen poly(acrylic acid) solution it forms a cement with suitable setting characteristics for dental use. The principle is based on the fact that zinc oxide is progressively deactivated as the temperature of its ignition is increased and prolonged.[38] Two methods of preparing a suitable zinc oxide powder are given below:[2]

(1) Analytical grade zinc oxide powder, or zinc oxide powder prepared by igniting zinc carbonate at 350–600°C for 30 min, is heated in an electric furnace at 900–1000°C until sufficently deactivated so that when combined with the chosen liquid a cement is formed with a clinically suitable setting

time. The period of heating is 12–24 hr. The oxide so produced is yellow, probably because zinc exceeds the stoichiometric proportion, and can be used without colouring.

(2) An intimate mixture of pure zinc oxide and magnesium oxide, in the weight ratio of 90:10, is heated at 1000–1300°C for 8–12 hr. The resulting sintered mass is ball milled to pass a 300 mesh sieve and then reheated for 8–12 hr. The exact conditions chosen are those that will produce a cement which is clinically satisfactory, when the powdered oxide is mixed with the selected liquid. The oxide produced by this method is white, which can be coloured if necessary by the choice of appropriate pigments.

The role of the magnesium oxide addition has not been fully understood, but apart from its effect in producing a white rather than a yellow powder, it appears that it improves the rate of hardening of these cements. However, excessive additions result in cements which absorb too much water, possibly because of the tendency of Mg^{2+} to hydrate. These metal oxide powders are generally similar to those employed in the traditional dental zinc phosphate cement, and this is an example where there has been a carry over of technology from one material to another. The role of magnesium oxide in the metal oxide powders of the zinc phosphate has also never been fully explained, but apparently its usage constituted an important advance in cement technology, and Fleck in 1902[39] considered that it improved the quality of the cement. Zhuravlev and Vol'fson[40] state that magnesium oxide is added to slow the setting time and increase the strength in zinc phosphate cements. When zinc oxide is fused with a little magnesium oxide, a sintered mass is formed containing ZnO and a solid solution of ZnO in MgO.[41] This material has a higher apparent density than plain zinc oxide, and when ground its activity towards acids is reduced.[42] Zhuravlev and Vol'fson[41] studied the reaction between MgO and ZnO using X-ray diffraction. They found that when a 50:50 mixture was sintered, although the product exhibited both ZnO and MgO lines, closer inspection revealed that the MgO lines had shifted, indicating an increase in lattice constant, which was attributed to the inclusion of Zn^{2+} in the MgO cubic lattice to form a solid solution. The substitution of Zn^{2+} (ionic radius 0·87 Å) for Mg^{2+} (ionic radius 0·78 Å) involved an increase in the MgO lattice constant. The limit of solubility of ZnO in MgO is 35–40%.

Poly(acrylic acid) solutions are prepared by aqueous free radical polymerisation using ammonium persulphate as the initiator and isopropyl alcohol as a chain transfer agent.[2] Polymerisation must be carried out using aqueous solutions where the concentration of acrylic acid does not exceed

25%, otherwise the reaction may proceed with explosive violence. The final concentration is attained by vacuum distillation.

A typical laboratory preparative method is given. Three solutions are employed:

Solution I
200 ml pure water
0·5–2·5 g ammonium persulphate
Solution II
100 ml of pure water
100 g redistilled inhibitor free acrylic acid
20 g isopropyl alcohol.
Solution III
60 ml of pure water
0·5–2·5 g ammonium persulphate.

A round bottom flange type flask is charged with solution (I) which is purged with nitrogen and heated to 80–85°C maintaining continuous purging. Solutions (II) and (III) are then added in the ratio of 3·4:1 over a period of two hours with continuous stirring. Then heating is continued for a further 2 hr at 80–85°C. This solution is concentrated by vacuum distillation at 20 mm of mercury pressure at 40–45°C until the desired concentration is attained, generally about 42%.

The molecular weight of the product is between 20 000–55 000 depending on the amount of ammonium persulphate used. Decrease in the ammonium persulphate concentration results in an increase in the molecular weight of the polymer. Increase in the temperature of polymerisation results in a reduction in the average molecular weight.

For water setting materials, a 40% solution of poly(acrylic acid) is allowed to air dry on a plastic sheet to form a thin glassy film containing about 80% of poly(acrylic acid). This material is milled and mixed with zinc oxide powder.[43]

Setting mechanism
The dental zinc polycarboxylate cements are rapidly hardening cements and will set within a few minutes at oral temperatures. Smith[44] has shown that strength is developed rapidly and quotes an example where 75% of the 24 hr strength was attained within 15 min of preparation (Fig. 1). Such accounts that have been given of the setting reaction and cement structure have been based on reasonable inferences rather than experimental

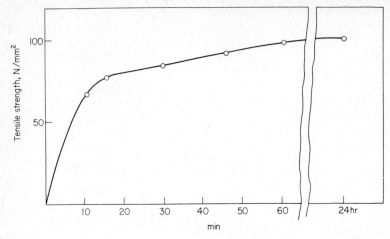

Fig. 1 Increase of the strength of a zinc polycarboxylate cement following preparation. (From Smith, D. C. (1971). J. Canad. Dent. Ass. 37, 22. Reprinted by permission.)

evidence. The description of the zinc polycarboxylate cement given here is largely based on analogies with the glass ionomer cement, and with the solid zinc polycarboxylate salts of Nielsen[45,46] where reaction mechanisms and structures have been the subject of experimental investigations reported elsewhere in this chapter.

On mixing the powder and liquid, an acid–base reaction undoubtedly occurs as would be expected, and the recorded pH of the cement has been observed by the authors to increase rapidly. It must be supposed that hydrated protons from the liquid rapidly penetrate the zinc oxide particles displacing zinc ions which migrate into the aqueous polyelectrolyte phase. Changes will occur in the configuration of the poly(acrylate) chain. Poly(acrylic acid) in aqueous solution is largely un-ionised and approaches a spherical ball configuration. However when the polyacid is partly neutralised, as hydrogen is replaced by zinc, then the chain will carry a corresponding proportion of negatively charged carboxylate ions and the resulting mutually repulsive forces will cause the coiled chains to unwind and expand and the cement paste to thicken. A more important effect is that of ion binding. The zinc cations migrating to the aqueous phase will become ionically bound to the polyacrylate chain, and will in fact progressively cross-link these chains rendering them insoluble, and causing the cement to set as a gel is formed.

The change in rheological properties observed as the cement ages is in accord with this outline mechanism for at a certain stage, the paste, originally a pliable plastic mass, becomes 'stringy' with the development of elastomeric qualities. This phenomenon is to be expected and is indicative both of the configurational change to linear chains and the partial cross-linking of polyanionic chains which occur at an early stage in the reaction. As more zinc ions migrate into the polyelectrolyte phase the degree of cross-linking increases and the 'stringy' elastomeric phase is replaced by a hard solid phase. A useful analogy may be drawn with the process of vulcanising rubber, where sulphur progressively cross-links the hydrocarbon chains causing the physical state to change from a rubbery state to a brittle mass.

A comparison between this cement and the zinc phosphate cement formed from zinc oxide and phosphoric acid is of interest. Both use forms of zinc oxide which have been deactivated to a similar extent and have a similar setting time. This is somewhat surprising, since the strengths of the two acids differ considerably: the pK_a for phosphoric acid is 2·2 while the apparent pK_a for poly(acrylic acid) lies between 5 and 6. Consequently it would be expected, at first sight, that phosphoric acid would react with zinc oxide powder and liberate zinc much more rapidly than poly(acrylic acid), so that the phosphate cement would set much more rapidly than the polyacrylic one. Since this is not so explanations have to be sought and two such are possible:

(1) The acidic function of the polyacids is supplemented by their ability to ionically bind or chelate zinc ions, an assumption which would indicate that the binding or chelating power of the polyacrylate chain is considerable.

(2) The zinc polycarboxylate cement sets at a much lower degree of neutralisation than does the zinc phosphate cement. The zinc phosphate cement matrix is bound by purely ionic interactions between zinc and phosphate ions and hence it is reasonable to suppose that a greater proportion of zinc ions are required to effect a set compared with the zinc polycarboxylate cement, as linear covalent bonded polymer chains are already present and only require the zinc ions to effect cross-linking.

However experimental evidence based on the increase of strength and pH of the cements with time[44,47] indicates that the cement forming reaction is complete in about the same time, from which it may be inferred that the explanation (1) is correct, and that the binding power of the polyacrylate chain towards zinc ions is considerable, suggesting that zinc indeed forms a complex as it is known to do with poly(methacrylic acid).[24]

Cement structure
The structure of the set cement consists of unconsumed zinc oxide particles embedded in a zinc poly(acrylate) matrix. Taking a simple combination of a zinc oxide powder with a 40% PAA solution, then calculation shows that if these components are mixed in the normal powder:liquid ratio of 1·5–3·0 about 8–15% of the powder is consumed in the reaction, the remainder acting as a particulate filler. The composition of such a range of cements will vary from 50–70% zinc oxide, 25–15% zinc poly(acrylate) and 25–15% water. In those cements where magnesium oxide is present in the powder as a minor constituent, then there will be a proportion of magnesium oxide and magnesium poly(acrylate) present also. The nature of the ionic linkages between cation and polyanion may be similar to the structures proposed by Fields and Nielsen[45,46] for solid zinc polyacrylate salt. They considered that there were three possible types of ionic linkage between the zinc ions and the polyacrylate chains:

(a) an intramolecular attachment of the cations to a single polyacrylate chain, *i.e.* internal salt formation,
(b) an intermolecular connection where zinc ions ionically cross-link adjacent chains,
(c) the formation of a pendant half-salt.

The formation of a chelate is another possibility. Although the formation of an 8-membered chelate ring is rare,[48] this cannot be excluded. Multidentate ligands are capable of forming large membered chelate rings[49] and the numerous functional groups on the poly(acrylate) chain make this possible.[50] Experimental evidence has been advanced for a Cu^{2+} polyacrylate chelate and Zn^{2+} poly(methacrylate) chelate.[21–24]

This model, derived from Nielsen's structures, explains certain properties of these cements.

(1) The increase of cement strength with poly(acrylic acid) concentration. Increase of poly(acrylic acid) concentration results in the decrease of average distance between poly(acrylate) chains, favouring the formation of intermolecular, *i.e.* cross-linking, bonds, over intramolecular bonds, with consequent increase in cement strength.
(2) There is an optimum powder:liquid ratio for maximum cement

strength (Fig. 2). As fast as the powder:liquid ratio increases, cement strength increases because the filler content increases. However, above the optimum powder:liquid ratio, excessive amounts of pendant half-salts may be formed, weakening the matrix.

Physical properties
The zinc polycarboxylate cement is one of a large family of zinc oxide cements. Traditionally these are represented by the zinc phosphate and zinc oxide eugenol (ZOE) cements, where phosphoric acid in aqueous solution and eugenol are the cement liquids. However, there are very many more zinc oxide cements than these: Brauer et al.[33] and Nielsen[34] reported a considerable number of liquid chelating agents which were effective cement formers.

The properties of the zinc polycarboxylate cement are compared with those of other dental zinc oxide cements in Table 2. The zinc polycarboxylate cement most closely resembles the zinc phosphate cement: both are bound by aqueous gels and the deactivated zinc oxide powders are interchangeable. The zinc polycarboxylate cement is weaker than the zinc phosphate cement in compression but stronger in tension. Both are much weaker in compression than the cement based on ion-leachable glasses

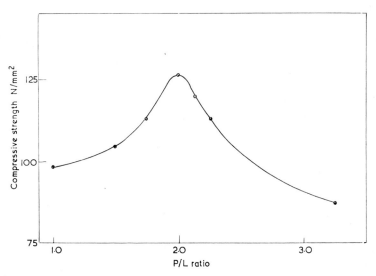

Fig. 2   Effect of the powder:liquid ratio on cement strength. (*From Smith, D. C. (1971). J. Canad. Dent. Ass.* **37**, *22. Reprinted by permission.*)

TABLE 2

PROPERTIES OF ZINC POLYCARBOXYLATE CEMENTS COMPARED WITH OTHER ZINC OXIDE DENTAL CEMENTS

	Poly-carboxylate	Phosphate	Eugenol[a]		EBA[b]
			Simple	Reinforced	
Powder:liquid ratio g/ml	3·6	4.2	2·6	2·35	6·2
Volume fraction of powder %	40	43	32	30	56
Setting time (37°C) mins	3·25	4·25	3·75	3·75	6·0
Strength $Nm^{-2} \times 10^{-6}$					
Compressive, 1 day	85	128	13	39	91
Tensile 1 day	12	8	1·6	3·5	7·6

[a] Eugenol 2-methoxy 4-allyl phenol ($C_6H_3(OH)(OCH_3)(CH_2$—CH=$CH_2)$), is the chief constituent of clove oil.
[b] EBA. The liquid consists of 37·5% eugenol and 62·5% 2-ethoxybenzoic acid ($C_6H_4COOH(OC_2H_5)$).

discussed below—the dental silicate and glass ionomer cements—so that it would appear that this property is a function of the filler rather than the chemistry of the system.

The enhanced tensile strength of the zinc polycarboxylate cement compared with that of zinc phosphate must be connected with the polymeric nature of the cement. Cement gel matrices, in general, consist of aggregates of particulate or plate-like structures bound together by weak physical forces[51] and thus they are weak in tension. Chatterji and Jeffery[52] consider that when subject to tension, the particulate matrix of hydraulic cements is held together by interstitial water so that the ultimate tensile strength is that of water. Probably in the zinc polycarboxylate cements, the physical forces binding the presumably particulate structure are supplemented by some long chain polyacrylate bridges.

The zinc polycarboxylate cement, like all other dental zinc oxide cements, is opaque and this property is considered to be a function of the opaque zinc oxide powders. By contrast cements based on ion-leachable glasses are translucent.

The most important properties that the zinc polycarboxylate cement has lie in its adhesive properties and its blandness towards living tissues. One of the most important qualities claimed for the zinc polycarboxylate cement is adhesion to enamel, dentine and orthodontic stainless steel. The importance of this quality is best illustrated by reference to the traditional dental cements; the zinc phosphate, zinc oxide–eugenol and silicate cements which lack it.[53] The retention of inlays, crowns and orthodontic appliances, when traditional cements are used for luting, is entirely due to mechanical interlocking, the cement paste flowing and setting in the gap between restoration and tooth material. Again, when these cements are used for filling, good tooth material has to be removed because undercutting is necessary for their mechanical retention. Another disadvantage arising from lack of adhesion is that the margin between restoration and cavity wall provides an avenue for the percolation of bacteria and a site for dental plaque which can cause erosion of the restoration, secondary caries and gingival inflammation.[54] The importance of the adhesive quality of the zinc polycarboxylate cement towards enamel, dentine protein and stainless steel cannot therefore be overestimated.

Experimental results have confirmed the superiority of the zinc polycarboxylate cement over the zinc phosphate cement in adhering to enamel, dentine and stainless steel.[3,44,55-58] Not only is the strength of the attachment much greater with the zinc polycarboxylate cement, but it is also little affected by thermal cycling[56] unlike that of the zinc phosphate cement where complete failure occurs under these conditions.[58] Moreover, whereas under test the zinc phosphate cement failed at the adhesive margin, the zinc polycarboxylate cement fractured cohesively indicating that the adhesive bond strength is somewhat greater than the tensile strength of the cement.[55] However, when mixed too thickly, so that the cement does not wet the substrate surface, adhesive failure of the latter cement is observed.[58] Proof that the cement is truly adhesive comes from tests which showed that bond strength was greatest on a substrate having a smooth clean surface.[56,58] Etching and mechanical roughing did not improve bond strength and sometimes decreased it. The zinc polycarboxylate cement shows no adhesion towards materials which are chemically inert towards acids for example dental gold and porcelain.

The mechanism of adhesion of the cement has been attributed by Smith[44] to chemical forces resulting from the chelation of calcium ions of the enamel by poly(acrylic acid). Doubt has been cast on this mechanism,[57,59] since 8-membered rings are relatively rare in chelate chemistry.[48] Beech favours electrostatic interaction as the adhesive mechanism because it has been shown that the binding of calcium to poly(acrylate) is electrostatic in nature. However, as Mellor[49] argues, 7-membered, 8-membered and larger chelate rings are possible with certain multidentate ligands and polyacids must fall within this class. Morawetz[50] considers that the numerous functional groups on the polyacrylate chain make chelation possible.

Despite these arguments the chelation theory must be considered inadequate for other reasons. Chelation could result in the complete removal of calcium from the enamel surface. A chelate bond has a strength that is far in excess of that required to account for the adhesive bond and the distance over which it acts is only of the order of 1 or 2 Å, an inadequate range of action to explain all aspects of adhesion. Physical forces acting over greater distances are of adequate strength to account for the adhesive bonds of the zinc polycarboxylate cement and polar forces are the most likely cause of adhesion. Initially when the paste is fluid it will contain many free carboxylic acid groups so that hydrogen bonding will play a role, since this bond is known to aid the wetting of surfaces, a prerequisite for adhesive bonding.

The polar nature of the poly(acrylate) chains in the cement (with —COOH, —COO⁻ functional groups) will result in dipole interaction between them and the polar surfaces of dentine collagen (—COOH, NH$_2$, functional groups), enamel apatite (O^{2-} ions at the surface) and oxide-coated metals (Fig. 3).

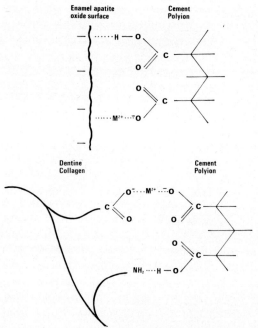

Fig. 3 *Probable mechanisms of adhesion of polycarboxylate cements to enamel apatite and dentine collagen. (Wilson, A. D. 11th Annual Conference on Adhesion and Adhesives, London, April 1973.)*

Ionic forces could also contribute to the adhesive forces. There could be direct ionic interaction between the carboxylate groups in the cement and amine groups in proteins. Metals from the cement or substrate could bridge carboxylate groups in the cement to other anionic groups, oxygen anions in enamel and base metal surfaces and carboxylate groups in proteins. In this connection it should be noted that this cement does not bond to the inert surfaces of noble metals and porcelains, which observations indicate that reactive surfaces are required for bonding. It follows that the diffusion of metal ions from the substrate stainless steel or enamel plays an essential role in adhesion and strongly suggests that ionic metal bridges are present. The long linear covalent bonded chains of the cement may also serve to bridge flaws between the body of the cement and the substrate, bearing in mind that when partially neutralised these chains expand from the initial random coil configuration. Undoubtedly these chains are physically absorbed by reactive oxide surfaces thus aiding adhesion. Eventually, as is well-known, some chemisorption would take place. Possible mechanisms of adhesion are shown in Fig. 3.

Biological properties
The biological effects of dental materials on living tissues are now recognised as being of equal importance to their physical and mechanical properties. The mouth is at once a hostile and sensitive environment. Traditional dental cements based on phosphoric acid—the zinc phosphate and dental silicate cements—are strong but irritant. The zinc oxide–eugenol cement on the other hand is bland, but too weak to resist mechanical and abrasive action in the mouth.

The response of living tissues towards zinc polycarboxylate cement has been the subject of a number of studies using human teeth [60,61] and monkey teeth.[62,63] Implants have been placed in the subcutaneous tissues of rats.[64] Evidence from these clinical trials show that this cement is compatible with dentine and dental pulp.[62] Although there is some initial response, tissues, whether dental pulp or subcutaneous tissue, soon recover and there is no permanent deleterious effect. This cement is as bland as zinc oxide–eugenol cement used as a control in standard biological tests.[65]

The blandness of the zinc polycarboxylate cement compared with the irritant action of the zinc phosphate is to be explained in terms of acidity. The weight of opinion is that the irritant action of the phosphate bonded cements results from their highly acidic reaction when freshly prepared.[66] Poly(acrylic acid) with an apparent pK_a of 5 compared with the pK_a of 2·2 for phosphoric acid is much weaker and much less irritant on this account. Perhaps more important is that it is much less mobile than phosphoric acid, for there is a M_w of say 20 000–50 000 to compare with one of about 100. Thus the free acid will show much less tendency to diffuse down dentinal tubules. In addition there is the restriction imposed by chain entanglement and partial cross-linking, anchoring even a predominantly acidic chain to the body of the cement.

4.1.3 Glass ionomer cements

New translucent surgical cements based on the hardening reaction that occurs between varieties of ion-leachable glasses and various polycarboxylic acids were patented by Wilson and Kent in 1969.[4] These ionic polymer cements, which they termed 'glass ionomer cements'[5,6] may be regarded as cold cured ceramic–organic polymer hybrids. They were developed in order to provide a dental filling and luting material which was translucent, bland towards tissues and adhesive towards tooth material. The first member of the glass ionomer cements, ASPA cement, utilised poly(acrylic acid) and was described by Wilson and Kent[5,6] in 1971–1972. This cement system was developed following the investigation of Wilson and his co-workers[51,67] on the nature of the dental silicate cement which is the product of the hardening reaction that occurs between special fluoro-alumino silicate glasses and liquids based on aqueous solutions of phosphoric acid. This dental silicate cement is manifestly superior to all dental zinc oxide cements in durability, strength and translucency and provides the best model on which to base future developments. It had been supposed that setting occurred as the result of the polymerisation of silicic acid to silica gel. However, the work of Wilson[51,67] established that setting occurred as the result of the liberation of cations from what were in fact species of ion-leachable glasses, by hydrated protons from the liquid.

These cations, principally the aluminium ion, take part with phosphate in a pH dependent polymerisation process to form a hard insoluble gel of aluminium phosphate. From this it was reasoned that other liquids might be substituted for phosphoric acid to yield cements; although at that time no others were known. This situation contrasted sharply with that found in the zinc oxide cements where numerous liquids are known to combine with zinc oxide to form useful cements. In search for alternative liquids, Wilson[68] found that certain criteria had to be satisfied.

The liquid had to have an acidic property in order to liberate ions from the glass, and the acid anion had to be one which would form complexes with di- and trivalent metal ions. Water was required as a reaction solvent and to hydrate the reaction product, and lastly the complex forming acids had to be very soluble in water so that 50% solutions could be prepared.

Few compounds were found to meet these requirements; they included various condensed phosphoric acids, fluoroboric acid and certain chelating organic acids: tartaric, citric, mellitic, pyruvic and tannic acids. None of these cements had any practical potential as they proved to be weak and hydrolytically unstable.

Later Wilson and Kent turned to investigating the use of aqueous solutions of various polycarboxylic acids, principally poly(acrylic acid) as cement-forming liquids. These liquids proved unsatisfactory cement formers with normal dental silicate cement glass powders on account of poor working qualities and sluggish setting and hardening characteristics. However, Wilson and Kent[69] discovered a series of fluoroaluminosilicate glasses, somewhat more basic than the normal dental silicate glass, which formed the basis of formulations for practical dental cements. These cements have obvious affinities both with the dental silicate and zinc polycarboxylate cements.

Materials

The ion-leachable glasses for the glass ionomer cements are prepared by fusing a mixture of quartz and alumina in a flux of sodium, calcium and aluminium fluorides and aluminium phosphate. All materials are finely ground and thoroughly mixed before fusion at 1050–1350°C for 40–150 minutes. The melt is shock cooled by first pouring on to a metal tray and then plunging the mass, when cooled to a dull red heat, into water. The glassy product, which may be transparent or opal, is finely ground to pass a 350 mesh sieve. Composition of two broad types of glass suitable for this application are given in Table 3, based on 100 parts by weight of Al_2O_3.

The microstructure of the glass has been shown, by transmission and scanning electron microscopy to be a complex multiphase glass[70] consisting of a continuous calcium aluminosilicate phase interspersed with massive inclusion of glass droplets occupying some 20% of the total volume (Fig. 4). The microstructure of these droplets has been shown by X-ray diffraction and non-dispersive X-ray analysis to be complex, and dependent on the temperature of glass fusion and apparently affects the quality of the cement paste. At 1150°C the droplets are 1·7 μm in diameter with a central crystalline core of fluorite containing no aluminium or silicon, sheathed by an amorphous layer which does, and which is preferentially attacked by acids (Fig. 5). At 1300°C the droplets are smaller and almost completely amorphous (Fig. 6). Glasses formed at this temperature are not of use in forming practical cements, since the pastes have unmanageable mixing

TABLE 3

ASPA GLASSES
COMPOSITION OF FRIT

	General formulation Type I	Type II
Al_2O_3	100	100
SiO_2	160–190	75–100
Total metal fluorides calculated as fluorine, F	105–150	50–150
AlF_3	0–100	0–100
$AlPO_4$	0–125	0–125
Na_3AlF_6	0–150	50–100

Specific examples

	(I)	(II)	(III)	(IV)	(V)	(VI)	(VII)	(VIII)	(IX)	(X)	(XI)
SiO_2	176	176	175	175	176	95	95	95	95	95	95
Al_2O_3	100	100	100	100	100	90	100	100	100	100	100
TiO_2	—	—	—	—	—	10	—	—	—	—	—
Na_3AlF_6	135	135	—	30	65	135	76	76	76	76	76
CaF_2	87	87	240	207	168	87	56	56	56	—	—
MgF_2	—	—	—	—	—	—	—	—	—	45	—
LaF_3	—	—	—	—	—	—	—	—	—	—	94
AlF_3	32	32	32	32	32	32	—	—	96	96	—
$AlPO_4$	56	100	60	60	60	56	73	121	73	73	73

After Wilson A. D. and Kent, B. E., *Ger. Offen.*, 2,061,513 (1971).

qualities which may be attributed to the largely amorphous nature of the droplets making the glass too reactive.

These glasses, whilst generally similar to dental silicate cement glasses, have to be made more reactive because the cement-forming liquids, poly(acrylic acid) and related poly(alkenylcarboxylic acids), are much weaker than the phosphoric acid which is used with dental silicate cements. The reactivity of these fluoroaluminosilicate glasses can be controlled by heat treatment. However unlike zinc oxide, the reactivity of these glasses is *increased* by increasing the temperature of ignition. As indicated above, high fusion temperatures tend to impart undesirable rheological properties into the cement pastes prepared from the glass, making them, at times, unworkable. This phenomenon has been investigated by Barry *et al.*[70] and will be discussed later. The prime method of controlling the reactivity of the glass is by adjusting its basicity. Experimental studies by Wilson and Kent[69] showed that the reactivity of these glasses could be increased by

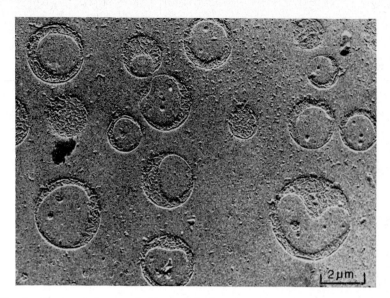

Fig. 4 Electron micrograph of a polished surface of G 200 glass prepared at 1150°C. (From Barry, T. I., Miller, R. P. and Wilson, A. D. 11th Conference of the Silicate Industry, Budapest, 1973.)

increasing the ratio of alumina (not total aluminium) to silica in the glass fusion mixture. These glasses have a higher $Al_2O_3:SiO_2$ ratio than dental silicate glasses (Table 3). Wilson and Kent found that increasing the fluoride content of the glass decreases the reactivity of the glass.[71] Remembering that the interaction between the proton-accepting ion-leachable glass and the proton-donating poly(acrylic acid) solution is essentially an acid-base one, glass reactivity can be explained in terms of the basicity of the glass.

The structure of the main phase of the glass following Zachariasen[72] may be regarded as a 3-dimensional random framework of linked $[SiO_4]$ and $[AlO_4]$ tetrahedra. In effect, this may be seen as a silica structure where Si^{4+} is partially replaced by Al^{3+} so that for valency reasons the network carries a negative charge and is thus protophilic and decomposable by acids. The extent of susceptibility to protonic attack will depend on the density of the negative charge on the network, *i.e.* on the Al:Si ratio. (This discussion, based on hypothetical ions, runs counter to Chapters 7 and 8 where these structures are treated as covalent networks with ionic sites. However, it has advantages for discussing polarisation phenomena—Ed.)

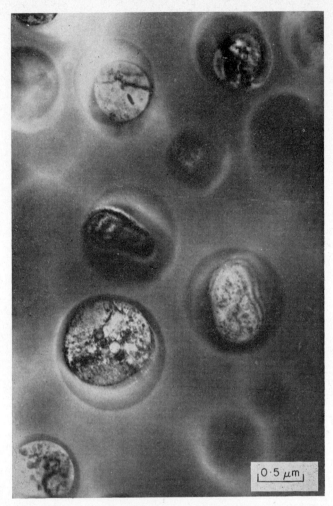

Fig. 5 Transmission electron micrograph of G 200 glass prepared at 1150°C. (From Barry, T. I., Miller, R. P. and Wilson, A. D. 11th Conference of the Silicate Industry, Budapest, 1973.)

In the theoretical treatment which follows, the aluminosilicate systems have been described on the basis of ions and their mutual polarisation. This concept, which stems from the rules of Fajans as applied by Weyl and Marboe,[73] enables us to describe in qualitative terms, relative reaction rates of these systems. This involves an extension of the concepts of acidity–basicity to condensed systems. These authors concluded that the basicity of

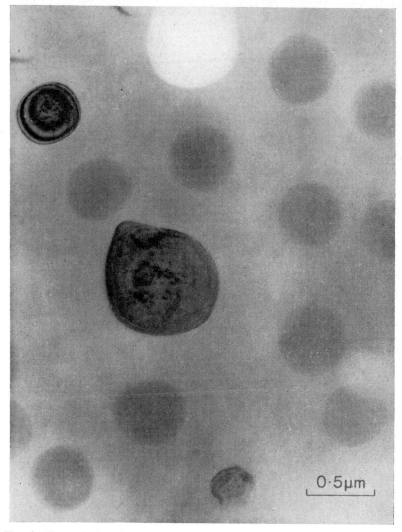

Fig. 6 Transmission electron micrograph of G 200 glass prepared at 1300°C. (From Barry, T. I., Miller, R. P. and Wilson, A. D. 11th Conference of the Silicate Industry, Budapest, 1973.)

a solid was proportional to the polarisability of its anions, which is minimal when the electron cloud of the anions is little affected by cations. The polarisability of an O^{2-} ion depends on its environment. According to Fajans' rules, an anion is highly polarised by cations of high field strength, *i.e.* multivalent cations of small ionic radius, and in this condition has little residual polarisability, *i.e.* it is acidic. In the case of aluminosilicate networks, the progressive replacement of Si^{4+} cations by Al^{3+} cations of lower field strength serves to loosen the electron cloud about the O^{2-} ions, increasing their polarisability and resulting in an increase in the basicity of the glass. Thus the glass network becomes increasingly vulnerable to the penetration of protons. This is the theoretical reason for increasing the $Al_2O_3:SiO_2$ ratio in the glass in order to increase its reactivity. Again, the progressive replacement of the O^{2-} ions by the F^- ions which have little polarisability tends to decrease the basicity of the glass, *i.e.* decrease its reactivity, a deduction which is in accord with experiment.

Structure

The setting mechanism and the structure of ASPA cement have been the subject of studies using chemical techniques, infra-red spectroscopy, and optical and electron-microscopy.[70]

The first step in the reaction is the liberation of ions from the glass[74] and their appearance in the aqueous phase of the cement paste; a process which has been shown by infra-red spectroscopy,[75] to be accompanied by the progressive ionisation of carboxylic acid groups to carboxylate groups. Apparently protons from the liquid penetrate into the powder particles, decomposing the aluminosilicate network of the main glass phase, exchanging with the network dwelling cations (Ca^{2+}, Na^+) and attacking the calcium and fluorine rich droplets of the disperse phase. Liberated ions migrate into the aqueous phase of the cement where salt formation takes place leading to gelation and setting (Fig. 7). The mechanism of gelation is

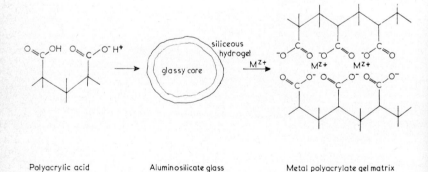

Fig. 7 *Proposed setting reaction of the ASPA cement.* (From Kent, B. E., Lewis, B. G. and Wilson, A. D. (1973). Brit. Dent. J., **135**, 322.)

connected with the phenomenon of ion-binding and has been discussed previously. Most likely, gelation results from the cross-linking of poly-(acrylate) chains by metal ions either acting as salt bridges or forming co-ordination complexes. The cement paste which initially has a putty-like consistency goes through an elastomeric stage and sets within a few minutes to a carvable solid mass when only a small fraction of these metal cross-links are formed. At this stage, the cement has unique rheological properties so that unlike other dental cements it can be shaped and contoured by a carving technique similar to that used for dental silver–tin amalgams. Subsequently, as more cross-links are formed the cement hardens into a stone-like mass. Calcium poly(acrylate) alone is found, by infra-red spectroscopy, in the fresh paste and is most probably the cause of the initial set since the formation of aluminium poly(acrylate) is delayed by one hour.

The fully hardened cement consists of partially reacted glass particles in a complex matrix (Fig. 8). Studies using a scanning electron microscope fitted with a non-dispersive analyser for elemental analysis indicate that the predominant species present are the metal salts of poly(acrylic acid). Minor amounts of silica are also found in the matrix, and the presence of fluoride salts is to be inferred from wet chemistry studies. Both aluminium

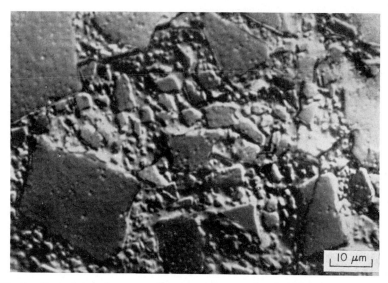

Fig. 8 Optical interference reflectance micrograph of ASPA cement. (From Barry, T. I., Miller, R. P. and Wilson, A. D. 11th Conference of the Silicate Industry, Budapest, 1973).

and calcium ionic links are possible but scanning electron microscopy and infra-red spectroscopy indicated that calcium links somewhat predominate.[75]

Properties
The glass ionomer cement, ASPA, has affinities both with the dental silicate cement and the zinc polycarboxylate cement. The powder component, a ground ion-leachable glass, is of the same general type as that used in the dental silicate cement, and the liquid used is similar to the poly(acrylic acid) solutions used in Smith's cement. It is therefore appropriate to compare the physical properties of ASPA cement with those of these two other cements (Table 4[76]).

In respect of high compressive strength (ca. 200×10^6 N/m^2) and translucency, ASPA cement follows the dental silicate cements. All zinc oxide cements, including the zinc polycarboxylate cement (80×10^6 N/m^2) are much weaker in the compressive mode and all are opaque. These properties

TABLE 4
COMPARATIVE PROPERTIES

	ASPA I	Dental silicate cement		Zinc polycarboxylate Durelon
		Super Syntrex	Others	
Powder:liquid ratio, g/ml	3·0–3·5	4·0	2·70–4·0	3·6
Volume fraction of powder (%)	52–56	62	52–62	40
Setting time 37°C, min	4–5	3·75	3–7	3·25
Strength $N\,mm^{-2} \times 10^{-6}$				
Compressive, 1 day	180–213	226	68–262	85
Compressive 7 days	222	246	—	—
Tensile 1 day	13–17	13	—	12
Tensile 7 days	16	14	—	—
Solubility 1 day %	0·2–0·6	0·5	0·5–3·8	0·04
Acid erosion %				
1 day	0·1	0·8	—	—
7 days	0·3	—	—	—
Opacity $C_{0.70}$	0·72	0·55	0·52–0·71	1·00
Initial (3 min) pH	2·5	1·3	—	2·6
Adhesion[a]	Adhesive	Non-adhesive	Non-adhesive	Adhesive

[a] To base metals and tooth material.

of ASPA cement appear to be a function of the filler, which is similar to that of the dental silicate cement, rather than those of the polyacrylate matrix which is similar to that of the zinc polycarboxylate cement. The point may be noted that, whereas zinc oxide particles decrease in size as they are consumed in the cement forming reaction, this is not the case with ion-leachable glass particles which maintain their size and shape—silica gel relicts replacing the glass network where attack has occurred. In consequence, the aluminosilicate cements are better space fillers than oxide cements, with a corresponding reduction in porosity and increase in strength.

ASPA cement is considerably more resistant to the effect of both aqueous and acid attack, an important quality in the oral situation, than the dental silicate cement. This point is illustrated by Fig. 9a and b, which shows the effect of dilute lactic acid/lactate buffer (0·01 N, pH 4·0) on the surface of both cements. The scanning electron micrographs (replica technique) show that in all cases the surface of ASPA cement is featureless *i.e.* the surface integrity has been maintained. By contrast, the surface of the dental silicate cement has disintegrated. These results are consistent with the acid erosion figures (in Table 4) obtained for both cements which show that the dental silicate cement disintegrates considerably in dilute acid solution. This effect is a mechanical one because the ion exchange that occurs is the same for both cements; it is clear that the removal of cations from the dental silicate cement structure has a far greater effect on cement cohesion than in the case of ASPA cement. This may be attributed to the nature of the linkages present in both cements. The aluminium and phosphate ions of the aluminium phosphate matrix of the dental silicate cement are linked by bonds, which are of a predominantly ionic character. Removal of cations from such a structure must result in the disintegration of the matrix. In ASPA cement only the cross-links have an ionic character, the linear polyacrylate chains being bound by covalent carbon–carbon linkages. Hence the removal of some of the bridging metal ions does not result in cement disintegration.

Another most important quality of the ASPA cement is its adhesion to enamel, dentine and orthodontic stainless steel. Preliminary findings indicate that the adhesive bond is somewhat better than with the zinc polycarboxylate cements and more resistant to abrasion[9] (theoretical aspects of adhesion of poly(carboxylic acid) cements have already been fully discussed in the section on zinc polycarboxylate cements).

Another expected quality of this cement is blandness towards living tissues, but at present no firm experimental data is available.

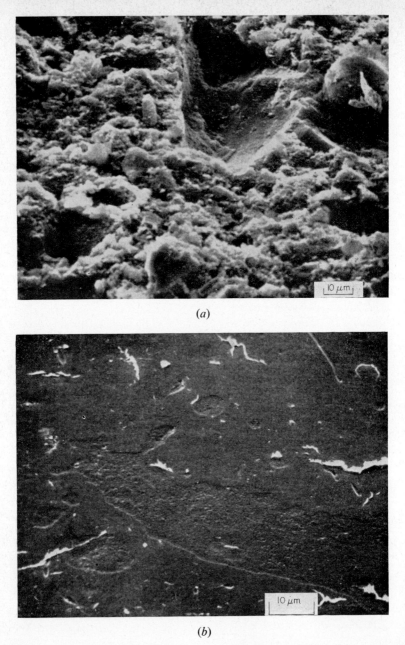

Fig. 9 The effect of acid attack on the surface of (a) ASPA and (b) a dental silicate cement. (From Kent, B. E., Lewis, B. G. and Wilson, A. D. (1973). Brit. Dent. J., **135**, 322.)

4.2 MONOLITHIC PLASTICS FORMED FROM METAL POLYACRYLATE SALTS

An interesting type of compound was developed in the 1960s by Nielsen[45,46] by reacting zinc oxide and other metal oxides and solid poly(acrylic acid). They possessed improved properties over normal plastics, characterised by higher shear moduli, lower thermal expansion, and good thermal stability. The main drawback of these materials for building applications is their brittleness, and to date they have found no practical application.

The fabrication technique resembles those of powder metallurgy. A copolymer of 94% acrylic acid and 6% 2-ethylhexyl acrylate in a finely powdered form is thoroughly mixed with powdered zinc oxide in a Mixer-Mill. The mixed powders are heated at 10 000 psi pressure at 300°C, the pressure being released at least six times before the temperature is attained, to expel water vapour. These conditions were varied for experimental purposes to elucidate the structure of the system.

4.2.1 Structure and properties

Nielsen has reviewed the effect of cross-linking on the physical properties of polymers.[77] He stated that rubber theory is not applicable to thermoset polymers of moduli greater than 10^8 dynes cm^{-2}. In this situation one has to resort to an empirical interpretation of data, and this approach was adopted by Nielsen in his study of the fused metal poly(acrylate) ionic polymers. The extent of reaction was assessed by X-ray diffraction for the presence of zinc oxide, and the absence of carboxyl carbonyl groups was confirmed by infra-red transmission spectroscopy. The shear modulus and mechanical damping were measured over a wide temperature range on a recording torsion pendulum.[78] An Instron tester was used to measure compressive strengths and flexural tests. A special apparatus was developed to measure the coefficient of thermal expansion.[79]

Five types of species can be present in the system *viz*, the pendant half-salt, the in-chain di-salt, the cross-chain di-salt, unreacted acid, and unreacted zinc oxide which acts as a filler (*see* Fig. 10). If excess acid is used, all five species are present except zinc oxide. The di-salts have more restricted rotation than the half-salt and will tend to produce a stiffer product. Although the higher metal content of the pendant half-salt ensures a higher stiffness per unit, the bridging bond in the di-salt O—Zn—O will be stronger than that of the half-salt and consequently the di-salts will be less likely to interchange sites. Overall it would seem that the presence of the

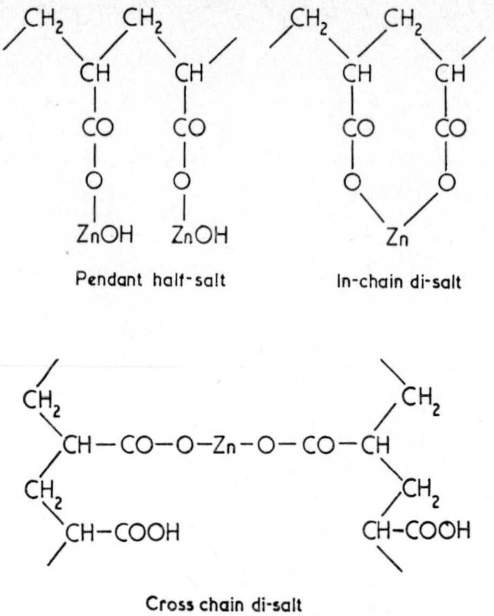

Fig. 10 Various salt types occurring in zinc poly(acrylate) melts. (S. Crisp and A. D. Wilson, Laboratory of the Government Chemist. Crown Copyright.)

half-salt will reduce the rate of increase in shear modulus, lead to higher damping, and lower the glass transition temperature.

These considerations were investigated by Nielsen[46] who prepared a series of salts of various compositions. The results obtained are shown in Table 5. The shear modulus of poly(acrylic acid) is about three times that of most common rigid organic polymers. This increased stiffness is accounted for by hydrogen bonding. As zinc oxide is introduced, the shear modulus increases but when unreacted zinc oxide is present there is a reduction in the glass transition temperature. Sample 5 in Table 5 corresponds to the di-salt and hence gives a measure of the shear modulus of that species.

The behaviour of the salts is shown in Fig. 11. Poly(acrylic acid) becomes much less stiff on heating (Sample 1) and has a low glass temperature of 127°C. Sample 2a contains only 25% of the theoretical amount of zinc oxide which is fully reacted to give the di-salt. Its behaviour on heating is similar to the acid, but the shear modulus is always higher, the damping loss and the glass temperature (210°C) considerably higher. Sample 2 is presumably a complex of all five species as only 24% of zinc oxide reacted. The behaviour of this sample is rather similar to that of Sample 2a up to

TABLE 5
COMPOSITION AND DYNAMIC PROPERTIES OF NIELSEN SALTS

Sample No.	Composition % of theory			Shear modulus (25°C) 10^{-10} dyne cm^{-2}	Glass temperature °C
	ZnO added	ZnO reacted	ZnO as filler		
1	0	0	0	2·97	127
2	100	24	76	4·40	200
2a	25	25	0	4·17	210
3	50	50	0	5·58	300
4	100	94	6	6·07	240
5	100	100	0	6·55	300
6	150	86	64	7·58	230
7	200	146	54	8·02	250

the glass temperature. At 200°C there is a discontinuity in the curve and with a further increase in temperature the shear modulus rises. When the glass temperature is reached, the unreacted zinc oxide is able to diffuse freely and further reaction occurs. Samples 3 and 5 are fully reacted and the system is strongly cross-linked. These samples are relatively unaffected by temperature and no glass temperature was observed below 300°C. In going from 25°C to 300°C the respective changes in logarithmic decrement are 0·03 to 0·15 (Sample 3) and 0·04 to 0·15 (Sample 5).

To assess the significance of ionic bonding in the system, the improvement in shear modulus was calculated on the assumption that zinc oxide acted simply as a filler in a matrix of poly(acrylic acid). Three theories were used and compared for the calculation viz Kerner,[80] Guth–Smallwood,[81,82] and Mooney.[83] The increase in shear moduli were 60–80% by the Kerner theory and 40–60% according to the other theories. The data are presented in Table 6, where the improvement ratio represents the ratio of measured modulus to calculated modulus based on the 'filler' theory.

When zinc oxide is added in excess of the theoretical amount, the shear modulus at 25°C increases. The change of shear modulus with temperature is very revealing. Sample 6 has not fully reacted and the modulus shows a marked decline between 150–200°C. A minimum occurs at about 200°C when the zinc oxide is able to react further and increase the cross-linking in the system. To account for the greater than theoretical reaction in Sample 7, the half-salt must have formed. There will be twice the theoretical excess, so the system will contain 92% half-salt, 54% di-salt, and 54% of zinc oxide. The presence of a high proportion of half-salt leads to the large decline in modulus with increasing temperature and a decrease in the improvement

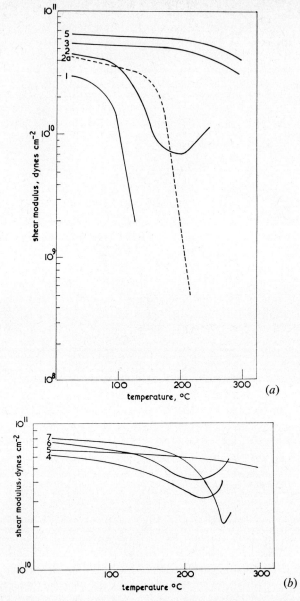

Fig. 11 Variation of shear modulus with temperature for zinc poly(acrylate) melts. (*Based on Fields, J. E. and Nielsen, L. E. (1968). J. Appl. Polym. Sci.* **12**, *1041. Reprinted by permission.*)

TABLE 6
EFFECT OF CROSS-LINKING ON SHEAR MODULUS OF NIELSEN SALTS

Sample No.	Relative modulus[a] (found)	Kerner theory	Mooney theory	Improvement ratio over: Kerner theory	Improvement ratio over: Mooney theory
2	1·48	1·29	1·42	1·15	1·05
2a	1·41	1·07	1·08	1·32	1·31
3	1·88	1·14	1·19	1·64	1·59
4	2·04	1·29	1·42	1·59	1·44
5	2·21	1·29	1·42	1·71	1·56
6	2·55	1·43	1·71	1·78	1·49
7	2·70	1·57	2·09	1·72	1·29

[a] Relative to Sample 1.

ratio (Table 6). The modulus of Sample 5 can be equated with the di-salt, viz. $6·55 \times 10^{10}$ dyne cm^{-2}. Calculations on Sample 7 using this figure lead to a value of $7·08 \times 10^{10}$ dyne cm^{-2} for the half-salt. As stated earlier, this higher figure reflects the higher metal content of the half-salt unit, and is more than compensated by the stronger bonding of the di-salt.

There is a progressive increase in shear modulus of polystyrene, poly-(acrylic acid), and the zinc salt in the ratio $1:2·3:5·2$. Polymers, such as polystyrene, derive their strength from weak intermolecular forces such as van der Waals, and London dispersion forces. Poly(carboxylic acids) are able to form hydrogen bonds to produce a more rigid structure as in poly(acrylic acid). The divalent salts, such as zinc, can form intermolecular ionic bonds which are stronger than hydrogen bonding. These three compounds illustrate an hierarchy of bonding forces. The importance of the bonding type is shown by other physical properties. Poly(acrylic acid) has a glass temperature of 127°C and the zinc salt does not have a glass temperature below 300°C. Other properties are collated in Table 7. The increase in

TABLE 7
COMPARISON OF PHYSICAL PROPERTIES OF SOME POLYMERS

Properties	Polystyrene	Poly(acrylic acid)	Zn poly(acrylate)
Compressive strength (Nm^{-2})	103×10^6	182×10^6	373×10^6
Coefficient of thermal linear expansion (per °C)	8×10^{-5}	$5·5 \times 10^{-5}$	$1·4 \times 10^{-5}$
Flexural strength (Nm^{-2})	—	71×10^6	68×10^6

compressive strength and decrease in the coefficient of thermal expansion are as expected in the sequence polystyrene to zinc poly(acrylate). The flexural strengths of all three compounds are about the same, but Nielsen observed that these measurements are very sensitive to sample imperfections.[78]

Nielsen prepared calcium and lead poly(acrylate) by the same method.[78] The products obtained were inferior to the zinc salt, particularly the compressive strength. The results are presented in Table 8.

TABLE 8
PROPERTIES OF FUSED METAL POLY(ACRYLATES)

Cation	Shear modulus 10^{10} dyne cm^{-2}	Compressive strengths $N\,m^{-2}$
Calcium	5·56	106 × 10^6
Lead	5·15	128 × 10^6
Zinc	6·20	320 × 10^6

In a study on the water stability of the salts, Nielsen found the properties of zinc poly(acrylate) were unaffected after treatment with water for 90 days at 22°C or 180 hr at boiling point.

The work of Nielsen included an investigation of introducing fillers into the system.[45] Fillers used included: mica, asbestos, zinc oxide, alumina, iron flake, aluminium flake, glass, stainless steel, boron, graphite, and molybdenum disulphide. When mica was used at a volume fraction of 40%, the shear modulus was as high as $1·8 \times 10^{11}$ dyne cm^{-2} (*i.e.* 15 times that of polystyrene). Oriented boron filler at 42% volume fraction produced a plastic with a shear modulus $1·7 \times 10^{11}$ dyne cm^{-2} and a flexural strength 780×10^6 Nm^{-2}. The highest compressive strength was obtained for a stainless steel filler at 20% volume fraction.

4.3 ACRYLIC SOIL CONDITIONERS

The use of salts of acrylic acid, its polymers and derivatives as soil conditioners was stimulated by the need to construct temporary roads and runways quickly in the 1939–45 war. A substantial amount of the literature on soil conditioners was published by American workers in the 1950s[84] and both engineering and agricultural applications of acrylate systems have been studied. Work in this subject appeared in Eastern European countries by the early 1960s; these appear to be mainly accounts of field studies.

There is a report that the Vietnam war has led to renewed interest in poly(acrylates) as soil conditioners.[85] A particular objective is to eliminate the nuisance of dust raised by aeroplanes on landing and taking off. Poly(acrylates) find widespread use in sedimentation techniques, and because silicate minerals are frequently involved in these procedures, some aspects of sedimentation theory have been considered in view of the similar mechanisms proposed for soil conditioning and sedimentation. The sedimentation of a silicate mineral can be regarded as either a model system, or an extreme case, of soil stabilisation.

Many of these applications are registered as inventions, and scant attention is often paid to chemical or physical mechanisms. By far the most thoroughly studied uses of poly(acrylates) are as dental cements, and in soil conditioning and sedimentation. In the former, as discussed elsewhere, the poly(acrylate) is cross-linked by polyvalent cations to form a hard setting matrix. In the later two cases it will be shown that interparticle bridging is an important mechanism, in addition to chelating reactions.

The various methods of treating soil to increase mechanical strength have been reviewed by Lambe.[86,87] The engineer can deal with unsuitable soil in three ways. It can be by-passed by driving piles to reach the rock stratum; or removed and replaced by a more suitable material; or lastly it can be treated. The first two methods are expensive, so there is a strong economic stimulus to find cheap soil treatments. Non-chemical treatments include mechanical compaction, heating to 600°C to irreversibly dehydrate the soil, freezing, drainage by electro-osmosis, or, in the case of slopes, planting grass. However, the most successful treatments are a combination of chemical with mechanical methods.

Calcium or sodium chlorides have been used as stabilisers. Their hydrated state maintains soil moisture, facilitates compaction, and prevents freezing but without greatly increasing soil strength. Portland cement (about 1%) has been added to soil. There is no evidence of a reaction between the cement and soil particles, but a rigid concrete-like mass is formed as the cement sets. This treatment has been successful but it can fail due to differential settlements if the soil is not on a uniform base.

Similar considerations are applicable to the use of bitumens. Various resins, natural and synthetic, have been used: their properties vary considerably. Sodium silicate/calcium chloride form a gel within the soil, but there is evidence of deterioration with ageing. Clay can increase soil strength. Other materials that have been used include tung oil, linseed oil, rubber latex, plasticised sulphur, sodium or calcium carbonate, long chain amine hydrochlorides, silicones, complex quaternary ammonium salts, starch and polyphosphates. More recently a number of these materials have been considered, together with poly(acrylates) for treatment of surface soil to suppress dust clouds.[85]

Hedrick and Mowry[88] compared the agricultural treatment of soil with poly(acrylates) with that of poly(uronic acid), its salts, and related polysaccharides. The latter compounds were found less desirable as large quantities were required (5–10 tons per acre) for effective treatment which had to be repeated at intervals because these polymers decomposed in 2 to 8 weeks. One consequence was that large amounts of cations were released into the soil particles and large amounts of carbon accumulated leading to soil denitrification. These are major disadvantages which were not obtained with poly(acrylate) formulations.

In using soil conditioners two major factors are very relevant: the cost and the problem of adding the agent to the soil. The first is somewhat ameliorated with poly(acrylates) as dilute solutions only are required.

4.3.1 Engineering treatments

In engineering applications, it has already been noted that it is often preferable to carry out the polymerisation *in situ*. A practical advantage is that problems of storing polymer and mixing it effectively with the soil are by-passed.[89] However, it has been suggested that a definite reaction with the soil is achieved by absorption of acrylate/metal ions. It has been suggested by de Mello[90] that lead and calcium acrylate form ions of the type $[Ca(CH_2CHCOO)]^+$. Absorbed sodium ions are exchanged at the soil particle surface. Consequently sites are formed on the soil with a vinyl group attached to the soil particle via a calcium ion bridge. As the catalyst is added these absorbed vinyl groups can polymerise with free acrylate ions to form covalent bridges between particles. Unabsorbed calcium cations are free to cross-link polymer chains, thus conferring additional rigidity on the system.

The reaction scheme and the type of idealised structure that is formed is illustrated in Fig. 12. The conditions of catalyst and water amounts will be discussed later with an indication of the effects of soil type.

Lambe[86] described the *in situ* treatment of a sandy soil with a mixture of calcium acrylate and a catalyst of 1:1 ammonium persulphate and sodium thiosulphate. The mixture was added as a dry powder to wet top soil to form a load bearing surface. To increase the strength or decrease the permeability of subsurface soil, the mixture was injected as a solution. This treatment was found to be effective when the amount of calcium acrylate added was between 4 to 25% by weight of the dry soil. Mixing was

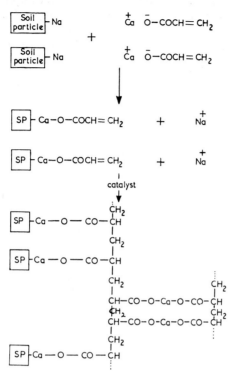

Fig. 12 Suggested mechanism of soil stabilisation by in situ *polymerisation of divalent metal acrylates. (S. Crisp and A. D. Wilson, Laboratory of Government Chemist. Crown Copyright.)*

difficult when less than 4% was added and the treatment became uneconomic if more than 10% was used. Meunier *et al.*[91] made a systematic study of the soil–water–calcium acrylate system of Fort Belvoir soil (20% clay, 40% silt, 40% sand). Their work was done with catalyst concentrations to give a constant gel time. The main conclusions of their work are presented in Fig. 13. The area 1 2 3 4 is the optimum compositional region and is bounded by four curves. The upper water limit is the maximum quantity of water to give a homogeneous mixture, the other curves are self-explanatory. The amount of calcium acrylate required is in substantial agreement with Lambe's conclusion. If a composition above line 1 4 is selected, the soil is easier to mix but the strength of the treated soil is below its maximum value. Greater strengths than those in the region 1 2 3 4 can be obtained for compositions to the left of line 1 2, but the treatment is

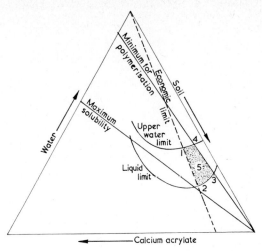

Fig. 13 Water–calcium acrylate, Fort Belvoir soil system. (From Meunier, V. C., Williamson, G. J. and Hopkins, R. P. (1955). Ind. Eng. Chem., 47, 2265. Reprinted by permission.)

uneconomic. To the right of line 3 4 the product is weakened due to incomplete polymerisation. Other workers have reported effective treatments with less than 1% of calcium acrylate.[92] The amount of polymer formed and its quality is unaffected by the proportion of catalyst added, provided there is sufficient for the reaction. However, the gel time, *i.e.* the time to form the gel after adding the catalyst, is affected by the concentration of the catalyst. Thus at ambient temperature, whilst the gel time is about 10 min with a catalyst to calcium acrylate ratio of 1:5, it is extended to 30 min when this ratio is reduced to 1:10. Various catalysts can be used provided they do not form insoluble salts with the cation moiety of the acrylate salt. Generally a stronger soil is formed if the minimum amount of water is used, and in fact, most soils contain sufficient water for the reaction. It is advantageous to flush some soils with a solution containing divalent metal cations before treatment with acrylates.[93]

For treatment of oil drill mud an acrylate copolymer has been found effective.[94] The copolymer was formed from sodium acrylate and acrylamide in a 7:3 mole ratio, with a molecular weight of 200 000. It is commercially available as Cypan (Cyanamide Co.) and used at 0·285% concentration of the total slurry. Davis describes an *in situ* polymerisation method for oil mud.[93] It seems that for engineering applications the recommended molecular weight range of polymers is 10^5–10^6.[95]

The result of soil treatment with poly(acrylates) depends on a variety of factors: soil type, polymer structure, molecular weight and method of preparation, cation, particle size distribution, and water content. The earliest work on this system was confined to studying the polymers of divalent metal acrylates, especially the calcium salt which was the subject of a review by Lambe.[86] For a sandy clay, consisting of 40% kaolinite, 40% quartz and 20% limonite, treatment with calcium acrylate alone markedly increased soil fluidity. Adding between 5 to 10% of calcium acrylate reduced the liquid limit from 50% to around 36% and the plastic limit from 24% to around 14%.† This effect of calcium acrylate on the soil aids the mixing of the monomer before polymerisation.

In engineering applications large volume changes accompany water absorption, and on the other hand water loss can lead to surface cracking. The extent of soil shrinkage on drying depends on the amount of water incorporated in the mixture. When the amount of water equals 60% of the soil weight, then shrinkage on drying is 43% by volume. With 20% water this shrinkage is reduced to 10%. Untreated soil has a shrinkage 30% greater than treated soil. Another feature is the time taken for the shrinkage to be completed. Using experimental samples, it has been found that while untreated soil completely shrinks within 2 hours, this period is extended to 4 or 5 days following treatment with calcium poly(acrylate). In field situations these effects would be partly ameliorated by the use of mechanical compaction and the availability of water in the subsurface soil. The amount of water absorbed by soil depends both on the amount of polymer used and the cation present.[96] These effects are illustrated by the data in Table 9. From this table it can be seen that the treatment is best done using a lower concentration of polymer. The variation of behaviour with cation provides a certain flexibility in the system so that the polymer may be used either for water proofing or water sensitising.

Tensile strength is increased considerably by treatment, according to one report[86] for a sandy clay from zero to around 170 psi. The tensile strength of treated soil depends on the amount of water used in mixing and the polymer concentration. Hopkins[96] found that as the amount of polymer increased, the soil took on the mechanical properties of the polymer. There was a broad maximum for tensile strength when the polymer concentration was 5 to 20% of dry soil weight. Optimum strength (43 psi) was obtained

† The liquid and plastic limits of a soil are the respective water contents, expressed as a percentage of dry soil weight, at which the soil becomes liquid and plastic. The procedures for measurement of these quantities are to be found in any standard work on soil mechanics.

TABLE 9
WATER ABSORPTION BY SOIL USING METAL POLYACRYLATES

	Soil shrinkage %		
Polymer concentration	3	5	10
Calcium poly(acrylate)	16	23	30
Zinc poly(acrylate)	16	16	19
Magnesium poly(acrylate)	18	—	30

at 10% polymer concentration. In some engineering situations high compressive strength is demanded. *In situ* polymerisation of calcium acrylate has been reported to increase the compressive strength of soil by as much as 1 500%.

The flexibility of treated soil depends on the amount of calcium poly(acrylate) and water used. In one case,[96] the percentage elongation for treatments with 3, 5, 10, 20, 50 and 100% calcium poly(acrylate) was 2, 5, 25, 100, 45 and 53 respectively. Lambe[86] found treatment with calcium poly(acrylate) successful for 20 different soils, the compositions including sandy, silty, and clayey soils. Greater strength was found with soils containing a greater proportion of fines. For soils with high liquid limits, more mixing water is required to be available for the reaction. Difficulties occur with soils containing few fine particles, as there is insufficient mixing water present. Soils with an expanded lattice, such as montmorillonite, present mixing problems since water is absorbed and is unavailable to participate in chemical reactions.

Poly(acrylates) have also been used to reduce fluid loss from drilling muds. The polyelectrolyte used has to be carefully selected to avoid increasing the viscosity and gel strength of the treated mud. Scanley[97] has described the use of copolymers of acrylic acid and acrylamide on three mud systems. The polymers used were of two types. Type I copolymers were copolymers of molecular weight 1·8, 2·5, and 3·9 × 10^5. The carboxylate:amide ratios were 58:42, 68:32, and 73:27, making a total of 9 copolymers. The mud systems studied were; attapulgite clay mud, low solids lime base mud and red mud containing 3·5% NaCl. It was found that higher molecular weight polymers are more effective fluid loss controllers, but also increase viscosity and gel strength. There was no specific effect with carboxylate:amide ratio: the behaviour was dependent on the ionic character of the mud. In the NaCl system, higher carboxylate:amide ratios were more effective, but worse in the lime base mud. Type II polymers were prepared by partial hydrolysis of polyacrylamide. Five polymers were prepared of molecular weight around 2.5×10^5 and carboxylate: amide ratios 9:91, 23:77, 37:63, 52:48, 66:34. Fluid loss was minimised at a carboxylated content of 50%. The viscosity and gel strength problems were minimised at around 23% carboxylate. At a carboxylate content of between 52 and 66% there was a sharp peak in viscosity and gel strength when 0·316 lb per barrel of polymer was used. This study illustrated that the effect of polymer treatment is variable, and optimum conditions can be obtained by paying attention to the salient factors in the system.

4.3.2 Agricultural treatments

Many acrylic polymers have been found effective for agricultural use. Hedrick and Mowry have published data for 61 polymers.[98] Claims have been made for other polymers and their salts.[99-101] For the calcium salt of partially hydrolysed poly(acrylamide), it is important to control the degree of hydrolysis. If the nitrogen content falls below 24% of its original value, according to Basdekis the calcium salt is too insoluble to achieve an effective soil treatment.[102] Yost found a preferred range for partially hydrolysed poly(acrylamide) of 65 to 85%, although treatment was possible over the wider range of 50–100% hydrolysis.[103]

In agricultural applications the soil is usually treated directly with a preformed polymer rather than carrying out *in situ* polymerisation. Hedrick and Mowry[104] have described the methods of application. For greenhouse experiments two techniques are available. The polymer can either be added in aqueous solution to dry or moist soils, using sufficient water to produce optimum workability, or alternatively added dry followed by the requisite amounts of water to give optimum workability. For field work, the dry polymer is spread over the prepared seed bed and mixed into the soil to the required depth by a rotary tiller, a technique which is not so effective with wet soils. Mixing is completed by rain or artificial watering. Formulations have also been patented in which the solid polymer is added with a carrier such as peat moss, limestone, sand, clay, mineral fertiliser, or silage.[98,105] In agricultural uses, the polymer is used as a flocculant and only small dosage is required; apparently an application of 0·05–0·1% is effective.[92]

The molecular weight of polymers should be greater than 10^4 according to Hedrick and Mowry, with an optimum value of 15 000 and maximum of 10^5.[98,106,107] However other workers put the lower limit at 5×10^4,[108] and for a copolymer of methacrylic and acrylic acids 3×10^5.[109] For poly(acrylamide) the molecular weight should be greater than 88 000 but reaches an economic limit at 5.3×10^5.[103]

Polycarboxylate electrolytes are undoubtedly effective for agricultural applications. A trial of 2–3 years has been reported by Hedrick and Mowry[88,104] where their work is summarised. The usefulness of poly-(acrylates) to agriculture arises from the formation of stable soil crumbs. Treated soil was evaluated for oxygen consumption by Warburg's method[110] and aggregate size by wet sieve analysis.[111] Materials used were the sodium salt of hydrolysed polyacrylonitrile (CRD 189) and a carboxylated polymer with a high calcium tolerance (CRD 186). The aeration factor, as measured by Warburg's method, increased consistently at four soil moisture contents. The effect was observed for fine soils, 0·1% of polyelectrolyte being sufficient. The results are presented in Table 10.

TABLE 10

AERATION FACTOR OF SOILS AT DIFFERENT MOISTURE CONTENTS

Moisture content	25%	37·5%	50%	62·5%
Soil				
Miami silt loam	82	68	8	0
Miami silt loam + 0·1% CRD 186	106	86	45	11
Paulding clay	94	92	64	0
Paulding clay + 0·1% CRD 186	105	102	85	57
Grenada silt loam	83	69	46	0
Grenada silt loam + 0·1% CRD 186	99	103	99	76
Sandy loam	83	55	0	0
Sandy loam + 1% CRD 186	130	99	63	30
Sandy loam + 0·1% CRD 186	128	92	66	40
Sandy loam + 0·1% CRD 189	113	99	72	35
Memphis silt loam	92	80	56	0
Memphis silt loam + 0·1% CRD 186	104	100	82	51

In the Warburg method, the soil is saturated with water to form a sticky mass and then allowed to air dry. In view of this treatment, it is also considered a measure of crumb stability to slaking by water. This is also measured by wet sieve analysis and results are shown in Table 11 for Miami silt loam. A similar increase in the percentage crumbs of this size was found for 22 soils treated similarly. The treatment has a negative effect, however, for Portsmouth fine sandy loam, but this was observed to be a difficult soil to wet.

Treated soil was found to have improved workability, being less sticky and having a lower plastic limit. Percolation rates were increased under laboratory tests from 100-fold for heavy soils to 4·4 times for soil with good natural percolation. The water equivalent of the soil increased by 20 to 70

Table 11
PERCENTAGE OF SOIL CRUMBS OF DIAMETER GREATER THAN 0·25 MM

Polyelectrolyte	0·1%	0·05%	0·02%	0·01%
CRD 186	96	84	70	50
CRD 189	92	72	42	16

times the weight of added polymer, which indicates that the effect is due to improved soil structure. As the wilting† percentage was unchanged, the additional water in the soil must be available for plant growth. Treatment of soil reduced rates of evaporation. Untreated soil lost half its water content in 15 days whilst it took 35 days to lose the corresponding quantity of water after treatment with 0·1% CRD 186, this being equivalent to treatment with 1 to 2% of methyl cellulose.

The effect of treatment of soil appears to be permanent. In the laboratory, soil crumbs remained stable to constant flow of water for 32 months. Under similar conditions, untreated soil was stable for a few minutes; for 2 weeks when treated with 1% sodium alginate and a month with 1% dried compost. In the field, no serious loss has been found after 2 to 3 years of normal working. In laboratory tests, heavy working of soil such as compression will reduce the amount of water-stable crumbs greater than 0–2·5 mm diameter to 21% and repeated treatment decreases the level to 16%. If this soil is treated with ammonium hydroxide (pH = 10·5), sodium hydroxide, or phosphates the original properties can be re-established.

These polymers have been found to be non-toxic to yeast, worms, rats, chickens, and the organisms responsible for nitrification. Vegetables grown in treated soil contain the usual nutrients and trace elements. Treated soil has led to improvements in plant growth. Germination and seedling emergence increases: germination of radishes is 68% in untreated soil, 77% in the presence of 0·02% CRD186 and 82% with 0·1% polymer. For carrots the germination is increased from 32% to 63% on using 0·05% CRD 186. Other advantages to agriculture are; faster growth, earlier maturity, increased yields, and increased root growth. For example, carrots grown in treated soil increased in weight compared to controls by a factor of 2·8. The improved performance of plant growth is attributed to the increased availability of air and water[112] due to the better soil structure.

† The wilting point is the soil moisture content below which plants are unable to draw moisture from the soil.

4.3.3 The mode of action of acrylate polymers on soil properties

Acrylate polymers can have a wide variety of effects on soil ranging from dispersion to flocculation. Relevant factors include soil aggregation, composition, structure and surface charges, polymer structure and molecular weight, and pH and the presence of metal cations.

One of the principal constituents of soil are the clay minerals which contain an aluminosilicate network, often in the form of layers. A few aluminium cations are substituted by divalent metal cations of similar ionic radius such as magnesium. The resultant unbalance of electrical charge produces a negative charge on the basal planes on clay particles such as montmorillonite. At the edge of particles, the aluminium cations have a residual positive charge. The resultant dipolar nature of these particles can lead to self-flocculation. Other cations can absorb on the particles and prevent flocculation. At high ionic strength, double layer effects lead to flocculation in accordance to the Schulze–Hardy rules. In view of the importance of surface charge, clayey soils are more susceptible to polymer treatments than sands. However several claims have been made for the treatment of sands by poly(acrylates).[113,114,95]

Kitchener[113] has discussed the nature of interparticle forces of silicates. The London–van der Waals attractive energy is given by Eq. (1) for spherical particles

$$V = \frac{Aa}{12h} \tag{1}$$

where A = constant (for silica in water $A = 10^{-13}$ erg), a = radius of particle and h = distance of separation.

For two particles of 0·1 μm radius and separated by 100 Å, the energy of interaction is -10^{-13} erg and there is an attractive force of 10^{-7} dyne which dominates inertial and gravitational forces (10^{-11} dyne). For particles of 10 μm the reverse situation obtains. For particles with charges of similar signs, the electrical double layer leads to a repulsive force dependent on ionic strength. Furthermore, surface solvation produces a force of repulsion.

The degree of aggregation has been shown to be dependent on the geological evolution of the soil.[116] If it is formed in the presence of aqueous sodium chloride it is highly dispersed, due to the absorption of highly hydrated cations and the development of high negative charges. Soils formed in the presence of alkaline earths, *e.g.* Mg and Ca, selectively absorbed these ions leading to lower electrostatic repulsion and higher aggregation. This type of consideration affords an explanation of electrokinetic dispersion and aggregation. The former is effected by treating soil with an alkali metal salt solution. The univalent cations displace any absorbed divalent cations which are precipitated as insoluble salts or complexes. On the other hand, aggregation can be achieved by adding divalent cations to the soil, which are selectively absorbed on to soil particles to displace univalent cations. The treatment of soil containing absorbed divalent cations with sodium poly(acrylate) leads to bridging by the divalent cation of the particle and the polyanion. This results in a negative charge on the particle and dispersion. If partially hydrolysed poly(acrylonitrile) is used in place of sodium poly(acrylate), aggregation occurs, presumably due to inter-chain association. No aggregation occurs for clays purified of absorbed divalent cations. These mechanisms are illustrated in Fig. 14.

The importance of surface charge on soil particles is further demonstrated by the effect of water on soil. In adding water, there is a decrease in cohesiveness and strength, the volume increases, and there is loss in permeability to air. This is attributable to the weakening of electrostatic forces in the higher dielectric medium which enables structural changes within the soil.

Turning now more specifically to the role of the polymer, unfortunately not all publications state important parameters such as molecular weight. The essential physico-chemical processes associated with polymer–soil

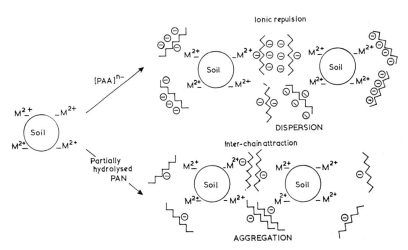

Fig. 14 The various modes of action of polyacrylates on soil. (S. Crisp and A. D. Wilson, Laboratory of the Government Chemist. Crown Copyright.)

interaction are ion absorption and bridging. Ion absorption alone results in dispersion, but when both processes occur then aggregation takes place. Aggregation, or flocculation, leads to an open structure if strong bonds are formed and it is this type of effect that produces the agricultural improvements in soil.

Ruehrwein and Ward[117] compared the action of sodium poly(methacrylate) (SPMA), a polyanion, and a poly-(β-dimethylaminoethyl methacrylate) (DMAEM), a polycation, on kaolinite. Both polymers consisted of around 15 000 units. X-ray studies demonstrated that the polycation alone was absorbed between layers. The polyanion was presumably absorbed at the particle edges. The absorption of SPMA is time dependent; full absorption takes a month, but is 75% complete after a day. The rate is determined by molecular diffusion and orientation. Absorption is accelerated by the addition of sodium chloride, due to ionic shielding reducing electrostatic repulsion. The amount of anion absorbed corresponds to the anion exchange capacity of the clay (ca. 2 meq per 100 g) and indicates that the process involved ion exchange. Colloidal clay is flocculated by cations in the following order: poly(vinylbutyl pyridinium) > DMAEM > Al^{3+} > Ca^{2+} > Na^+. Polyanions do not have a flocculating effect. If, however, the clay is first flocculated by sodium ions and SPMA then added, a stable floc is formed as polyanions are able to absorb over two particles and so this system is bound together and is stable to dilution. This work illustrates the importance of absorption and bridging in soil aggregation. Implicit is the importance of molecular weight, since this determines polymer length.

A thorough study of absorption of polyanions on kaolinite was made by Michaels and Morelos using simple techniques.[118] They made the observations that flocculation occurs with polyanions which have hydroxyl or amide groups or carboxylate salts. Hydrolysed poly(acrylonitrile) and poly(acrylamide) flocculate clays at neutral or slightly acidic pH values. No flocculation occurs in alkaline solution if the alkali is sodium hydroxide, but it is found for calcium hydroxide solutions. Three possible mechanisms of flocculation are considered:

(1) $Clay^+—A^- + RCOO^-M^+ \rightleftharpoons Clay^+—{}^-OOCR + M^+A^-$
(2) $Clay—OH + HOOCR \rightleftharpoons Clay—OOCR + H_2O$
(3) $Clay^-—M^{2+} + RCOO^- \rightleftharpoons Clay^-—M^{2+}—{}^-OOCR$

Mechanism (1) is favoured by $RCOO^-$ and will increase with pH, whilst the opposite effect is predicted for (2). In their study, the authors used clay in the hydrogen form, so their work is inapplicable to the third mechanism.

The polymers used in this study were sodium poly(acrylate), [SPA] and 20% hydrolysed poly(acrylamide) [PAM] and the clay was kaolinite. SPA is un-ionised and coiled by intermolecular hydrogen bonding at low pH, leading to colloidal aggregation, low viscosity, and turbid solutions. At high pH the polymer is ionised and uncoiled and solutions are optically clear and viscous. Comparison of viscosities suggests that PAM is less coiled than SPA. Titration against sodium hydroxide of the clay treated with the polymers shows a strong buffer action by SPA at about pH 7·8 and a smaller buffering effect by PAM at pH 6·9.

The maximum absorption of SPA occurs at a pH of about 6, when two-thirds of the polymer is absorbed. No absorption of SPA was found at pH 7·7. With PAM, absorption increased as the pH decreased from 6·7 with zero absorption, to 4·6 when absorption was complete. Because of charge repulsion, absorption increases with decreasing ionisation, *i.e.* lower pH which suggests mechanism (2). Michaels and Morelos experimentally estimated the acid constants of the polymers and calculated the ratio of anion to un-ionised polymer and also the ions absorbed per polymer unit. The results are given in Table 12. The explanation of absorption at pH 6 is that the ionic double layer around the clay particles is more hydrogen ion rich than the bulk of the solution, reducing the pH by about 2·2 units. Revised calculations are shown in Table 13. Only about two-thirds of SPA is absorbed due to intramolecular hydrogen bonding, which is less effective in PAM.

TABLE 12
ABSORPTION OF POLYMERS ON KAOLINITE

pH	SPA		PAM (20% hydrolysed)	
	Ratio of ionised to un-ionised acid groups	Ionised acid groups per monomer unit	Ratio of ionised to un-ionised acid groups	Ionised acid groups per monomer unit
8	1 500	1·0	2 900	0·20
7	150	0·99	290	0·20
6	15	0·94	29	0·19
5	1·5	0·60	2·9	0·15
4	0·15	0·13	0·29	0·04

Flocculating behaviour is deduced from sediment volumes, *see* Fig. 15. At low pH values, SPA does not have a flocculating effect as the molecular coiling prevents bridge formation. At pH 6·1 to 8 the slight increase in

Table 13
ABSORPTION DATA CORRECTED FOR SURFACE EFFECTS

pH of solution	pH at clay	SPA	PAM
8·2	6	0·94	0·19
7·2	5	0·60	0·13
6·2	4	0·13	0·03
5·2	3	0·02	0·004

volume suggests slight flocculation, but this may be caused by increased surface charge density due to ionised carboxylate polymer groups. With PAM there is a sharp volume increase at pH 6–7·5. Absorption occurs in this pH region, and the polymer coils must be sufficiently long to form

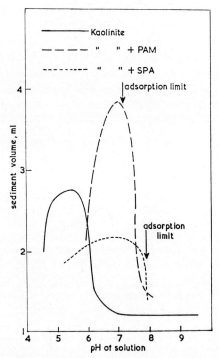

Fig. 15 Sediment volume of kaolinite suspension. (From Michaels, A. S., and Morelos, O. (1955). Ind. Eng. Chem. **47**, *1801. Copyright by the American Chemical Society. Reprinted by their permission.)*

bridges. At pH 4·6, PAM is ineffective presumably due to coil formation. Both polymers flocculate the clay at pH values where absorption does not occur. This is due to a combination of the polymer buffer action and the presence of simple cations e.g. sodium. At pH 8·5, the electrokinetic potential is sufficiently high to stabilise the clay suspension to either polymer.

This study illustrates both the role of absorption and inter-particle bridges in flocculation by poly(acrylates). It also brings to light many nuances of the system due to pH variations and polymer structure.

Kitchener[115] and others, [95,98,103,106–109] have drawn attention to the significance of the polymer molecular weight. The optimum dosage for a polymer of molecular weight 10^6 is less than that of molecular weight 10^5 and at a value of 10^4 the polymer cannot be used to flocculate, but may be a good dispersing agent. Poly(acrylamide) molecules of molecular weight 10^6 have an average length of 0·1 μm and are probably used on particles of 1 μm. The formation of inter-particle bridges has not been observed by electron-microscopy but evidence of their existence can be summarised. The absorbed polymer cannot be leached out, due to its multi-point attachment. Optimum flocculation occurs at less than full surface coverage: in the latter condition the suspension is reestablished. Flocs are large, elastic bodies capable of withstanding a moderate shear force and have a large sediment volume. They may be regarded as a spring connected system.

REFERENCES

1. McLean, J. W. (1965). *Rev. Belge Med. Dent.*, **20**, 277.
2. Smith, D. C. (1969). British Patent 1,139,430.
3. Smith, D. C. (1968). *Brit. Dent. J.*, **125**, 381.
4. Wilson, A. D. and Kent, B. E.(1973). British Patent 1,316,129 (application 1969).
5. Wilson, A. D. and Kent, B. E. (1971). *J. Appl. Chem. Biotech.*, **21**, 313.
6. Wilson, A. D. and Kent, B. E. (1972). *Brit. Dent. J.*, **132**, 133.
7. Guinard, S., Boyer-Kawenoki, F., Dobry, A. and Tonnelat, J. (1949). *Comp. rend.*, **229**, 143.
8. Hermans, J. J. and Overbeek, J. Th. G., (1948). *Rec. trav. chim.*, **67**, 761.
9. Flory, P. J. (1953). *J. Chem. Phys.*, **21**, 162.
10. Katchalsky, A. and Eisenberg, H. (1951). *J. Polym. Sci.*, **6**, 145.
11. Huizenga, J. R., Greiger, P. F. and Wall, F. T. (1950). *JACS*, **72**, 2636.
12. Manning, G. S. (1967). *J. Chem. Phys.*, **47**, 2010.
13. Bjerrum, N. (1926). *Proc. Danish Roy. Soc., Math-Phys Comm.*, **7**, No. 9.
14. Fuoss, R. M. (1958). *JACS*, **80**, 5059; Fuoss, R. M., Onsager, L. and Skinner J. F. (1965). *J. Phys. Chem.*, **69**, 2581.
15. Huizenga, J. R., Greiger, P. F. and Wall, F. T. (1950). *JACS*, **72**, 4228.
16. Ferry, G. V. and Gill, S. J. (1962). *J. Phys. Chem.*, **66**, 999.

17. Nagasawa, M., Takahashi, A., Izumi, M. and Kagawa, I. (1959). *J. Polym. Sci.*, **38**, 213.
18. Crescenzi, V. *et al.* (1959). *Chem. Abs.*, **54**, 4109.
19. Crisp, S. (1967). Ph.D. Thesis, London University.
20. Ikegami, A. and Imai, N. (1962). *J. Polym. Sci.*, **56**, 133.
21. Gregor, H. P., Luttinger, L. B. and Loebl, E. M. (1955). *J. Phys. Chem.*, **59**, 34.
22. Mandel, M. and Leyte, J. C. (1964). *J. Polym. Sci.*, **A2**, 2883.
23. Idem, *ibid*, 3771.
24. Morawetz, H. (1955). *J. Polym. Sci.*, **17**, 442.
25. Wall, F. T. and Gill, S. J. (1954). *J. Phys. Chem.*, **58**, 1128.
26. Jacobson, A. L. (1962). *J. Polym. Sci.*, **57**, 321.
27. Ikegami, A. (1964). *J. Polym. Sci.*, **A2**, 907.
28. Wall, F. T. and Drennan, J. W. (1951). *J. Polym. Sci.*, **7**, 83.
29. Rees, D. A. (1972). *Chem. Ind.*, 630.
30. Michaeli, I. (1960). *J. Polym. Sci.*, **48**, 291.
31. Jurecic, A. (1971). *Ger. Offen.* no. 2,110,665; (1973). *US Pat.* 1,304,987.
32. Skinner, E. W. and Phillips, R. W. (1967). *The Science of Dental Materials*, 6th edn, W. B. Saunders, Philadelphia and London.
33. Brauer, G. M., White, E. E. and Mashonas, M. G. (1958). *J. Dent. Res.*, **37**, 547.
34. Nielsen, T. H. (1963). *Acta Odont. Scand.*, **21**, 159.
35. Bertenshaw, B. W. and Combe, E. C. (1972). *J. Dentistry*, **1**, 13.
36. Idem (1972). *J. Dentistry*, **1**, 62.
37. Private communication, 1972.
38. Smith, D. C. (1958). *Brit. Dent. J.*, **105**, 313.
39. Fleck, H. (1902). *Dent Items of Interest*, **47**, 906.
40. Zhuravlev, V. F. and Vol'fson, S. L. (1946). *Trudy Leningrad Tekhnol Inst in Leingrad Sovela*, no. 12, 134.
41. Zhuravlev, V. F., Vol'fson, S. L. and Sheveleva, B. I. (1950). *J. Appl. Chem. USSR*, **23**, 121.
42. Crowell, W. S. (1927). *J. Am. Dent. Ass.*, **14**, 1030.
43. Baumann, E. and Gerhard, G. (1970). *Ger. Offen.* 1,903,087.
44. Smith, D. C. (1971). *J. Canad. Dent. Ass.*, **37**, 22.
45. Nielsen, L. E. (1968). *Polymer Preprints*, **9**, 596.
46. Fields, J. E. and Nielsen, L. E. (1968). *J. Appl. Polym. Sci.*, **12**, 1041.
47. Swartz, M. L., Phillips, R. W., Norman, R. D. and Oldham, D. F. (1963). *J. Am. Dent. Ass.*, **67**, 367.
48. Martell, A. E. and Calvin, M. (1952). *Chemistry of the Metal Chelate Compounds*, Prentice-Hall, London.
49. Mellor, D. P. (1964). In *Chelating Agents and Metal Chelates*, Chap II. Ed. F. D. Dwyer and D. P. Mellor. Academic Press, New York, London.
50. Morawetz, H. (1965). *Macromolecules in Solution*, Chap. VII. Interscience, New York.
51. Wilson, A. D., Kent, B. E., Clinton, D. and Miller, R. P. (1972). *J. Mat. Sci.*, **7**, 220.
52. Chatterji, S. and Jeffery, J. W. (1967). *Nature*, **214**, 559.
53. Swartz, M. L. and Phillips, R. W. (1955). *J. Am. Dent. Ass.*, **50**, 172.
54. Going, R. E. (1972). *J. Am. Dent. Ass.*, **84**, 1349.
55. Mizrah, E. and Smith, D. C. (1969). *Brit. Dent. J.*, **127**, 371.
56. Mizrah, E. and Smith, D. C. (1969). *Brit. Dent. J.*, **127**, 410.
57. Mortimer, K. V. and Tranter, T. C. (1969). *Brit. Dent. J.*, **127**, 365.
58. Phillips, R. W., Swartz, M. L. and Rhodes, B. (1970). *J. Am. Dent. Ass.*, **81**, 1353.
59. Beech, D. R. (1972). *Arch. Oral. Biol.*, **17**, 907.

60. Plant, C. G. (1970). *Brit. Dent. J.*, **129**, 424.
61. Barnes, D. S. and Turner, E. P. (1971). *J. Canad. Dent. Ass.*, **37**, 265.
62. Klotzer, W. T., Tronstad, L., Dowden, W. E. and Langeland, K. (1970). *Deutsch. Zahnarti Zeit.*, **25**, 877.
63. Truelove, E. J., Mitchell, D. F. and Phillips, R. W. (1971). *J. Dent. Res.*, **50**, 166.
64. Beagrie, G. S., Main, J. H. and Smith, D. C. (1972). *Brit. Dent. J.*, **132**, 351.
65. American Dental Association, Interim Recommended Standard Practices for Toxicity Tests on Dental Materials, *Guide to Dental Materials and Devices*, 6th edn, pp. 158–167 (1972–1973).
66. Zander, H. A. (1946). *J. Am. Dent. Ass.*, **33**, 1233.
67. Wilson, A. D. (1972). *Nat. Bur. Std. Special Publ.* 354, Dental Materials Research, Proc. 50th Ann. Symp., Gaithersburg (1969), p. 85.
68. Wilson, A. D. (1968). *J. Dent. Res.*, **47**, 1133.
69. Wilson, A. D. and Kent, B. E. unpublished data.
70. Barry, T. L., Miller, R. P. and Wilson, A. D. 11th Conference of the Silicate Industry, Budapest (1973).
71. Kent, B. E. and Wilson, A. D. unpublished data.
72. Zachariasen, W. H. (1932). *J. Am. Ceramic Soc.*, **54**, 3841.
73. Weyl, W. A. and Marboe, E. C. (1962). *The Constitution of Glasses*, Vol. 1, Chap. III, V, Interscience, New York.
74. Crisp, S. and Wilson, A. D. (1973). IADR, British Division, 21st Meeting.
75. Crisp, S. and Wilson, A. D. (1973). British Ceramic Society Meeting: Aberdeen.
76. Kent, B. E., Lewis, B. G. and Wilson, A. D. (1973). *Brit. Dent. J.*, **135**, 222.
77. Nielsen, L. E. (1969). *J. Macromol. Sci. Rev. Macromol. Chem.*, **C3**(1), 69.
78. Nielsen, L. E. (1951). *Rev. Sci. Instr.*, **22**, 690.
79. Nielsen, L. E. (1965). *Trans. Soc. Rheology*, **9**, 243.
80. Kerner, E. H. (1956). *Proc. Phys. Soc.*, **69B**, 808.
81. Guth, E. (1945). *J. Appl. Phys.*, **16**, 20.
82. Smallwood, H. M. (1944). *ibid.*, **15**, 758.
83. Mooney, M. (1951). *Colloid Sci.*, **6**, 162.
84. Anon, (1950). *Roads and Street*, **93**, 51.
85. Anon, (Oct. 17th 1966). *Chem. Eng. News*, 80.
86. Lambe, T. W. (1951). *Boston Soc. Civil Engineers*, 127.
87. Lambe, T. W. (1955). *Eng. News–Rec.*, **155**, 41.
88. Hedrick, R. M. and Mowry, D. T. (1952). *Chem. Ind.*, 652.
89. Hauser, E. A. and Mercer, E. (1948). US Patent 2,401,348.
90. de Mello, V. F. B., Hauser, E. A. and Lambe, T. W. (1953). US Patent 2,651,619.
91. Meunier, V. C., Williamson, G. J. and Hopkins, R. P. (1955). *Ind. Eng. Chem.*, **47**, 2265.
92. Lambe, T. W. and Michaels, A. S. (1954). *Chem. Eng. News*, **32**, 488.
93. Davis, R. W. (1959). US Patent 2,842,338.
94. Scanley, C. S. (1962). *Amer. Chem. Soc., Div. Petrol Chem. Preprints*, **7**, 65.
95. Fischer, R. F. (1966). US Patent 3,268,002.
96. Hopkins, R. P. (1955). *Ind. Eng. Chem.*, **47**, 2258.
97. Scanley, C. S. (1959). *World Oil*, 122.
98. Hedrick, R. M. and Mowry, D. T. (1952). US Patent 2,625,529.
99. Morrill, H. L. (1955). US Patent 2,717,884.
100. Boethner, F. E. and Niederhauser, W. D. (1959). US Patent 2,892,823.
101. Basdekis, C. H. (1957). US Patent 2,812,314.
102. Basdekis, C. H. (1953). US Patent 2,652,381.
103. Yost, J. F. (1956). US Patent 2,751,367.
104. Hedrick, R. M. and Mowry, D. T. (1952). *Soil Sci.*, **73**, 427.
105. Daline, G. A. (1956). US Patent 2,741,551.

106. Hedrick, R. M. and Mowry, D. T. (1953). US Patent 2,625,471.
107. Hedrick, R. M. and Mowry, D. T. (1953). US Patent 2,651,885.
108. Ziegler, G. E. (1956). US Patent 2,765,290.
109. Trommsdorff, E., Abel, G. and Volker, T. (1956). US Patent 2,763,961.
110. Quastel, J. H. and Webley, D. M. (1943). *J. Ag. Sci.*, **37**, 257.
111. Yoder, R. (1936). *J. Amer. Chem. Soc. Agron.*, **28**, 337.
112. Baver, L. D. (1951). *Agron. J.*, **43**, 359.
113. Hopkins, R. P. (1959). US Patent 2,871,204.
114. Burge, L. L. (1965). US Patent 3,223,161.
115. Kitchener, J. A. (1969). *Filtration and Separation*, **6**, 553.
116. Michaels, A. S. Proc. of Conf. on Soil Stabilisation, MIT, 59 (June 18–20, 1952).
117. Ruehrwein, R. A. and Ward, D. W. (1952). *Soil Sci.*, **73**, 485.
118. Michaels, A. S. and Morelos, O. (1955). *Ind. Eng. Chem.*, **47**, 1801.

CHAPTER 5

METAL DICARBOXYLATES—HALATOPOLYMERS

J. ECONOMY AND J. H. MASON

5.1 INTRODUCTION

In this chapter, attention is focused on the polymeric behaviour of metal dicarboxylates. The study of these materials is of considerable importance, since their structure is relatively uncomplicated and permits a more meaningful assessment of the effect of ionic bonding on polymeric properties. Obviously, the presence of highly ionic bonds in the main chain of a polymer-like structure should greatly influence the rheological properties of the material as well as the physical and mechanical behaviour.

The behaviour of these materials both in solution and in the melt has been sufficiently characterised to conclude that the metal dicarboxylates are indeed polymers.[1-3] On the other hand, it has been pointed out that metal dicarboxylates such as the metal sebacates tend to crystallise in a salt structure. The ability of these metal dicarboxylates to display both polymeric and salt-like properties appears to be unique. The term 'halatopolymers' has been introduced to describe that group of materials which can display salt and polymer properties. The transition of a metal sebacate salt into a polymer is referred to as a 'halatopolymeric transition'.

The feature which distinguishes the metal dicarboxylates from other polymers is the presence of the highly ionic metal bonds in the main chain of the polymer. For example, in calcium sebacate it has been estimated that the Ca—O bond may have as much as 80% ionic character. The presence of such bonds in the main chain results in a strong dipole interaction between chains and a corresponding brittle behaviour.

Of considerable interest is the fact that the halatopolymers lie at the borderline between polymers and inorganic salts. Hence, their characterisation permits a more meaningful differentiation between the fields of polymers and non-polymeric materials. In fact, the degree of ionic character of bonds in the main chain would appear to provide a method for differentiating between inorganic polymers and other inorganic compounds. For example, in the extreme case such as sodium chloride, polymer-like behaviour would not be expected since on melting, such structures lose any continuity of interunit linkage and then recrystallise into highly ordered structures. Unfortunately, the degree of ionic character does not provide a clear cut differentiation for polymers since there is no sharp demarcation between ionic and covalent bonds. A better differentiation is obtained if one assumes that polymers should display viscoelastic or plastic behaviour and should not crystallise into highly ordered salt structures. This type of crystallisation would be impeded by a highly viscous melt and relatively long chain lengths. More pertinent to the materials discussed in this chapter, this approach does not require that a continuity of interunit linkage be retained during melting but only that sufficient association of units occurs to permit the feature of melt extensibility. In these cases where highly ionic bonds are widely separated by stable covalent structures, one might expect an equilibrium association leading to polymer-like behaviour. This is actually observed with the metal salts of long chain dicarboxylic acids. For example, calcium sebacate salt at elevated temperatures forms a highly viscous melt consisting of 10–15 associated repeat units:[2]

$$Ca^{2+} \; Sebacate^{2-} \; \xrightarrow{t > 325°C} \; (CaSeb)_{10-15} \; \underset{t > 325°C}{\overset{cooling}{\rightleftharpoons}} \; (CaSeb)_{n \to \infty} \quad (1)$$

(salt) (viscous melt) (amorphous)

The polymer in the melt form is characterised by the absence of traditional end groups, and its molecular weight is best described in terms of a dissociation constant. Solidification of the melt leads to an amorphous material which behaves like a highly cross-linked thermoplastic.

In the following sections the available data on the preparation, structure, properties and uses of the metal dicarboxylates are discussed.

5.2 PREPARATION

The best method for preparing metal dicarboxylates is by the double decomposition reaction between a soluble alkali metal salt of the dibasic acid and a soluble salt of the cation as illustrated for the preparation of calcium sebacate from sodium sebacate and calcium chloride. All of the metal dicarboxylates prepared by this method from pure, monomeric

$$\text{NaOOC(CH}_2)_8\text{COONa} + \text{CaCl}_2 \longrightarrow$$
$$\bigl[\text{CaOOC(CH}_2)_8\text{COO}\bigr] + \text{NaCl} \qquad (2)$$

dicarboxylic acids have a crystalline salt structure as illustrated by the X-ray diffraction pattern of calcium sebacate (Fig. 1a). The ionic, crystalline nature of these salts is further demonstrated by the fact that colours associated with ions, such as those of copper and cobalt, are retained in the metal dicarboxylates. Generally, the only limitation to this method of preparation of the crystalline metal dicarboxylates is that the cation should be sufficiently electropositive to form a metal carboxylate salt.

Fusion of the crystalline metal dicarboxylates leads to a viscoelastic melt which, on cooling, forms amorphous, polymer-like solids as illustrated by the X-ray diffraction pattern of amorphous calcium sebacate (Fig. 1b). The requirement for melting below the decomposition temperature puts severe practical limits on the number of polymeric metal dicarboxylates which may be formed. Thus, all of the aromatic dicarboxylic acids and the aliphatic dicarboxylic acids with less than eight carbon atoms yield non-melting dicarboxylates. The readily available aliphatic dicarboxylic acids with more than eight carbon atoms are limited to azelaic, sebacic, dodecanedioic and the dimer acids. The requirement for melting also puts

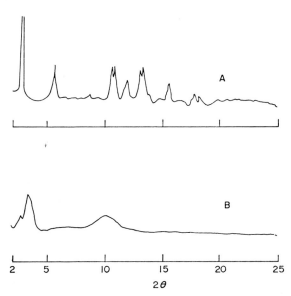

Fig. 1 X-Ray powder diagrams of calcium sebacate: (A) crystalline form precipitated from water; (B) amorphous form obtained after fusion.

limitations on the metal ions used. Thus, trivalent metals might tend to form cross-linked structures similar to those obtained with trifunctional acids, such as trimesic acid, which are discussed in the section on the structure of metal dicarboxylates. The divalent transition metals also afford a problem in that they generally tend to oxidise and become infusible. This problem can be circumvented by the use of organo derivatives of the highest oxidation state. Thus, Andrews et al.[4] have reported the preparation of dibutyl tin dicarboxylates by the condensation of dibutyl tin acetate with dicarboxylic acids. These materials are more covalent in nature and have much lower melting points, e.g. dibutyl tin sebacate is reported to have a molecular weight of 3000 and a melting point of 78–82°C.

The crystalline metal dicarboxylate salts formed by the double decomposition reaction may contain water of crystallisation. Thus, the crystalline magnesium sebacate may contain as many as three molecules of water per magnesium atom. The water of crystallisation can be removed by heating to ca. 175°C. In one instance, magnesium isophthalate, the water of crystallisation, or rather its loss, provides a mechanism for forming a polymeric aromatic dicarboxylate. Thus, if magnesium isophthalate containing water of crystallisation is moulded at 350°C a hard, tough amorphous piece is obtained. On the other hand, if anhydrous magnesium isophthalate is similarly moulded the piece readily crumbles and exhibits an unchanged crystalline structure. Presumably the water of hydration acts to plasticise the structure.

The polymeric form of the metal dicarboxylates can be prepared directly by the melt polymerisation of the dicarboxylic acid with the metal acetate or oxide. Temperatures in excess of 150°C are required to effect condensation and usually the condensation must be halted after approximately 90% of the theoretical acetic acid or water is evolved because of high viscosities. If the halatopolymer is pure, it is extremely difficult to effect recrystallisation either by very slow cooling of the melt or annealing at elevated temperatures. Exposure of pure amorphous calcium sebacate to water for 68 hr induces only partial recrystallisation. Recrystallisation of amorphous, halatopolymer is enhanced by the presence of impurities such as unreacted dicarboxylic acid or moisture.

A number of attempts have been made to prepare metal dicarboxylates containing longer chain aliphatic units. The fusion polymerisation of carboxyl terminated polyesters with metal acetates or oxides gave ester interchange which resulted in mixtures of pure polyester and metal dicarboxylate. Under the alkaline conditions necessary to dissolve the polyester in water for the double decomposition reaction, severe hydrolysis occurred. Interfacial polymerisation avoided hydrolysis but resulted in a great deal of unreacted polyester in the polymer. Attempts to polymerise barium and calcium di(aminocaproates) with sebacic acid, sodium sebacate or sebacoyl chloride gave similar results. Metal dicarboxylates containing hydroxyl and amino groups have been prepared using 12-hydroxystearic acid and 11-amino-undecanoic acid. However, attempts to react these groups with diepoxides and diisocyanates were unsuccessful. Prasad and Kacker[5] have reported the preparation of phenoxides of manganese. Since diphenoxides such as resorcinol might provide polymers similar to the halatopolymers, the reactions of zinc, magnesium and manganese acetate with resorcinol have been investigated and shown to produce polymeric materials. However, they are thermally less stable than the metal dicarboxylates and are also decomposed by carbon dioxide.

5.3 STRUCTURE

The structure of metal dicarboxylates has been studied by means of chemical analyses, infra-red spectra, differential thermal analyses, X-ray diffraction patterns, solution viscosities and melt viscosities. This work has shown that the metal dicarboxylates can exist either as salts or as amorphous polymer-like structures, depending on the method of preparation, the chain length and/or branching of the dicarboxylic acid as well as the nature of the metal ion. The polymeric structure is the result of ionic association of the metal carboxylate bonds along the dicarboxylic acid chains. The molecular weight of the polymeric structure is a function of the degree of this association.

5.3.1 Crystalline structure

X-ray diffraction studies have shown that when the metal dicarboxylates described in Table 1 are prepared by precipitation from water they all exhibit a high degree of crystallinity similar to that shown for crystalline calcium sebacate in Fig. 1(A). Crystallographic data obtained on single crystals of barium terephthalate and barium sebacate prepared by precipitation from water are given in Table 1. These unit cell dimensions correspond approximately to repeat units of the dibasic acid.

Paquot *et al.*[6] have ascribed polymeric structures containing 13–31 repeat units to the crystalline Mg, Cl and Pb sebacates prepared by the reaction of the divalent metal sulphate with sodium sebacate in water. This was done on the basis of metal analyses (Mg, Pb and Cd) which were

TABLE 1

CRYSTALLOGRAPHIC DATA OF SINGLE CRYSTAL METAL DICARBOXYLATES

Compound	System	a Å	b Å	c Å
Barium terephthalate	Monoclinic	9·65	5·89	6·90
Barium sebacate	Tetragonal	8·64		15·42

slightly lower than theoretical. They also analysed for sodium and sulphur which could be present in end groups from these reactions. They found insufficient sodium or sulphur to account for the low metal analyses and therefore assumed a carboxylic acid end group and calculated molecular weights of 5000–6000 based on the low metal analyses. However, studies on carefully purified calcium sebacate showed that there were negligible quantities of sodium or chlorine ($< 0.05\%$ from the divalent metal chloride) and that there were no carboxylic acid or —CaOH (from hydrolysis) groups detectable by infra-red analysis. This data confirms the results from the X-ray diffraction analyses and clearly indicates that the metal dicarboxylates prepared by precipitation from water are true salts and have no end groups in the conventional polymer sense.

Those metal dicarboxylates which can be fused, form extremely viscous melts which on cooling exhibit a much more amorphous X-ray diffraction pattern as illustrated by that of calcium sebacate in Fig. 1(B). The diffraction patterns obtained from these fused salts are essentially identical to those obtained from the same metal dicarboxylates by melt polymerisation. Differential thermal analyses of crystalline and amorphous calcium sebacate show an almost identical melting region (*see* Fig. 2) with the endotherm for the crystalline form being somewhat sharper and stronger. Presumably, if the calcium sebacate were completely amorphous, then one should observe a differential thermogram with a minor transition spread out over a broad range. The relative sharpness of the DTA indicates retention of a localised degree of order which X-rays cannot detect. The pure amorphous calcium sebacate could not be crystallised either by very slow cooling of the melt or annealing at elevated temperatures. However, some crystallisation could be induced by exposing the amorphous solid to water for 68 hours.

5.3.2 Polymeric structure

The idea that polymer-like structures could occur with highly ionic units spaced regularly along the chain is most intriguing, particularly in view of

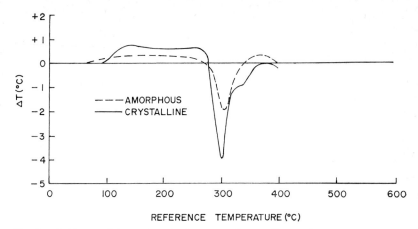

Fig. 2 *Differential thermal analysis curves of crystalline and amorphous calcium sebacate.*

the high bond energy and thermal stability of such links. The resinous nature of certain salts of organic acids, such as the calcium and zinc salts of abietic acid have been recognised for some time.[7] However, these and other similar materials are salts of monobasic acids and are characterised by their ready solubility in certain organic solvents. The resinous or amorphous nature of these salts is probably best explained by the inability of the relatively large negative ion to locate itself in a crystal lattice. The polymeric nature of the divalent metal salts of polybasic acids such as polyacrylic acid and alkyd resins of high acid number have also been reported,[8] and are discussed elsewhere in this volume. However, in these materials, the negative ions are pendant to the long chain polymeric structures so that it is impossible to attribute their properties solely to the possibility of ionic association.

Cowan and Teeter[1] were the first to clearly demonstrate, by means of solution viscosity studies, the polymeric nature of materials containing a divalent metal ion in the main chain. Although most of the metal dicarboxylates are essentially insoluble, zinc dimerate† is soluble in amines such as pyridine, quinoline and piperidine. Cowan and Teeter determined the viscosities of zinc dimerate in pyridine over the range of 0–10% concentration. Since independent means of evaluating molecular weight were not

† The dimerates mentioned here are metal salts of dimerised fatty acids and are essentially dimers of linoleic acid. They thus have a molecular weight of 500–600 and contain two COOH groups.

available, it was necessary to obtain approximation by comparison with zinc stearate which is chemically similar and has almost the same molecular weight as the zinc dimerate repeat unit but differs in its ability to associate into chains. The results which are summarised in Fig. 3 show that a plot of viscosity versus concentration for zinc stearate falls on a straight line. On the other hand the data for zinc dimerate lie on a curve which asymptotically approaches the line for zinc stearate at zero concentration. With increasing concentration this curve becomes increasingly steeper until at concentrations near 10% it becomes essentially linear. Since the slope of this curve should be proportional to the molecular weight, it appears that

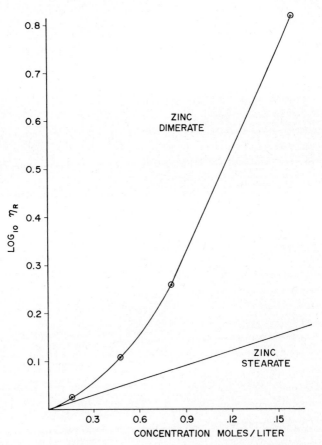

Fig. 3 Relative viscosities of zinc stearate and zinc dimerate in pyridine.

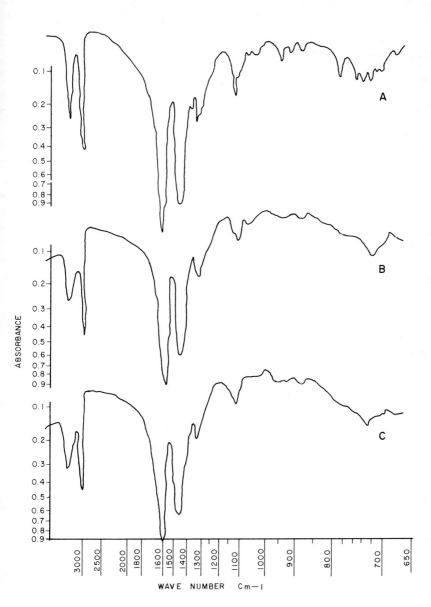

Fig. 4 Infra-red absorbances of: (A) *fused calcium azelate;* (B) *fused calcium dodecanoate;* (C) *fused 50–50 physical mixture of materials used for spectra A and B.*

in very dilute solution, the molecular weight of zinc dimerate approaches that of zinc stearate. However, in more concentrated solutions there is more and more association of the zinc ion–dimerate ion combination leading to a greater molecular weight. If the value of K obtained for zinc stearate is considered applicable, a molecular weight of 16 000 representing twenty-five repeat units is obtained for zinc dimerate.

Economy et al.[2] further extended the knowledge of the structure of metal dicarboxylates with their investigations of the melt phase of calcium sebacate. The metal carboxylate bond has been estimated to have 80% ionic character and might be expected to undergo rapid interchange reactions in the melt phase. Two approaches were used to examine this phenomenon, namely, infra-red analyses of mixtures before and after melting and melt flow measurements in the presence of chain modifiers.

The infra-red evidence for the interchange reaction is presented in Fig. 4. Spectra (A) and (B) are those of fused calcium azelate and calcium dodecanedioate, respectively. Spectrum (C) is that of a 50–50 physical mixture of the materials used to prepare Spectra (A) and (B) after re-fusing at the same temperature. As can be seen, absorptions at 735, 742, 780 and 910 cm^{-1} which are attributable to calcium azelate can no longer be detected. The spectrum closely resembles that of the calcium dodecanedioate, except that the absorptions at 720, 1110, 1310 and 1580 cm^{-1} are shifted toward the slightly lower wavelengths at which similar absorptions are observed for calcium azelate. Spectra identical to (C) can also be obtained by fusing a 50–50 physical mixture of crystalline calcium azelate and crystalline dodecanedioate or by fusing the coprecipitated calcium salt of a 50–50 mixture of sodium azelate and sodium dodecanedioate. This

TABLE 2

EFFECT OF 'MONOMERS ON THE MELT FLOW[a] OF CALCIUM SEBACATE'

Additive	Concentration mole/mole Ca Seb	Melt flow dg/min
None	—	0·88
CaCl$_2$	0·020	0·91
Ca(OH)$_2$	0·020	0·92
NaCl	0·026	1·10
Sodium Sebacate	0·010	1·04
Sebacic Acid	0·020	1·30

[a] Determined at 386 ± 1°C and 6·5 psi.

confirms that the calcium carboxylate bonds and probably the other metal dicarboxylate bonds undergo relatively rapid interchange on melting.

The rapid interchange of the metal carboxylate bonds in the molten state suggests that the polymeric behaviour of the melt is due to an association equilibrium of linear dicarboxylate units. Thus, at any given time there is a finite concentration of dissociated metal carboxylate end groups. If this is the case, materials such as $CaCl_2$, $Ca(OH)_2$, NaCl, sodium sebacate and sebacic acid might also be expected to interchange with the molten calcium sebacate and act as chain modifiers. Such chain modifiers, as indicated schematically in eqs. (3)–(7) might have an effect similar to that of adding end groups and result in marked increases in the melt flow.

$$—COOCaOOC— + NaCl = —COONa + ClCaOOC— \quad (3)$$

$$—COOCaOOC— + CaCl_2 = —COOCaCl + ClCaOOC— \quad (4)$$

$$—COOCaOOC— + Ca(OH)_2 = —COOCaOH + HOCaOOC— \quad (5)$$

$$—COOCaOOC— + NaOOC(CH_2)_8COONa$$
$$= —COONa + NaOOC(CH_2)_8COOCaOOC— \quad (6)$$

$$—COOCaOOC— + HOOC(CH_2)_8COOH$$
$$= —COOH + HOOC(CH_2)_8COOCaOOC— \quad (7)$$

The effects of these materials on the melt flow of calcium sebacate at 386°C are summarised in Table 2. From this it can be seen that the calcium hydroxide and calcium chloride have essentially no effect on the melt flow, whereas sodium chloride, sodium sebacate and sebacic acid markedly increase the melt flow. Although the calcium hydroxide and calcium chloride may not react, it seems more likely that the —COOCaCl and —COOCaOH groups formed by interchange are highly reactive and that the equilibria illustrated by eqs. (4) and (5) lie far to the left. This would also account for the sodium sebacate being about twice as effective a chain modifier as sodium chloride on a molar basis. In other words, the —COOCaCl groups or half the groups formed by the interchange of NaCl tend to react with each other according to eq. (4) and be ineffective, whereas the —COONa group is stable and acts as a modifier or chain terminator. On the other hand, each molecule of sodium sebacate affords two stable —COONa groups and is accordingly twice as effective.

Another aspect of the rapid interchange of chain modifiers is seen from the effect of adding a trifunctional acid such as trimesic acid (*see* Table 3). The trimesic acid effectively affords a branching point and the pressure sensitivity of the melt is increased. Also, the trimesic acid is less effective

TABLE 3
EFFECT OF BRANCHING ON MELT FLOW[a] AND PRESSURE SENSITIVITY OF CALCIUM SEBACATE

Material	Melt flow, dg/min		Pressure sensitivity factor (P)[b]
	325 g wt	2150 g wt	
Calcium Sebacate	0·88	8·57	1·47
Calcium Sebacate + 1·92 × 10⁻² mole fraction trimesic acid	1·10	14·44	1·98

[a] Determined at 386 ± 1°C.
[b] $P = 2150 \text{ gMF}/6.62 \times 325 \text{ gMF}$.

than the sebacic acid in increasing the melt flow at the same molar concentration. Presumably the branching resulting from the trimesic acid tends to offset the higher melt flow expected on the basis of chain termination.

The rapid interchange of calcium sebacate with chain modifiers provides a novel method of approximating the molecular weight of calcium sebacate in the melt. Since there is a rapid equilibrium interchange between the metal carboxylate groups, the polymer chains are present on a time-average basis and cannot be thought of in terms of discrete polymer molecules with ionic end groups. The time-average molecular weight, however, can be expressed in terms of an association or dissociation constant.

The relationship between viscosity and molecular weight of a polymer can be described by the equation:

$$\log \eta = 3.4 \log \overline{M}_w + K_0 \tag{8}$$

Since melt flow (MF) is inversely proportional to the viscosity, this equation may be restated:

$$\log (\text{MF}) = 3.4 \log M_e/M_p + K_1 \tag{9}$$

where M_e denotes moles of endgroups and M_p is moles of polymer repeat units. Expressing the molecular weight in terms of endgroups gives a number-average molecular weight. However, this is approximately equal to $\overline{M}_w$ if there is a normal distribution as might be expected for an equilibrium reaction.

Since the ionic endgroups in pure metal dicarboxylate are the result of an equilibrium dissociation reaction, their concentration may be expressed by the dissociation constant D:

$$D = (M_e/M_p)^2 \tag{10}$$

or, rearranging, the ionic endgroups resulting from the dissociation can be written as

$$M_e = M_p\sqrt{D} \quad (11)$$

When a small amount of a chain modifier capable of interchanging with the metal dicarboxylate, as indicated in eqs. (3)–(7), is added to the metal dicarboxylate, the concentration of endgroups is increased by an amount proportional to the concentration of the chain modifier. Thus eq. (9) may be rewritten:

$$\log (MF) = 3{\cdot}4 \log [(M_p\sqrt{D} + K_2 M_m)/M_p] + K_1 \quad (12)$$

where M_m is the moles of chain modifier. If it is postulated that in the case of sodium chloride, the interchange equilibrium shown in eq. (3) lies far to the right and that of eqn. (4) far to the left, K_2 becomes essentially equal to unity. This simplification allows the determination of D for calcium sebacate by determining the melt flow of calcium sebacate containing various quantities of sodium chloride as summarised in Table 4. The value of D which yields a slope of 3·4 for a plot of log MF versus log $[\sqrt{D} + (M_{NaCl}/M_p)]$ can then be determined. Plots of log MF versus log $[\sqrt{D} + (M_{NaCl}/M_p)]$ for various values of $\sqrt{D}$ are given in Fig. 5. From this it can be seen that the $\sqrt{D}$ for calcium sebacate equals approximately 0.2 (slope = 3·0). On this basis the calcium sebacate consists of approximately 10 repeat units or has a time-average molecular weight of 2000 at 386°C.

Similar data were obtained for the calcium sebacate–sodium sebacate system and are summarised in Table 5 and Fig. 6. In this case the equilibrium shown by eqn. (6) is assumed to proceed completely to the right and each mole of sodium sebacate is twice as effective as sodium chloride.

TABLE 4

EFFECT OF SODIUM CHLORIDE CONCENTRATION ON THE MELT FLOW[a] OF CALCIUM SEBACATE

$M_{NaCl}/M_p \times 10^2$	Melt flow (MF), dg/min	log MF
0	0·85	−0·071
2·16	1·1	0·041
4·57	1·6	0·20
10·74	2·2	0·34

[a] Determined at 386 ± 1°C and 6·5 psi.

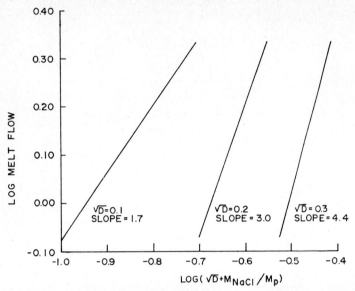

Fig. 5 Effect of sodium chloride concentration on the melt flow of calcium sebacate assuming various values for the dissociation constant of calcium sebacate.

Again a value of approximately 0·2 (slope = 3·5) is obtained for the $\sqrt{D}$ for calcium sebacate at 386°C.

The determination of an equilibrium association of approximately 10 repeat units in the calcium sebacate melt is in reasonable agreement with the 15–30 repeat units determined for zinc dimerate in concentrated pyridine solution. It can be concluded that the polymeric behaviour of the metal carboxylates is due to a dynamic association equilibrium through the metal carboxylate bond along the dicarboxylic acid chains. Association

TABLE 5

EFFECT OF SODIUM SEBACATE CONCENTRATION ON THE MELT FLOW OF CALCIUM SEBACATE

$M_{NaSeb}/M_p \times 10^2$	Melt flow (MF), dg/min	log MF
0	0·85	−0·071
1·015	1·0	0·000
5·268	3·1	0·491
10·030	8·5	0·929
11·120	16·4	1·046

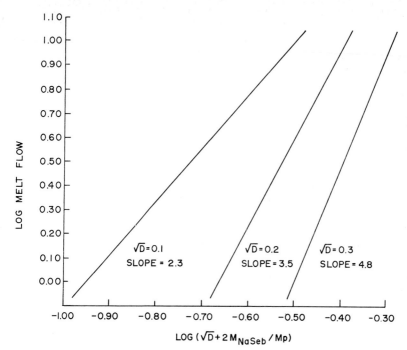

Fig. 6 Effect of sodium sebacate concentration on the melt flow of calcium sebacate assuming various values for the dissociation constant of calcium sebacate.

between metal dicarboxylate groups of neighbouring chains does not appear to contribute significantly to the viscosity. This is evidenced by the fact that the viscosity versus concentration plot for zinc stearate in pyridine is a straight line and that chain modifiers such as sodium chloride have no effect on the melt flow of calcium decanoate as shown in Table 6.

TABLE 6

EFFECT OF NaCl ON THE MELT FLOW OF CALCIUM DECANOATE

M_{NaCl}/M_{CaDec}	Melt flow,[a] dg/min
0	1·63
1·26 × 10^{-2}	1·66
2·62 × 10^{-2}	1·59

[a] Determined at 205°C and 6·5 psi.

However, a pressure sensitivity factor which decreases from 2·21 at 205°C to 1·49 at 213°C suggests there may be a minor association directly between the metal carboxylate groups which becomes increasingly important at lower temperatures. It is this type of secondary association that would account for the localised degree of order suggested by the DTA data in Fig. 2.

5.4 PROPERTIES AND USES

The metal dicarboxylates have been characterised with respect to their thermal stability, melting points, solubility, fibre forming character and mechanical properties. To date no commercial uses have emerged for this group of materials. In the following sections are summarised the reported properties and possible uses of the metal dicarboxylates.

5.4.1 Properties

The melting points and thermal stabilities in air of a series of metal dicarboxylates prepared by precipitation from water are summarised in Table 7. As expected, the relatively high ionic nature of the metal dicarboxylate bond leads to excellent thermal stabilities.

Of the metal sebacates, calcium sebacate possessed the highest melting point, except for barium sebacate which did not melt below its decomposition temperature of 350°C. In all cases where fusion occurred, highly viscous melt phases were obtained. Substituting terephthalate for sebacate led to much more thermally stable structures, e.g. barium terephthalate was stable in air up to 590°C. The effect of different dicarboxylic acids on melting point was evaluated. It was found that increasing the number of methylene groups in the dicarboxylate led to an approximately 12–20°C decrease in melting points of the polymer per methylene unit added. In the work on the polymeric salts of dimer acid, zinc dimerate was reported to melt at 130°C while calcium and magnesium dimerates melted in excess of 200°C.[1] Hence the decrease in melting point with increasing number of methylene units appears to slope off since the dimer acid has approximately 33–35 methylene units calculated on the basis of a dibasic acid.

Thermal gravimetric analysis curves both in air and argon are shown for calcium sebacate and calcium terephthalate in Fig. 7. Calcium sebacate appears to possess substantially higher thermal stability in argon up to 445°C. The plateau observed at 58% weight loss corresponds to the formation of $CaCO_3$ and the second plateau to CaO. No ketone formation

TABLE 7
MELTING POINTS AND THERMAL STABILITIES OF METAL CARBOXYLATES

(OOCRCOOM) M	R	Melting point, °C	Degradation temperature in air, °C[a]
Mg	$(CH_2)_8$	240	350
Ca	$(CH_2)_8$	320	380
Ba	$(CH_2)_8$	None	350
Zn	$(CH_2)_8$	315	330
Cd	$(CH_2)_8$	280	310
Pb(+2)	$(CH_2)_8$	275	320
Sn(+2)	$(CH_2)_8$	170	320
Mn(+2)	$(CH_2)_8$	275	305
Mg	C_6H_4	None	495
Ca	C_6H_4	None	525
Ba	C_6H_4	None	590
Sn(+2)	C_6H_4	None	400
Sn(+2)	C_6H_4	None	400
Ca	$(CH_2)_4$	None	360
Ca	$(CH_2)_7$	340	340
Ca	$(CH_2)_8$	320	380
Ca	$(CH_2)_{10}$	296	420

[a] Temperature at which 5% weight loss occurs.

could be detected during the decomposition of calcium sebacate. On the other hand, cadmium sebacate yielded a ketone-containing product. In the case of calcium terephthalate, thermal gravimetric analysis indicates a stability in argon up to 600°C. Again, the plateaus correspond to the formation of the carbonate and oxide. Most of the metal carboxylates supported combustion in a Bunsen flame except for magnesium sebacate and zinc sebacate. From an examination of the thermal gravimetric analysis curve for magnesium sebacate, it is apparent that magnesium carbonate does not form as a stable intermediate but probably decomposes, producing CO_2 which acts as a flame retardant.

One interesting feature of the metal dicarboxylates is the ease with which they can be fiberised from the melt. The fibres are unfortunately weak and brittle. In the case of zinc dimerate a critical temperature at about 125°C was noted above which birefringence in the fibre was weak and disappeared within a few hours. If fibres were drawn below this temperature the birefringence was pronounced and permanent. X-ray diffraction of the oriented fibres indicated a low degree of crystalline order. The brittleness

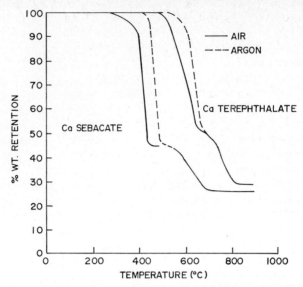

Fig. 7   Thermal gravimetric analysis curves of calcium sebacate and calcium terephthalate in air and argon.

and weakness of these fibres is undoubtedly due to the ionic linkages. No improvement would be anticipated in highly oriented filaments since any planes of ions in the oriented structure would serve as planes of cleavage as in ionic crystals.

Practically all the metal dicarboxylates are insoluble in organic solvents. Zinc dimerate, however, is soluble in various amines as well as methylcyclohexanone. As little as 5% butyl amine will act to solubilise zinc dimerate in non-solvents. All of the metal dimerates dissolve in vegetable oil at 300°C but separate after cooling. Such mixtures show considerable tolerance for non-solvents such as xylene.

The mechanical properties of the amorphous metal dicarboxylates were examined. Test specimens of calcium sebacate containing from 50–75% asbestos filler were formed by compression moulding at 350°C and 7000 psi. Tensile strengths ranged from 5000 to 7000 psi. Undoubtedly the asbestos filler acted as a reinforcing agent and also tended to reduce the thermal stresses resulting from the high moulding temperature. Pure calcium sebacate could be moulded but the samples contained numerous microcracks resulting from the thermal stresses induced in cooling from the high temperatures.

5.4.2 Uses

One potential use of the metal dicarboxylates is in the preparation of polyketones. Thus, Paquot and Perron have reported[9] that the cadmium dicarboxylate on heating to 300°C converts to the corresponding ketone. For example, cadmium sebacate on heating gave a 76% yield of the polyketone with a molecular weight of 1547.

$$\left[\begin{array}{c} O \\ \parallel \\ OC-(CH_2)_n-CO \end{array} \right] Cd \xrightarrow{300°C} \left[(CH_2)_n-\overset{O}{\underset{\parallel}{C}} \right] + CdO + CO_2 \quad (13)$$

Use of metal carboxylates as an additive to phenolic resins significantly improved the tensile strength of moulded samples by about 30%, and the strength after heating to 285°C for 24 hr by 2–3 times the value for the control.[10] In these blends, there is a possibility that the metal dicarboxylate group could interact with the phenolic hydroxyl; however, the far greater chemical stability of the metal dicarboxylate bond argues against this

TABLE 8

EFFECT OF METAL DICARBOXYLATES ON THE STRENGTH OF PHENOLIC RESIN[a]

Additive[b]	Original tensile strength[c]		Tensile strength after 24 hrs. at 285°C	
	psi	As % of control	psi	As % of control
Control	9 300	100	2 000	100
Calcium carbonate	9 200	99	900	45
Calcium valerate	5 600	60	800	40
Calcium hexanoate	5 600	60	200	10
Sebacic acid	6 100	66	700	35
Calcium sebacate	8 800	95	5 900	295
Calcium isophthalate	12 400	133	7 800	390
Calcium terephthalate	12 000	129	3 200	160
Barium sebacate	9 500	102	3 600	180
Zinc sebacate	6 300	68	3 400	170
Calcium azelate	8 400	90	4 200	210

[a] An unmodified novolak with a molecular weight of 839, m.p. 102–108°C and containing 9·3% hexamethylene tetramine was used in all samples.
[b] Control contains 20% by weight calcium oxide. All other samples contain 18% calcium oxide and 10% additive by weight.
[c] Samples were compression moulded for ½ hour at 171°C and 1800 psi, ASTM D-638 specimens, Type 1 were used for testing.

possibility. Presumably the role of the metal dicarboxylate is primarily that of a very efficient stress relaxant during thermal cycling. The effect of these additives on mechanical properties is shown in Table 8.

Use of the metal dimerates as varnish resins has also been explored.[1] It has been reported that the zinc and calcium dimerate varnishes are superior to varnishes from ester gum in drying time and in hardness of baked films. On the other hand, because of their chemical nature, it is not surprising that metal dimerate films possess almost no resistance to alkali. Similar problems were noted in evaluating zinc dimerate as a shellac substitute.

REFERENCES

1. Cowan, J. C. and Teeter, H. M. (1944). *Ind. Eng. Chem.*, **36**, 148.
2. Economy, J., Mason, J. H. and Wohrer, L. C. (1970). *J. Polym. Sci.*, A1, **8**, 2231.
3. Economy, J. and Mason, J. H. (1970). *Thermal Stability of Polymers, Vol. I*, Ed. R. T. Conley, Marcel Dekker, New York, p. 590.
4. Andrews, T. M., Bower, F. A., LaLiberte, B. R. and Montermoso, J. C. (1958). *J. Am. Chem. Soc.*, **80**, 4102.
5. Prasad, S. and Kacker, K. P. (1958). *J. Ind. Chem. Soc.*, **35**, 890.
6. Paquot, C., Perron, R. and Vassillieres, C. (1959). *Bull. Soc. Chim. France*, 317.
7. Clark, G. L. (1940). *Applied X-Rays*, 3rd edn, McGraw-Hill, New York, p. 474.
8. Hagedorn, M. (1936). US Patent 2,045,080.
9. Paquot, C. and Perron, R. (1962). US Patent 3,061,639.
10. Economy, J., Wohrer, L. C. and Mason, J. H. (1968). US Patent 3,418,273.

CHAPTER 6

TECHNOLOGY OF POLYELECTROLYTE COMPLEXES

M. J. LYSAGHT

6.1 SUMMARY AND INTRODUCTION

Polyelectrolyte complex resins, also called polysalts or polyion complexes, are the reaction product of two oppositely charged, strong polyelectrolytes. They are formed via the following reaction, also illustrated in Fig. 1:

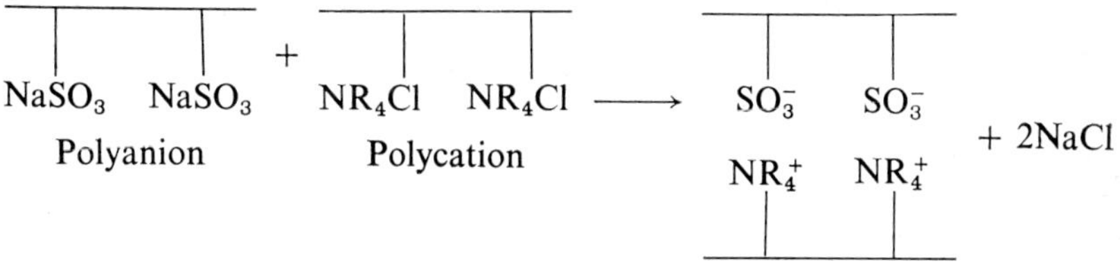

$$\begin{array}{c|c|c} | & | \\ NaSO_3 & NaSO_3 \\ \text{Polyanion} \end{array} + \begin{array}{c|c} | & | \\ NR_4Cl & NR_4Cl \\ \text{Polycation} \end{array} \longrightarrow \begin{array}{c|c} | & | \\ SO_3^- & SO_3^- \\ | & | \\ NR_4^+ & NR_4^+ \\ | & | \end{array} + 2NaCl$$

The polyelectrolyte resin structure is especially interesting because ionic and covalent forces contribute equally to the network structure.

Polyelectrolyte complex resins containing equimolar quantities of polyanion and polycation are described as 'neutral'. These are strongest, most easily prepared, and generally preferred. Other formulations containing up to 0·1 milli-equivalent excess of polyanion or polycation per gram of resin are described as 'non-stoichiometric'. As would be expected, the non-stoichiometric materials display an enhanced capacity to sorb water and electrolytes and have found favour in biological and biomedical applications.

Alone, polyelectrolyte complex resins are hard, brittle, readily ground, and preferably handled as fine powders. Plasticised with 1–10 percent salts,

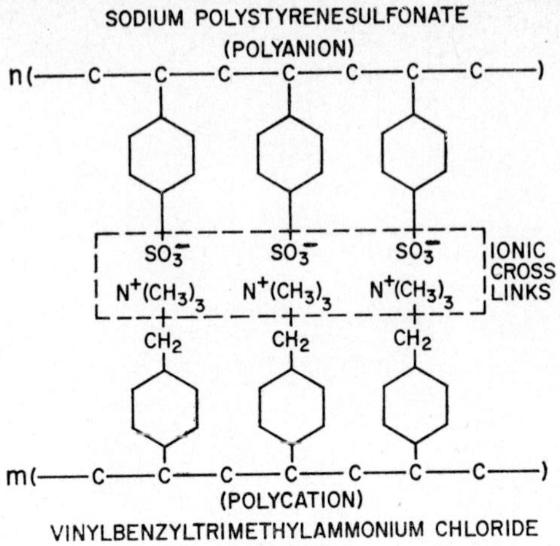

Fig. 1 Schematic structure of typical polyelectrolyte complex.

certain acids, and oxy-sulphurous compounds, the resins become tough, horn-like, and leathery. Water-equilibrated, the same materials behave as hydrogels with unusually good chemical stability and strength. Versatility is thus a key feature of the resins.

Polyelectrolyte complex resins appear slightly amber and are completely odourless and tasteless. They are insoluble in water, acids, bases, and all organic solvents. They are soluble in a few three-component solvents comprising water, an electrolyte, and a high polarity organic solvent. The oppositely-charged resins were originally co-reacted in a salt-free, very dilute aqueous solution ($<0.1\%$ solids). A neutral product was always obtained regardless of the ratio of reactants. Other reaction methods capable of reproducibly yielding non-stoichiometric products have subsequently been developed.

It is important to clearly distinguish polysalts from other classes of materials having superficially similar structures. The polyelectrolyte complex resins differ from ionomers (and their urethane and carboxylated rubber analogues) in that the latter contain only a slight ionic functionality. Also, ionomers contain a single, weak polyelectrolyte which is cross-linked by a divalent counterion. PEC resins are also distinguished from ion-exchange resins in that the latter contain covalent cross-linking and are thus completely insoluble and non-thermoformable. These relations are expanded in Table 1.

TABLE 1

COMPARISON OF POLYELECTROLYTE COMPLEX RESINS WITH RELATED MATERIALS

Ionomers	Ion-exchange resins		Polyelectrolyte complexes	
C=O–O–Ca–O–C=O	SO_3Na	SO_3Na	SO_3^-	SO_3^-
	SO_3Na	SO_3Na	$N(R_4)^+$	$N(R_4)^+$
Readily thermo-formed	Completely insoluble		Soluble in select solvents	
Hydrophobic	Permanently cross-linked; not thermoformable		Hydrophilic, forms good, volumetrically stable gels	
Quite insoluble	Volume in gels may vary with pH		Thermoformable (when plasticised)	
Not gel forming				

6.2 HISTORY

The reported study of precipitation reactions between oppositely-charged macromolecules dates back to 1896 when Kossell precipitated egg albumen with protamine.[19] Similar work with naturally occurring, weak, and often amphoteric complexes, has continued intermittently ever since and is well summarised by Doty and Ehrlich.[8] Perhaps the best systematic work was undertaken by Bungenburg de Jong[20] while the most advanced practical developments can be attributed to Green and his colleagues at National Cash Register.[16]

The literature on complexes formed from *strong* polyelectrolytes is scant. In the late forties, Fouss and Sadek reported on the reaction of polyvinylbutylpyridinium bromide with a polystyrenesulphonate, but the complexes themselves were not studied.[13] Jackson and his co-workers at DuPont synthesised several polyelectrolytes and subsequently formed and studied innumerable complexes.[17,18] However, little attention was paid to stoichiometry and the complexes were evaluated specifically as 'breathing enhancers' for synthetic leather. In the work reported in the patents, the complexes were rarely isolated and their properties never measured in the absence of carriers. Beginning in the late fifties and continuing well into the sixties, Michaels and his students at MIT published a series of theses and papers dealing with the formation, isolation, and evaluation of strong polyelectrolyte complexes.[11,12,15,21,24,25,33,34,46] These publications elucidated the structure of the complexes, the reaction mechanism, and—most importantly—the methods for dissolving, shaping, and manipulating the resins. Subsequent commercial exploitation of these resins has been carried out largely at Amicon Corporation of Lexington, Massachusetts. More recently, commercial interest has been seen elsewhere.[40,42] Over two-score patents have been issued and several excellent review articles have been published.[23,37,38]

6.3 SYNTHESIS

6.3.1 Starting materials

As a last resort, the starting resins, *i.e.* polyanion and polycation, can be synthesised (and the patent literature is replete with recipes), but it is far more practical to use commercially available materials. Some representative commercial products are:

Polyanion—sodium polystyrene sulphonate NaSS (Dow SA-1291; Enjay RS-781)
—polyvinylsulphonate (Hercules)
—polytak RNA (Peninsular Chemicals)
Polycation—polyvinylbenzyltrimethylammonium chloride (Dow QT-2781)
—polyvinylpyridinium bromide (Ionac PP2025-X)
—polydiallyldimethylammonium chloride (Calgon)

At present, commercially available, strong polyanions are limited to sulphonates; polycations may be quaternary ammonium pyridiniums, and—if needs be—polysulphoniums.

The starting resins are generally highly contaminated with salt and can be purified by passage through ion-exchange column or by fractional precipitation. Activity (ionic equivalence per unit weight of polymer) can be determined via titration; conductive titration against an acid or base is preferred. Relative activity can be obtained by titrating a dilute, purified solution of polycation versus polyanion and determining the end point by (a) observing flocculation (good to $\pm 10\%$), (b) monitoring supernatant conductivity (good to $\simeq 1$ or 2%), (c) analysing the resulting polyelectrolyte complexes (good to 1%).

6.3.2 Fabrication of neutral resins
Once in hand and characterised, the oppositely-charged polyelectrolytes may be reacted to a neutral product (*i.e.* one containing equimolar quantities of polyanion and polycation) by at least four methods:

(1) A dilute (<0·1%) aqueous solution of polycation may be added slowly to a well stirred, equally dilute aqueous solution of polyanion. A small floc forms which precipitates at the stoichiometric point. If the starting materials are pure, solids concentration may be as high as 0·5%.

(2) In a slight variation on the above procedure, two concentrated aqueous polyelectrolyte solutions may simultaneously be added to a larger mass of water. The concentration of free polyelectrolytes in the reaction bath should not be allowed to exceed 0·1%. If addition rates are carefully controlled, a stoichiometric product can be obtained from impure starting materials.

(3) In the most popular synthesis, appropriate quantities of each resin are separately dissolved in a ternary solvent for the complex (*e.g.*, 60% water, 20% NaBr, 20% acetone *or* 35% parts $Ca(NO_3)_2$, 35% dioxane, 30% water). The two solutions are combined without precipitation. The solution is subsequently drowned in water to precipitate the complex and to extract salts, organics, and impurities.

(4) Finally, quite concentrated (up to 40%) aqueous solutions of the polyelectrolytes may be combined forcefully on a rubber mill or in a high shear mixer. The product is then ground up, washed, and dried.

These processes are summarised in Table 2.

6.3.3 Non-stoichiometric resins
In principle, any of the preparative methods for neutral resins can be modified to yield non-stoichiometric products. In methods (1) and (2), for example, the employment of somewhat more concentrated polyelectrolyte solutions (>0·5%) or the presence of dissociated salts will lead to a resin containing either excess polyanion or polycation. Unfortunately, the final composition of products produced by this route is neither predictable nor reproducible.

Accordingly, aqueous precipitation from a ternary solvent is the synthesis of choice for non-stoichiometric resins. However, the ratio of polyanion to

TABLE 2
SUMMARY OF RESIN PREPARATION METHODS

Method	Literature description	Advantages	Disadvantages
A. Titrate solution of polyelectrolyte into solution of other	Ref. 26	Yields stoichiometric product regardless of mix ratio; requires no purification of starting materials; no solvents wasted	Dilute solutions ($<0.1\%$) require excessive water; filtration may be difficult
B. Add solutions of polyelectrolytes simultaneously to drowning bath	Ref. 32	Yields stoichiometric product regardless of mix ratio; requires no purification of starting materials; no solvent waste	Control of product addition rates difficult. Reactor mixing and product filtration also troublesome
C. Co-precipitate complex from ternary solution	Ref. 30	Allows working with concentrated polyelectrolytes (10% vs. 1%) pure compact product	Product not necessarily stoichiometric. Solvent handling and disposal can be difficult, solvents costly
D. Force blend on rubber mill; Ross Mixer, etc.	Ref. 31	Eliminates large volume of water and/or undesirable solvents	Requires expensive equipment and skilled operators; product not necessarily stoichiometric; filtration may be difficult

polycation in the product varies from that in the ternary solution since some of the resins are washed out before fully reacting, especially at highly non-stoichiometric ranges. The magnitude of this effect is illustrated in Fig. 2 which plots the composition of the casting solution against the composition of the product. Once such a curve is obtained for a given system, it is highly reproducible and a resin of any composition can be obtained with relative ease.

6.3.4 *In-situ* polymerisation

In-situ polymerisation is an entirely different alternative procedure for preparing polyelectrolyte complex structures.[6,42] An aqueous solution (up to 30% solids) is prepared containing a single strong polyelectrolyte and a water-soluble monomer of a strong polyelectrolyte of opposite charge. The monomer is polymerised in place by addition of a catalyst and an insoluble polyelectrolyte complex rapidly forms. With sodium polystyrenesulphonate as the polyanion, appropriate monomers for the polycation might be

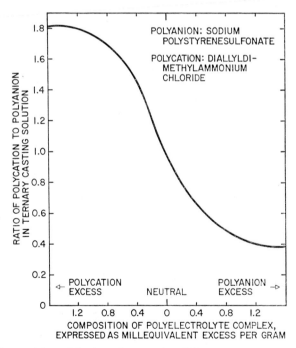

Fig. 2 Composition of final product produced by aqueous precipitation versus composition of ternary casting solution.

2-hydroxy-3-methacrylyloxy propyltrimethylammonium chloride (available from Shell as 'G-methyl methacrylate') or vinylbenzyltrimethylammonium chloride. Free radical generators (cumene hydroperoxide) or redox pairs (sodium metabisulphite/ammonium persulphate) are suitable catalysts. After formation, if desired, the polyelectrolyte complexes may be masticated, washed, dried, and redissolved in a conventional ternary solvent. The *in-situ* polymerisation procedure is useful for forming both neutral and non-stoichiometric resins.

6.4 PROPERTIES AND PHYSICAL CHEMISTRY

In pure, dry form, polysalt resins are brittle, friable plastic powders. Useful and interesting properties emerge only when the PEC resins are formed into hydrogels or blended with electrolyte plasticisers.

The resins are insoluble in all common solvents. However, they may be dissolved over narrow ranges of composition in special ternary solvents comprising water, a polar organic (acetone, ethanol, dimethyl formamide, γ-butacryl-lactone, etc.) and an electrolyte (NaCl, NaBr, HCl, H_2SO_4, NH_4Cl, etc.). A typical solubility phase diagram is shown in Fig. 3.

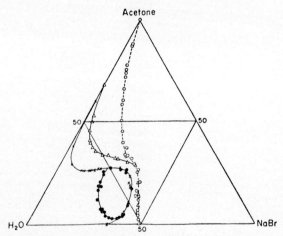

Fig. 3 Phase diagram of polyion complex (●), sodiumpolystyrene sulphonate (×), and polyvinyl pyridinium bromide (△) in acetone–water–NaBr system at 30°C. Broken line (○) denotes the boundary for miscibility among solvent components, in the absence of polymer.

Hydrogels are generally formed by precipitation into water of the preformed resins from a ternary casting solution. Hydrogels may be equilibrium or two-phase, depending upon the fabrication process:

(1) Equilibrium gels have a specific water content which varies only with the nature of the constituent polyelectrolytes and the degree of nonstoichiometry. As would be expected, the water content is lowest for neutral gels. An equilibrium gel is always clear. If desiccated, it will return to its original water level. Figure 4 demonstrates the effect of stoichiometry on equilibrium water content for two different polycations with the same polyanion.

At low water content ($< 10\%$), obtainable by desiccation, the PEC resins are glassy solids which are plasticised into strong, leather-like materials by salts and electrolytes. At intermediate water contents (20–60%), the resins form unusually tough and highly transparent hydrogels. Much above 60% the gels are weak (like wet cellophane).

(2) While single-phase 'thermodynamic' gels are easiest to envision, polyelectrolyte complex resins often form two-phase gels in which micro

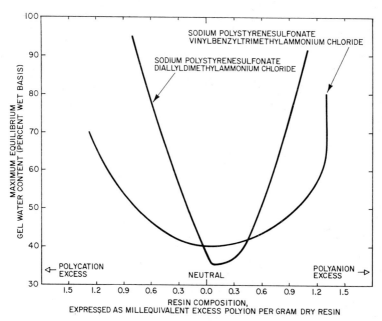

Fig. 4 Effect of composition on equilibrium water content for two polyelectrolyte complex systems.

droplets of water are surrounded by a PEC–water gel phase. When the dispersed phase is in the half-micron size range, the gels become hazy instead of clear. In more minute form, the presence of the second phase can be determined only by complicated and somewhat conjectural procedures.

Many of the properties of hydrogels depend upon their water content. Figure 5 is a composite plot of hydraulic permeability, dialytic permeability for 1000 M_w polysaccharide, oxygen permeability, and conductivity versus equilibrium water content for polyelectrolyte complexes formed from sodium polystyrenesulphonate and polyvinylbenzyltrimethylammonium chloride.

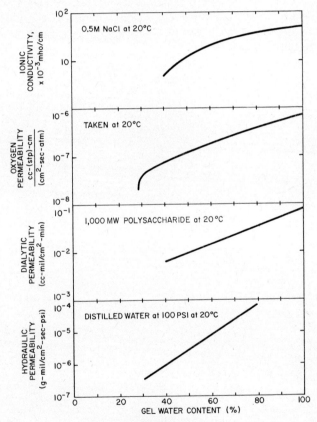

Fig. 5 Properties of polyelectrolyte complex hydrogels as a function of gel water content system: sodium polystyrenesulphonate; polyvinylbenzyltrimethylammonium chloride.

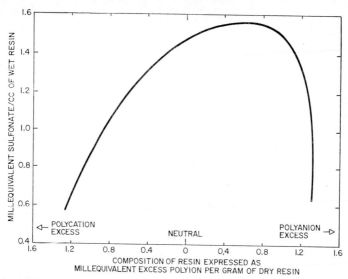

Fig. 6 Effect of composition on sulphonate ion density of polyelectrolyte complex hydrogel. System: Sodium polystyrenesulphonate/polyvinylbenzyltrimethylammonium chloride.

Because the water content of equilibrium hydrogels increases with the extent of non-stoichiometry, density of a particular ionic group in the hydrogel may be decreasing even as density in the corresponding dry resin is increasing. This effect is illustrated in Fig. 6 for the sulphonate ion content of an equilibrated hydrogel.

Polyelectrolyte complexes are stable in acids, weak bases, organic solvents. They are compatible with olefins, nylons, vinyls and most other thermoplastics. Of special interest is the fact that polyion complexes have sufficient chemical stability to be immersed in or even gelled with 35% sulphuric or phosphoric acid up to 200°F. Contact or gelation with 50% or 80% acid is possible indefinitely at room temperature. Such chemical stability is very impressive for any materials and truly unique for a polymeric gellant. Conventional gel-forming polymers work by virtue of their polyhydroxyl functionality (—OH, C—O—C, —C=O) and are thus inherently unstable at low pH.

6.5 APPLICATIONS

Although the literature contains many suggestions for potential applications of polyelectrolyte complexes, successful commercialisation has been quite limited to date. This is partly because the most immediate applications for polyelectrolyte complexes involve their utilisation as hydrogels, and products based on such materials ('soft' contact lenses, non-fog sports goggles, and the like) have just begun to emerge into the marketplace in the seventies. A second factor is the high cost of raw materials and processing. Finally, and as has been abundantly illustrated in the fields of plastics and other polymers, the transition of novel technology from a laboratory curiosity to a commodity product is always arduous and time-consuming.

Since 1965, Amicon Corporation, Lexington, Massachusetts, has been marketing a grade of Diaflo® ultrafiltration membranes based upon polyelectrolyte complex resins.[39] These membranes are 'anisotropic' filters with an extremely fine pore structure and are capable of passing water, salts, and low molecular weight sugars while holding back proteins, polysaccharides, and other macromolecules. The molecular weight cutoff ranges from 500 daltons for the tightest polyelectrolyte membranes up to 10 000 daltons for the loosest. Figure 7 is a plot of solute molecular weight versus percentage rejection for the polyelectrolyte membrane series.

In operation, a membrane disc is placed in a pressurised cell which contains a magnetic stirring bar (*see* Fig. 8). The solution to be ultrafiltered is placed in a chamber upstream of the membrane and a pressure up to 50 psi is applied. The purified filtrate, containing nothing above the specified

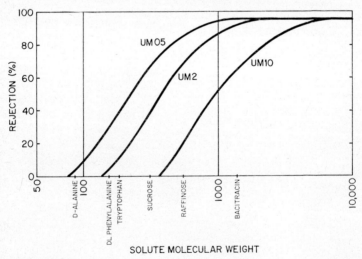

Fig. 7   Percentage rejection as function of solute molecular weight for Diaflo® ultrafiltration membranes fabricated from polyelectrolyte complex resins.

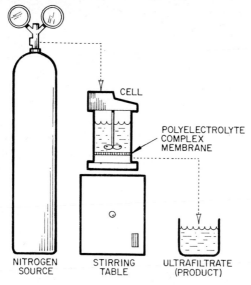

Fig. 8   Equipment for use of ultrafiltration membrane.

membrane cutoff, passes through the membrane and is collected in a beaker, graduate cylinder, or the like. Even at very modest pressures (10 to 50 psi) ultrafiltrate is produced at a surprisingly high rate (a disc with a 2 in diameter is capable of ultrafiltering upwards of 50 ml per hour). Polyelectrolyte complex membranes have found wide acceptance by biochemical investigators who are interested in desalting and concentrating proteins, enzymes, or other delicate mixtures.[5] Membrane ultrafiltration has replaced, in many instances, the use of sausage-casing dialysis or complex chemical fractionations. This application enjoys substantial and growing utilisation on a worldwide basis.

Polyelectrolyte complexes have also been evaluated for other membrane applications, in particular for desalination[12] and dialysis.[26] Reverse osmosis grade membranes contain no pores at all and separate similarly sized species (*e.g.* water and salt) on the basis of chemical solubility. Reverse osmosis membranes prepared from polyelectrolyte complex resins were found to display transmembrane fluxes about three orders of magnitude higher than those made from cellulose acetate.[12] However, cellulose acetate membranes had a markedly enhanced sieving coefficient for sodium chloride and were still preferred for desalination purposes. Suitably prepared polyelectrolyte complex reverse osmosis membranes were observed to exhibit a striking rejection capacity for sulphonate, calcium, and other divalent ions and have been evaluated under contract to the United States Office of Saline Water as a pretreatment filter in the membrane desalination.[41,43] The purpose of the pretreatment was to prevent any build-up of a fouling layer of multivalent ions and their salts on the surface of the cellulose acetate membrane.

Dialysis membranes were offered as a commercial product and enjoyed some popularity because, unlike cellophane and other commercially available materials, they could be made available in a variety of water contents and surface charges. The product has been discontinued because of problems with mechanical strength.

Polyelectrolyte complex hydrogels have demonstrable compatibility with sensitive areas of bodily tissue and have been evaluated successfully both as corneal and external 'soft' contact lenses.[6] The corneal compatibility has suggested an evaluation as antithrombogenic materials (*i.e.* surfaces which do not cause clotting in contact with blood *in vivo*), and two US patents have been issued on the blood compatible formulations of polyelectrolyte complex resins.[3,4] Cross and others have reported the results of a series of tests in which small grafts—9 mm × 8 mm I.D.— were inserted into the vena cava of dogs and after two hours and again after two weeks examined for clots.[2,4,36,44] The results of these so-called Gott Ring Tests are summarised in Fig. 9. It can be seen that the moderately anionic form of the resins are virtually non-clotting. Such results are highly significant in as much as standard plastics, the olefins, teflon, and the like, clot up immediately and even heparinised surfaces fail within two weeks. It is interesting to note that the most thrombo-resistant polyelectrolyte resins were the moderately anionic versions which also have the highest sulphonate ion density (refer back to Fig. 6). This has led to the intriguing hypothesis that polyion surfaces are mimicking synthetic heparin which also contains a high sulphonate ion density and may actually be similar to the surfaces of living biogels.

In a related area, Friedman has published results of studies using polyelectrolyte complex resins as percutaneous tissue adhesives.[14] Here, the *in-situ* polymerisation method is employed and highly purified pre-resin is applied to a wound or surgical incision along with a catalyst and allowed to harden in place.

It has been noted that polyelectrolyte complex resins form gels with strong acids just as well as with water.[22] Gels containing 40 to 60% phosphoric acid and sulphuric acid were evaluated by the United States Army as matrixes for hydrogen/oxygen fuel cells and were found to provide

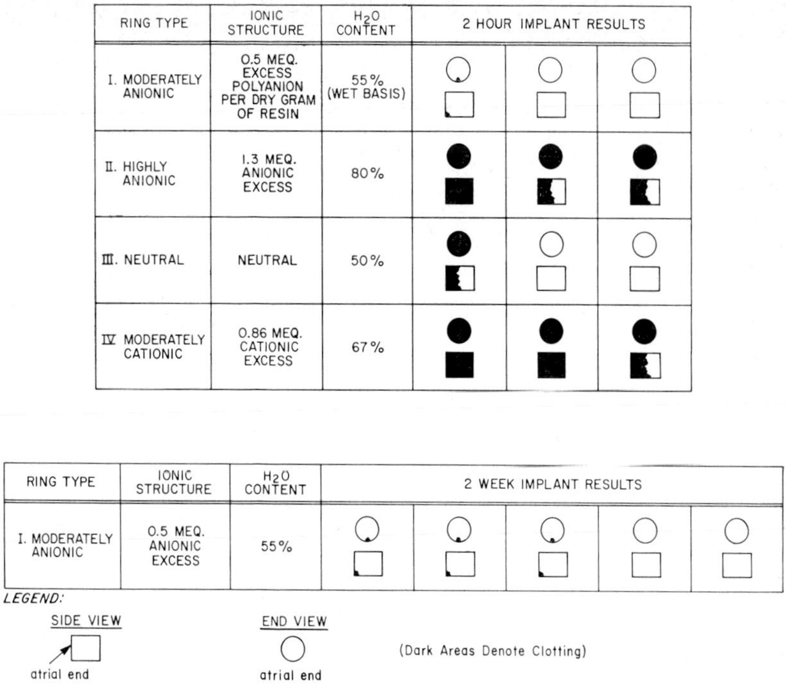

RING TYPE	IONIC STRUCTURE	H₂O CONTENT	2 HOUR IMPLANT RESULTS		
I. MODERATELY ANIONIC	0.5 MEQ. EXCESS POLYANION PER DRY GRAM OF RESIN	55% (WET BASIS)	◔ ▫	○ ▫	○ ▫
II. HIGHLY ANIONIC	1.3 MEQ. ANIONIC EXCESS	80%	● ◪	● ◪	● ◪
III. NEUTRAL	NEUTRAL	50%	● ◩	○ ▫	○ ▫
IV. MODERATELY CATIONIC	0.86 MEQ. CATIONIC EXCESS	67%	● ■	● ■	● ◪

RING TYPE	IONIC STRUCTURE	H₂O CONTENT	2 WEEK IMPLANT RESULTS				
I. MODERATELY ANIONIC	0.5 MEQ. ANIONIC EXCESS	55%	◔ ▫	◔ ▫	◔ ▫	○ ▫	○ ▫

LEGEND:
SIDE VIEW — atrial end
END VIEW — atrial end
(Dark Areas Denote Clotting)

Fig. 9 Results of Gott Ring tests with polyion hydrogels. (Reprinted from Ref. 44 with permission of the publisher.)

low electrical resistivity while forming excellent barriers to gas blowby.[7] The gels were not stable in alkaline solutions; and as fuel cell technology moved toward high temperature alkaline fuel cells, further work in this area was discontinued.

Patents have been issued to Douglas of Gould, Inc., which teach the use of high acid content gels, first as acid 'sponges' and subsequently as separators in a water-activated, dry-charged automotive battery.[1,9,10] In essence, the concentrated sulphuric acid which is stored in the gels during the shelf life of the battery is leached out upon activation through the simple addition of water. The leaching process reduces the concentration of the electrolyte to 35% (specific gravity 1·280) which is appropriate for battery operation. The polyelectrolyte complex resin stays behind in sheet form to serve the role conventionally played by a resin-bonded paper separator during the conventional life of the battery.[27,28] It is believed that further commercial exploitation of this scheme awaits development of lower cost polyelectrolyte complex resins.

In another application, Michaels[29] has developed 'dry hydrogels' in which water is removed from the gel by rapid desiccation—so rapid that the gel network cannot collapse. The highly open, macroreticulated structure is then finely ground and dispersed into a vinyl plastic—markedly increasing its breathability and moisture permeability.

Taylor of Polaroid Corporation, Cambridge, Massachusetts, has received a patent for the use of the *in-situ* polymerisation method described earlier to form photoresistant sheets from polyelectrolyte complex resins.[42] He proposes applying a liquid, water-soluble mixture of a strong polyelectrolyte and a photo-catalysable monomer of oppositely charged polyelectrolyte on to a surface and subsequently exposing a portion of the mixture to visible ultraviolet light thereby catalysing local polymerisation. In this manner an insoluble polymer will be formed in those regions which have been exposed whilst the unreacted materials may be simply washed away with water.

In an exciting new area, Wallis and Melnick[45] have recently prepared products similar to polyelectrolyte complexes by reacting viruses with polycations. The amphoteric reaction products are subsequently injected into living animal bodies to engender antibody response and eventual immunity against the viral strain.

Polyelectrolytes have been considered for innumerable other purposes, including the following intriguing possibilities where no application data is available:

(a) A transparent, non-fogging coating for automobile windows.
(b) An electroconductive, transparent inner liner capable of heating glass laminates.
(c) A dielectric sensor capable of monitoring temperature, relative humidity, etc.
(d) An extremely lossy dielectric filler for dielectric heated plastics.
(e) A temporary 'artificial skin' for severe burn victims.
(f) A gellant for the free acid in H_2SO_4, H_3PO_4 capacitors.
(g) A slow-release matrix for implantable drugs, suppositories, etc.

Interestingly, none of the application areas thus far considered capitalise on the specific nature of the polyanion/polycation reaction or the demonstrated ability—in pure, dilute solutions—of the substituent ionic groups to seek out and react with oppositely charged moieties on a stoichiometric basis. The governing principles of these phenomena must closely resemble those currently being elucidated for immune and genetic reactions in molecular biology, and the next phase in polyelectrolyte complex research will almost certainly involve studies with polyions substituted with mixed ionic species (possibly even biologically significant groups) spaced at regular repeating intervals along the backbone chain. The prospects of highly specific reactions, with important relevance to immunochemistry or chemical information storage, are not too remote. Clearly, the fullest potential of this field of technology has yet to be exploited.

REFERENCES

1. Biddick, R. E., Douglas, D. L. and Ockerman, J. B. *Water-activated dry-charged lead–acid batteries*, Sixth International Power Source Symposium held at Brighton, Sussex, (23–25 September 1968), pp. 85–92. In preprint—available from Pergamon Press, Ltd.
2. Bixler, H. J., Markley, L. M. and Cross, R. A. (1968). Utilisation of polyelectrolyte complexes in biology and medicine. *J. Biomed. Res.*, **2**, pp. 145–155.
3. Bixler, *et al.* US Patent 3,475,358. 'Antithrombogenic material,' to Amicon Corp. (28 November 1969).
4. Bixler, *et al.* US Patent 3,514,438. 'Antithrombogenic materials,' to Amicon Corp. (6 June 1969).
5. Blatt, W. F., *et al.* (October 8, 1965). Protein solutions: concentration by a rapid method, *Science*, **150**, No. 3693, pp. 224–226.
6. Cross, Robert A. and Michaels, Alan S. Structure, properties, and biocompatibility of polyelectrolyte complexes. Presented at the International Congress of Pure and Applied Chemistry, Boston, Massachusetts (July 1971).
7. Dankese, J. P. and Massucco, A. A. 'Evaluation of polyelectrolyte complex as a fuel cell electrolyte, Final Technical Report from Contract DA-44-009-AMC-1580 (T) under ERDL at Ft. Belvoir, Virginia (1 February 1967).
8. Doty, P. and Ehrlich, G. (1952). Polymeric electrolytes, *Ann. Rev. Phys. Chem.*, **3**, 108.
9. Douglas, D. L., *et al.* US Patent 3,556,850. Lead–acid, polyelectrolyte complex containing storage battery and method of storing and handling same, to Gould, Inc. (19 January 1971).
10. Douglas, *et al.* US Patent 3,556,851. 'Water activatable dry-charged lead–acid storage cells,' to Gould, Inc. (19 January 1971).
11. Falkenstein, G. 'The dielectric properties of polysalt films.' M.I.T. Thesis, (August 1963).
12. Fleming, S. 'The reverse osmosis performance of a polyelectrolyte complex membrane.' M.I.T. Thesis (September 1970).
13. Fouss, R. M. and Sadek, H. (1949). Mutual interaction of polyelectrolytes. *Science*, **110**, 552–554.
14. Friedman, E., *et al.* 'Biologically compatible polyelectrolyte complex hydrogels for use as tissue/lead adhesives for long-term percutaneous leads.' Proceedings—Artificial Heart Program Conference, Washington, D.C. (9–13 June 1969).
15. Gray, C. 'The mechanical properties of polyelectrolyte complexes.' M.I.T. Thesis (September 1965).

16. Green, B. K., et al. U.S. Patent 2,800,457 to National Cash Register (23 July 1957).
17. Jackson, H. L. US Patent 2,832,746. 'Water-insoluble homogeneous polymer blends of a water-insoluble, water-dispersible, non-electrolyte film-forming polymer with water-soluble linear organic polymers united by ionic cross-linkages, and their preparation, and fabrics coated therewith,' to DuPont (18 February 1953).
18. Jackson, H. L. US Patent 2,832,747. 'Polymeric compositions comprising linear polymers united through ionic cross-linkages and through covalent cross-linkages, process for their preparation, and fabrics coated therewith,' to DuPont (18 February 1953).
19. Kossell, A., (1896). *J. Physiol. Chem.*, **22**.
20. Kruyt, H. R. (1962). Ed. *Colloid Sciences II* Chapter 10, Elsevier, Amsterdam.
21. Lewis, R. W. 'Mechanical and swelling properties of polyanion–polycation complexes.' M.I.T. Thesis, Department of Chemical Engineering (1970).
22. Lysaght, M. J., (September 1970). Electrochemical properties of polyelectrolyte complex resins. *ACS Reprints* (*Division of Organic Coatings and Plastics Chemistry*), **30**, No. 2.
23. Michaels, A. S. (October 1965). Polyelectrolyte complexes. *Ind. Eng. Chem.*, 32–40.
24. Michaels, A. S., Falkenstein, G. L. and Schneider, N. S. (May 1965). Dielectric properties of polyanion–polycation complexes. *J. Phys. Chem.*, **69**, No. 5, 1456–1465.
25. Michaels, A. S. and Miekka, R. G. (1961). Polycation–Polyanion complexes: preparation and properties of poly(vinylbenzyltrimethylammonium)–poly(styrene sulphonate). *J. Phys. Chem.*, **65**, 1765.
26. Michaels, A. S. et al., US Patent 3,276,598. (Polyelectrolyte films, to Dow Chemical (October 4, 1966).
27. Michaels, A. S. US Patent 3,419,430. 'Electrical energy device containing polyelectrolyte gel separator,' to Amicon Corp. (31 December 1968).
28. Michaels, A. S. US Patent 3,419,431. 'Polyelectrolyte gel separator and battery therewith,' to Amicon Corp. (31 December 1968.)
29. Michaels, A. S., US Patent 3,467,604. 'Moisture permeable polyion complex resinous composites,' to Amicon Corp. (19 September 1969).
30. Michaels, A. S. et al., US Patent 3,546,142. 'Polyelectrolyte structures,' to Amicon Corp. (8 December 1970).
31. Michaels, Alan S. et al., US Patent 3,558,744. 'Process for making polyelectrolyte complex resin,' to Amicon Corp. (26 January 1971).
32. Michaels, Alan S., US Patent 3,565,973. 'Purifying cross-linked polyelectrolytes,' to Amicon Corp. (23 February 1971).
33. Miekka, Richard. 'Polycation–polyanion complexes: preparation and properties of poly(vinylbenzyltrimethylammonium)–poly(styrene sulphonate). M.I.T. Thesis, (June 1961).
34. Mir, L. 'Interactions of polyelectrolyte in dilute aqueous solutions.' M.I.T. Thesis (March 1963).
35. Najajima, Akio and Sato, Hiroko. (1969). Phase relationships of polyion complex composed of sodium–polystyrene sulphonate and polyvinyl pyridinium bromide in three-component solvent systems. *Bull. Inst. Chem. Res.*, **47**, No. 3, Kyoto University, 177–183.
36. Nelsen, L. et al. (May 1970). Synthetic thromboresistant surfaces from sulphonated polyelectrolyte complexes. *Surgery*, **67**, No. 5, 826–830.
37. Polyelectrolyte complexes. *Encyclopedia of Polymer Science and Technology* Interscience, New York, Vol. 10, 765–780 (1969).
38. Polyelectrolyte complexes. *Kirk Othmer Encyclopedia of Chemical Technology*, 2nd Ed. Interscience, New York, Vol. 19, 117–133 (1968).

39. Rigopulos, P. N., US Patent 3,549,016. 'Permeable membrane and method of making and using same,' to Amicon Corp. (12 December 1970).
40. Schaper, R. J. et al., US Patent 3,579,613. 'Polysalts containing sulphonated acrylics,' to Calgon (18 May 1971).
41. Strathmann, H. and Cross, R. A. 'Evaluation of anisotropic, ionogenic membranes (Diaflo UM-Series) for selective removal of divalent cations and anions from brackish water by reverse osmosis.' Final Report submitted to Office of Saline Water, for the period 1 April 1970 to 1 April 1971, Contract No. 14-30-2640.
42. Taylor, Lloyd D. US Patent 3,578,458. 'Ionically cross-linked photopolymerised addition polymers,' to Polaroid Corp. (11 May 1971).
43. Testa, A. J., Cross, R. A. and Strathmann, H. 'Ultrafiltration membranes for prefiltration of brackish feed waters.' Presented at the Third OSW Conference on Reverse Osmosis, Palm Springs, California (January 1972).
44. Vogel, Mary K., Cross, Robert A., Bixler, Harris J. and Guzman, Ruben J. (May 1970). Medical uses for polyelectrolyte complexes. *J. Macromolec. Sci.— Chemistry*, **A4** (3), 675–692.
45. Wallis, Craig and Melnick, Joseph, L. US Patent 3,651,213. 'Method for the immunisation of a living animal body against viral disease,' to Monsanto (21 March 1972).
46. Wu, S. H. 'Viscoelastic properties of new polyelectrolyte complexes.' M.I.T. Thesis, Department of Chemical Engineering (1969).

CHAPTER 7

CRYSTALLINE SILICATES AND PHOSPHATES AS IONIC POLYMERS

J. H. ELLIOTT

7.1 INTRODUCTION

Several excellent general reviews[1,2,3] have been published during the past two decades on silicate and phosphate science and there have been many more which deal with the more specialised geological,[4,5] mineralogical,[6,7,8] physicochemical[9,10,11] and technological[12,13,14] aspects of both the naturally occurring and the synthetic crystalline materials.

A growing insight into the structure[15,16] of the naturally occurring metallic silicates and of certain synthetically prepared phosphates has revealed the presence of polymeric rings, chains, sheets and three-dimensional networks and therefore these crystalline materials may be properly regarded as inorganic polymers having certain features in common with the organic polymers as well as with the inorganic glasses dealt with elsewhere in this book.

Although many of the known crystalline silicates and phosphates of the metals may be regarded as polymers, not all of them are ionic polymers; some contain monomeric anions whilst others consist of completely covalent networks. Between these two extremes there exists a range of polymers which contain both covalent and ionic bonds and which therefore are ionic polymers as defined earlier in this book.

The sections in this chapter deal only with the ionic crystalline silicate and phosphate polymers and cover only those aspects which are of particular interest from a materials point of view. Fundamental to an appreciation of the contents of these sections is an understanding of the way in which interatomic bonding leads to both covalent and ionic bond formation; this is therefore dealt with first.

7.2 INTERATOMIC BONDING IN IONIC SILICATES AND PHOSPHATES

The polymeric chains which constitute the backbones of the ionic silicate and phosphate polymers are formed by the sharing of common oxygen atoms between adjacent SiO_4 and PO_4 groups respectively. In order to understand how such units arise and why they should link together in this way it is necessary to refer back to the electronic structures of the atoms concerned.

7.2.1 Electronic structure

The ground state electronic configurations for the Si, P and O atoms are given in Table 1. Silicon has four electrons in its outermost quantum shell and oxygen has six. These outer quantum shells are depicted by a Kossel–Lewis diagram in Fig. 1 (a) to (c). On the basis of the 'octet rule'[17] silicon requires an extra four electrons to complete its octet and thus to reach electronic stability. In the silicates, the silicon atom achieves this stability by sharing its four electrons (covalent bond formation) with one of the six electrons on each of four oxygen atoms as depicted in Fig. 1 (d). Such an arrangement does not, however, confer electronic stability on the SiO_4 group as a whole because each of the four oxygen atoms linked to the central silicon atom has only seven electrons in its outer shell and still needs one more electron to complete its own octet. This additional electron may be acquired in one of two ways: (a) by sharing its seventh electron with an electron from some other atom or group or (b) by accepting an electron from some other atom or group. The former of these two processes leads

TABLE 1
ELECTRONIC CONFIGURATIONS AND ELECTRONEGATIVITIES

Element	Atomic number	Ground state electronic configuration	Electronegativity (after Pauling[18])
Silicon	14	$1s^2 2s^2 2p^6 3s^2 3p^2$	1·8
Phosphorus	15	$1s^2 2s^2 2p^6 3s^2 3p^3$	2·1
Oxygen	8	$1s^2 2s^2 2p^4$	3·5

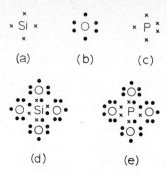

Fig. 1 Electrons in outer shells of free and combined atoms: (a) free silicon atom; (b) free oxygen atom; (c) free phosphorus atom; (d) neutral SiO_4 group; (e) neutral PO_4 group.

to covalent bond formation, the latter to ionic bond formation. This dichotomy of behaviour is the key to all silicate structures and is fundamental to the formation of ionic polymers not only in silicate systems but also in phosphate systems as well.

One important difference between the bonding in silicates and phosphates should however be noted. In the silicates, all four oxygen atoms of the SiO_4 group are capable of forming covalent bonds or ionic bonds or a mixture of covalent and ionic bonds; in phosphates, however, only three of the four oxygen atoms present have this choice. The reason for this may be seen by reference to Fig. 1 (c) and (e); the free phosphorus atom has five electrons in its outermost shell and three of these form covalent bonds with oxygen by electron sharing in exactly the same way that silicon forms covalent bonds with the same element. Once these three covalent bonds have been formed, however, the phosphorus atom is surrounded by a stable outer shell of eight electrons and can only enter into further chemical combination by donating its remaining electron pair to a fourth oxygen atom, with the formation of a co-ordinate bond. The important point to note here is that by accepting this electron pair from phosphorus, the fourth oxygen atom acquires the two electrons it needs to form its own stable octet of outer electrons. The fourth oxygen atom in the PO_4 group is therefore quite different from the other three because unlike them (and unlike the oxygen atoms in the SiO_4 group) it is unreactive and unable to take part in further covalent or ionic bond formation. It is often convenient to distinguish the co-ordinate bond by means of an arrow as shown in Fig. 3. (b)

It is worth noting at this point that although the bonds between silicon and oxygen and between phosphorus and oxygen in the SiO_4 and PO_4

groups respectively have been referred to as covalent, both bonds do in fact have some ionic character. The percentage of ionic character in the bonds may be calculated from Pauling's electronegativity values (Table 1) using the formula:[18]

$$\text{Amount of ionic character} = 1 - \exp - (\tfrac{1}{4}(X_A - X_B)^2)$$

This gives values of 53·5% ionic character for the Si—O bond and 39% ionic character for the P—O bond. This means therefore that the bonding in the PO_4 group has a lot of ionic character, and in silicates the bond is approximately equally ionic and covalent. In view of the intermediate nature of the Si—O bond the stereochemistry of the SiO_4 group may be considered using either a completely ionic or a completely covalent model as an approximation. The ionic model based on radius ratio considerations [19] leads to a tetrahedral structure for the SiO_4 group and a similar conclusion is reached if a covalent model is used as the starting point. We shall now consider this latter approach in more detail since it is applicable not only to silicates but to phosphates as well.

7.2.2 Tetrahedral symmetry of the SiO_4 and PO_4 groups

The orbital representation of the outer electronic structures of the silicon, phosphorus and oxygen atoms is shown in Fig. 2. In the ground state, the silicon atom has only two unpaired electrons in its outer quantum shell but in the excited state there are four unpaired electrons in the 3s and 3p orbitals which can pair with similar electrons in other atoms. In the silicates, one unpaired electron from each of four oxygen atoms couples with the four unpaired electrons on silicon to form four σ-bonds. The energy required for the promotion of the electron from the 3s orbital of the ground state to the 3p orbital of the excited state, which makes the formation of four σ-bonds possible, comes from hybridisation; since the four sp³ hybrid bonds formed have tetrahedral symmetry, the SiO_4 group itself has a tetrahedral configuration. However, each oxygen atom has still one unpaired electron remaining, which it can use for either covalent or ionic bond formation.

In a similar way, phosphorus in its excited state is capable of forming four σ-bonds with oxygen and since they are once again sp³ hybrid bonds, the PO_4 group will have a tetrahedral configuration. It should be noted, however, that the excitation in this case has led to one electron in a 3d orbital which does not take part in the hybridisation and therefore does not contribute to the symmetry of the arrangement. This electron can form a

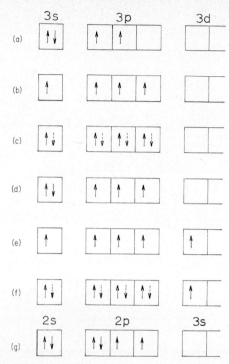

Fig. 2   Electronic structures. (↑ and ↓ *denote electrons of opposite spin*).
  (a) *Silicon atom in ground state.*
  (b) *Silicon atom in excited state.*
  (c) *Bond formation of silicon with oxygen; 4 sp^3 hybrid bonds.*
  (d) *Phosphorus in ground state.*
  (e) *Phosphorus in excited state.*
  (f) *Bond formation of phosphorus with oxygen; 4 sp^3 hybrid bonds.*
  (g) *Oxygen atom in ground state.*

π-bond by coupling with an unpaired electron on one of the four oxygen atoms to give the arrangement shown in Fig. 3 (a). Since this oxygen atom has now 'used up' its valency electrons, one in σ-bond formation and one in π-bond formation with phosphorus, it is incapable of forming a normal covalent or ionic bond with other atoms or groups. In this respect it differs from the other three oxygen atoms of the PO_4 group. For comparison Fig. 3 (b) shows the alternative co-ordinate bond representation based on the simpler Kossel–Lewis model. For most purposes this simpler model is quite satisfactory for an understanding of the principles of ionic polymer

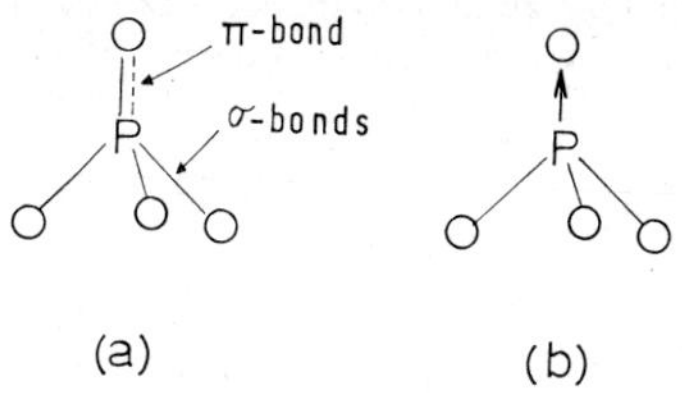

Fig. 3 Tetrahedral PO_4 group: (a) localised π-bond representation; (b) co-ordinate bond representation.

formation from the basic tetrahedral units and will therefore be used in the remainder of this section.

7.2.3 Silicate polymers

As mentioned earlier, there are two ways in which each oxygen atom of the SiO_4 tetrahedral group having seven electrons in its valency shell can achieve a stable electronic configuration: (a) by covalent bond formation or (b) by ionic bond formation. Various possibilities arise according to the number of covalent and ionic bonds formed by the four oxygen atoms present. These are set out in Table 2. Clearly, covalent bond formation could take place with any suitable atom or group of atoms, but in the silicate polymers the mutual electronic requirements of the oxygen atoms in the tetrahedra are met by one oxygen atom of each SiO_4 group sharing its seventh electron with an unpaired electron on the silicon atom of a neighbouring pseudo-tetrahedral SiO_3 group. In this way both of the silicon atoms are tetrahedrally co-ordinated by oxygen and the central (shared)

TABLE 2

RELATIONSHIP BETWEEN COVALENT/IONIC BOND FORMATION AND CONNECTIVITY IN POLYMERIC SILICATES

Number of bonds between tetrahedra		Example of polymeric unit	Connectivity
Covalent bonds	Ionic bonds		
0	4	$(SiO_4)^{4-}$	0
1	3	$(Si_2O_7)^{6-}$	1
2	2	$(SiO_3)_n^{2n-}$	2
3	1	$(Si_2O_5)_n^{2n-}$	3
4	0	$(SiO_2)_n$	4

oxygen atom achieves electronic stability; the first stage of polymer formation has taken place (Fig. 4). This, however, still leaves three electronically unstable oxygen atoms attached to each silicon atom and each of these has the twofold choice of (a) covalent or (b) ionic bond formation. If all three oxygen atoms on each tetrahedral unit go ionic then we are left with a pyrosilicate ion $(Si_2O_7)^{6-}$ but if one of the three oxygen atoms forms a covalent bond with the silicon atom of a further pseudo-tetrahedral SiO_3 group, then a trimer is formed. It will be clear that if the oxygen atoms of the adjoining groups and of those subsequently added on follow a similar course then a polymeric chain of composition:

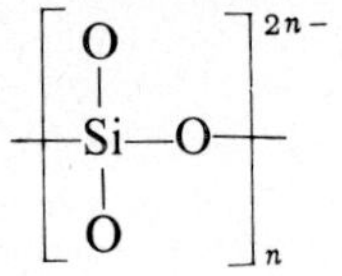

$$\left[\begin{array}{c} O \\ | \\ -Si-O- \\ | \\ O \end{array} \right]_n^{2n-}$$

will result if the two remaining oxygen atoms attached to each silicon atom form ionic bonds with neighbouring cations.

If, on the other hand, one of the two remaining oxygen atoms forms a covalent bond with the silicon atom of an adjacent pseudo-tetrahedral SiO_3 group rather than an ionic bond with a cation, then a sheet structure of composition $(Si_2O_5)_n^{2n-}$ is formed provided the fourth oxygen atom of every tetrahedral unit in the chain forms an ionic bond with neighbouring cations.

Finally, if all four oxygen atoms of the SiO_4 groups form covalent bonds, then we have a network structure of composition $(SiO_2)_n$ which is neutral because none of the oxygen atoms has taken part in ionic bond formation. We see, therefore, that polymer formation in silicates depends on the formation of covalent bonds, the degree of polymerisation increasing with the number of covalent bonds, that is with the connectivity[20] of the structure.

Of course not all polymeric silicates have simple connectivity numbers of 0, 1, 2, 3 and 4 since it is possible to have some structures in which one SiO_4 group is covalently bound through one oxygen atom (1-connective)

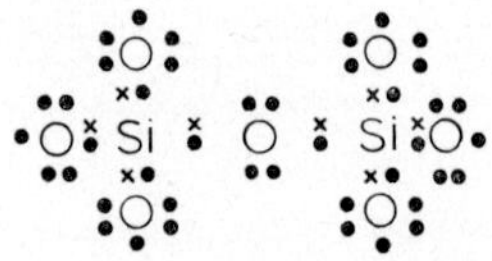

Fig. 4 Linking together of SiO_4 tetrahedra by covalent bond formation.

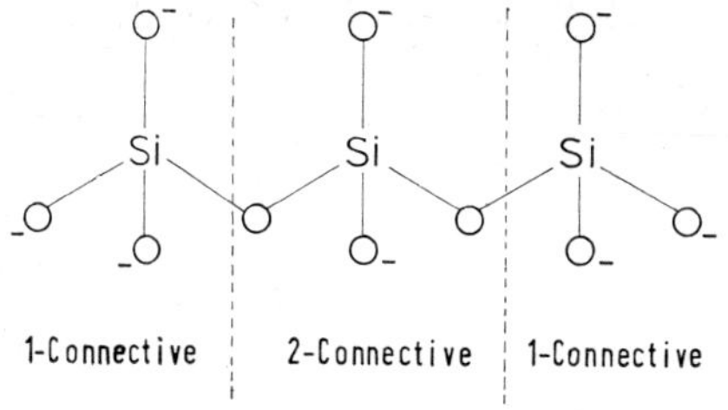

Fig. 5   Linear silicate polymer showing 1- and 2-connective groups.

and a neighbouring group is covalently bound through two oxygen atoms (2-connective). This occurs for example in linear silicates, where the end groups are 1-connective and the middle groups are 2-connective (Fig. 5). Similarly, there are structures in which some tetrahedra are 2-connective and others 3-connective, as in the double chain silicates of formula $[Si_4O_{11}]_n^{6n-}$ (Fig. 6).

All of the structures described above are formed by the linking together of silicon–oxygen tetrahedra and it will be clear that a wide variety of structures is possible; in the case of metallic phosphates the number of possible structures is somewhat more restricted for reasons which are set out below.

7.2.4 Phosphate polymers

It has already been shown that the fourth oxygen atom in the tetrahedral PO_4 group differs from the other three in that it already possesses a stable

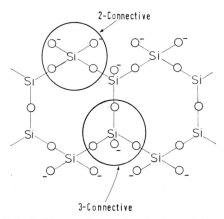

Fig. 6   Cross-linked silicate polymer showing 2- and 3-connective groups.

TABLE 3

RELATIONSHIP BETWEEN COVALENT/IONIC BOND FORMATION AND CONNECTIVITY IN POLYMERIC PHOSPHATES

Number of bonds between tetrahedra		Example of polymeric unit	Connectivity
Covalent bonds	Ionic bonds		
0	3	$(PO_4)^{3-}$	0
1	2	$(P_2O_7)^{4-}$	1
2	1	$(PO_3)_n^{n-}$	2
3	0	$(P_2O_5)_n$	3

octet of electrons and therefore does not take part in covalent or ionic bond formation of the type described above for the oxygen atoms of SiO_4. The other three oxygen atoms of the PO_4 group, like those of the SiO_4 group, have only seven electrons in their outer shells and are therefore capable of forming (a) covalent or (b) ionic bonds with other atoms or groups of atoms in just the same way as the oxygen atoms of the SiO_4 group. The connectivities in the case of the phosphates range from zero to a maximum of three instead of from zero to four, as occurred in the silicates. Some examples are given in Table 3. As with silicates, structures of mixed connectivity are possible; for example connectivities of 1 and 2 occur in the linear phosphate polymers and connectivities of 2 and 3 in cross-linked polymer chains (ultraphosphates) as illustrated in Fig. 7.

7.2.5 Copolymers

Although 4-connective homopolymer networks cannot be obtained by the linking up of PO_4 tetrahedra alone, 4-connective heteropolymer networks may be obtained if the phosphate group is copolymerised with pseudo-tetrahedral BO_3 or AlO_3 groups in a 1:1 ratio. In such cases, the labile oxygen atoms of the PO_4 tetrahedra form covalent links with neighbouring BO_3 or AlO_3 groups in the usual way and the fourth 'inert' oxygen atom, acting as a Lewis base (electron pair donor) combines with a fourth BO_3 or AlO_3 group acting as a Lewis acid (electron pair acceptor) to form a 4-connective framework (Fig. 8). The resulting network structures of boron phosphate $(BPO_4)_n$ and aluminium phosphate $(AlPO_4)_n$ are isostructural with $(SiO_2)_n$; aluminium phosphate in fact exists in seven crystalline modifications, whose conversions and inversions correspond closely with those

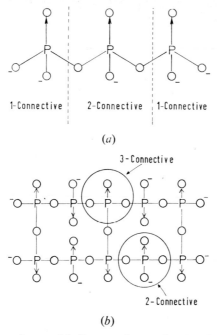

Fig. 7 Phosphate polymers (a) linear polymer showing 1- and 2-connective groups; (b) cross-linked polymer showing 2- and 3-connective groups.

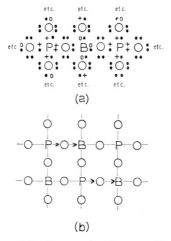

Fig. 8 Boron phosphate (a) electron distribution; (b) 2-D representation of 4-connective network.

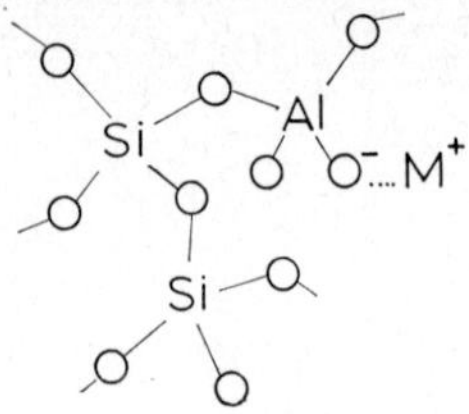

Fig. 9 Introduction of ionic bonds into a covalent silica network as a result of isomorphous substitution of aluminium for silicon.

of the seven crystalline modifications of silica.[21] The parallelism is particularly interesting in that the respective means of the radii and charges of the (hypothetical) Al^{3+} and P^{5+} ions are almost identical with the radius and charge of the (hypothetical) Si^{4+} ion. It is therefore possible to envisage the formation of $(AlPO_4)_n$ for example, as the result of the complete isomorphous substitution of the silicon atoms of silica by alternate phosphorus and aluminium atoms.

Such isomorphous substitution of aluminium (and more rarely boron) for silicon is very common in silicate structures; in some cases the extent of substitution is quite small but in others it may be as high as 50%. An example of a 1:4 copolymer is shown in Fig. 9; here one-quarter of the silicon atoms in a $(SiO_2)_n$ network are replaced by aluminium and this leads to a network of overall composition $(AlSi_3O_8)_n^{n-}$, the negative charge arising because the aluminium ion has a formal charge of only $3+$ whereas the silicon ion which it replaces in the neutral silica structure has a formal charge of $4+$. The negative charge is balanced in these copolymers by the presence of positively charged ions (cations) and since the centres of negative charge are the oxygen atoms of the lattice, bonds of the type $Al-O^-...M^+$ occur as in other ionic silicate polymers.

7.3 SOURCES AND SYNTHESIS OF CRYSTALLINE IONIC SILICATES AND PHOSPHATES

7.3.1 Naturally occurring silicate polymers

Many of the copolymers of silicon and aluminium mentioned above occur widely distributed in nature as do the non-substituted polymeric silicates themselves. Figure 10 gives some idea of the distribution of the polymeric silicates in igneous rocks of the earth's crust. The diagram is divided into seven areas representing the seven principal mineral groups in the earth's

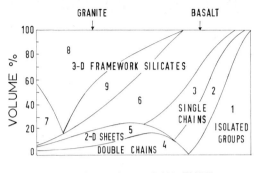

Fig. 10   Distribution of polymeric silicates in nature.

crust and these may be further grouped into the five main structural types:

Group 1. Isolated SiO_4 tetrahedra (area 1); the orthosilicates, *e.g.* olivine.

Group 2. Single chain silicates; the pyroxene minerals represented by orthopyroxenes (area 2) and clinopyroxenes (area 3).

Group 3. Double chain silicates; the amphibole minerals, *e.g.* amosite asbestos (area 4).

Group 4. 2-D Sheet silicates (area 5), *e.g.* biotite, mica.

Group 5. 3-D Framework silicates; this class includes quartz (area 9), the feldspathoids (area 7), the orthoclase feldspars (area 8) and the plagioclase feldspars (area 6).

All of these minerals with the exception of the orthosilicates (group 1) and quartz (group 5) are ionic silicate polymers as previously defined, and their structures and properties will be described later. It will be clear from Fig. 10, however, that of all the ionic silicate polymers in nature the feldspars are the most widespread; the commonest of the igneous rocks, granite and basalt contain about 45 and 50 volume percent of feldspar respectively and other types of igneous rock intermediate between these in composition contain even more.

The feldspars are important, however, not only on account of their incidence in nature in the form of igneous rocks but also because they are the parent minerals from which in the course of time 2-D sheet polymers and other breakdown products characteristic of sedimentary deposits have been formed. One possible breakdown scheme involving acid leaching, desilication and condensation is given below (Fig. 11). The main breakdown products on weathering are the layer lattice minerals (2-D sheet polymers) of the kaolin and montmorillonite groups which occur in clay deposits alongside unchanged feldspar, quartz and mica. The micaceous minerals are closely related to the montmorillonites and are degraded into montmorillonites and subsequently into minerals of the kaolin group by successive weathering.

7.3.2 Synthetic silicate and phosphate polymers

Although naturally occurring mineral deposits are the main commercial source of insoluble ionic silicate polymers, many can also be made on a small scale synthetically. Alkali metal silicates which are soluble and therefore do not exist in nature have to be prepared synthetically, as have all the phosphate polymers, since only the non-polymeric orthophosphates, e.g. apatite ($Ca_{10}F_2(PO_4)_6$) containing isolated PO_4 groups in the structure are stable in a natural environment.

Five general methods have been used for the synthesis of polymeric silicates and phosphates: (a) condensation polymerisation at elevated temperatures; (b) solid state reaction; (c) hydrothermal reaction; (d) crystallisation from the melt and (e) precipitation from aqueous solution. Gimblett[22] has given an excellent review of these methods; here we shall merely summarise the main points.

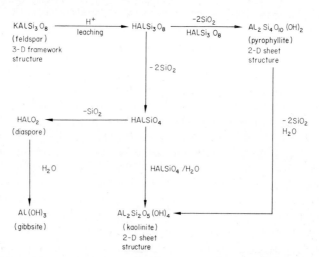

Fig. 11   Sedimentary deposits from the weathering of feldspar.

Condensation polymerisation at elevated temperatures
Although some polymeric silicates can be made by condensation polymerisation, the method is most extensively used for the preparation and manufacture of phosphate polymers. It involves the dehydration of acid salts such as sodium dihydrogen orthophosphate; in the first stage the dimer is formed:

$$\text{HO}-\underset{\underset{\text{ONa}}{|}}{\overset{\overset{\text{O}}{\|}}{\text{P}}}-\text{OH} + \text{H}-\text{O}-\underset{\underset{\text{ONa}}{|}}{\overset{\overset{\text{O}}{\|}}{\text{P}}}-\text{OH} \longrightarrow \text{HO}-\underset{\underset{\text{ONa}}{|}}{\overset{\overset{\text{O}}{\|}}{\text{P}}}-\text{O}-\underset{\underset{\text{ONa}}{|}}{\overset{\overset{\text{O}}{\|}}{\text{P}}}-\text{OH} + \text{H}_2\text{O}$$

monomer   dimer

and successive condensation reactions of the same kind lead to polyphosphates of increasing molecular weight. In practice, the length of the chain is governed by a number of factors including (a) the molar base/acid ratio (B/A); (b) heat treatment, (c) the cation present in the acid salt and (d) the humidity.

All of the polymeric phosphates may be represented stoichiometrically as a combination of an acidic oxide (P_2O_5) with a basic oxide (Na_2O, K_2O, etc.) and the composition then expressed by means of the base/acid (B/A) ratio; in the case of acid salts, water is included along with the basic oxides, e.g. for sodium dihydrogen phosphate ($NaH_2PO_4 = 1Na_2O \cdot 2H_2O \cdot 1P_2O_5$) the B/A value is 3. The relationship between the B/A ratio of the various polymeric phosphates and the structural type is shown in Table 4. It will be

TABLE 4

RELATIONSHIP BETWEEN STRUCTURAL TYPE AND B/A RATIO FOR POLYMERIC PHOSPHATES

B/A Ratio	Anion	Structural type	Common name
3	$(PO_4)^{3-}$	monomer monomer/dimer mixtures	orthophosphate
2	$(P_2O_7)^{4-}$	dimer	pyrophosphate
	$(P_nO_{3n+1})^{(n+2)-}$	chains	polyphosphate
1	$(PO_3)_n^{n-}$	rings or very long chains	metaphosphate
0	$(P_2O_5)_n$	cross-linked chains rings 3-D network	ultraphosphate phosphorus pentoxide

clear that the degree of polymerisation and the structural type of polymeric phosphate obtained depend strongly on the B/A ratio; the reactants used and their proportions must therefore be carefully chosen if the structure of the product is to be controlled. In practice, products with B/A = 1 may be prepared by heating sodium dihydrogen phosphate; those with B/A from 1 to 3 may be prepared by heating mixtures of sodium dihydrogen orthophosphate with sodium carbonate; products for which B/A is between 1 and 0 (the ultraphosphates) are normally obtained by heating $(P_2O_5)_n/(NaPO_3)_n$ or NaH_2PO_4/H_3PO_4 mixtures.

The heat treatment to which the reactants are subjected has a very important influence on the structure of the polymer formed at any given B/A ratio. Figure 12 shows the effect of heat treatment on disodium dihydrogen pyrophosphate (B/A = 2); loss of water as a result of condensa-

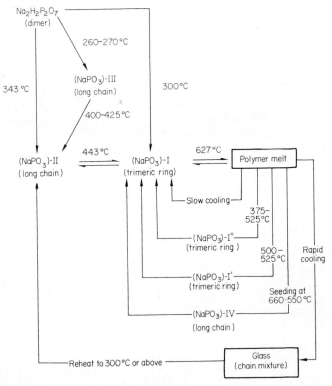

Fig. 12 *Effect of heat-treatment on the formation and structure of polymeric phosphates.*

tion polymerisation causes a decrease in the B/A ratio towards a minimum value of 1 so that the polymers expected on complete condensation will consist of rings and/or chains. However the degree of polymerisation and the structure of the product is largely determined by the temperature level and the nature of the thermal treatment (*e.g.* slow/rapid cooling, seeding, etc.). The structural differences between the trimeric rings in $NaPO_3$-I, -I′ and -I″ and between the long chains in $NaPO_3$-II, -III and -IV are discussed later.

In addition to the effect of B/A ratio and heat treatment, the nature of the product resulting from condensation polymerisation is also influenced by the cations (M^{n+}) present in the starting material. In general, it appears that (except for Li^+ and H^+) only those univalent ions having radii greater than 0·1 nm and those polyvalent ions having radii greater than 0·08 nm give rise to polyphosphates; other univalent ions give essentially trimers $(MPO_3)_3$ and other polyvalent ions tetramers $(MPO_3)_4$ as products of reaction. For example in the series of acid salts of composition MH_2PO_4 where $M = Li^+$, Na^+, K^+ or NH_4^+ the lithium and ammonium salts behave similarly on dehydration (in the absence of excessive water vapour) to give $(LiPO_3)_n \cdot H_2O$ and $((NH_4)PO_3)_n \cdot H_2O$ without difficulty above 247°C and below 200°C respectively. The reaction in fact passes through a number of intermediate stages, but the lithium salt intermediates are unstable above 100°C and therefore break down to give the long chain polymeric product. The intermediates formed in the dehydration of the potassium and sodium salts, however, are much more stable; the trimer $(KPO_3)_3$ is stable up to 245°C, and as may be seen in Fig. 12 the trimeric ring formed by the sodium salt, *i.e.* $(NaPO_3)_3$ is much more stable than the long chain polymer as the temperature nears the melting point.

Further complications arise if excess moisture is present in the atmosphere in which the condensation polymerisation is carried out. Although the precise mechanism involved is not clear, it appears that the presence of water vapour enhances the stability of oligophosphate intermediates and thereby reduces the degree of polymerisation. Whatever the precise mechanism may be, it is quite clear that the humidity of the environment is an important variable which must be taken into account if successful synthesis is to be achieved.

By taking account of the four factors, B/A ratio, heat treatment, nature of the cation and humidity, considerable control can be exercised on the type of structure of the polyphosphate resulting from the condensation polymerisation reaction. Of these four factors the B/A ratio is obviously of fundamental importance.

Solid state reaction

The B/A ratio is also of great importance when considering the synthesis of silicate and phosphate polymers by solid state reaction. The method has not been extensively used for the preparation of polymeric phosphates because the condensation polymerisation method is so suitable for this class of compounds; however, the method is an important one for the synthesis of silicate polymers. Silicate chemists and technologists usually refer to the acid/base (A/B) rather than the base/acid (B/A) ratio [23] but the latter will be used in the present review for ease of comparison with the phosphates. The relationship between the B/A ratio for the various polymeric silicates and the structural type is shown in Table 5. It will be evident that (excluding the oligomers) the polymeric silicates all have a B/A ratio less than unity. The raw material used as the source of SiO_2, which appears as the acid component in the denominator of the B/A ratio (Na_2O + K_2O + $\cdots$ etc./SiO_2) is almost invariably one of the forms of silica and the base may be a metal oxide, but more often than not is the corresponding carbonate or hydroxide particularly when alkali or alkaline earth salts are being prepared. The two components are usually mixed together and heated to temperatures within the range 720°C to 1520°C, when reactions such as the following occur:

(a) *univalent cations*

$$M_2O + SiO_2 \longrightarrow M_2(SiO_3)$$

(b) *divalent cations*

$$MO + SiO_2 \longrightarrow M(SiO_3)$$

Since such reactions depend on solid state diffusion followed by chemical reaction at the reactant/product interface,[24] temperature will be an important variable (volume diffusion usually occurs at temperatures above the Tammann temperature *i.e.* $0.5T_m$, where T_m is the melting point (°K)) as will the structural features of the reactants. Since the mechanism is thought to involve the diffusion of the base into 3-D silica network, the form of silica used must be particularly important. Of the three crystalline polymorphs of silica, quartz is the most densely packed and the least reactive; tridymite and cristobalite have much more open structures and therefore react rather more readily. Vitreous silica glass has also been used as a source of silica in experimental work and amorphous silicas prepared by precipitation from solution may also be used. Irrespective of the form of silica employed, the mechanism of the reaction may be regarded as a

TABLE 5
RELATIONSHIP BETWEEN STRUCTURAL TYPE AND B/A RATIO FOR POLYMERIC SILICATES

B/A ratio	Anion	Structural type	Common name
2	$(SiO_4)^{4-}$	monomer	orthosilicate
1·5	$(Si_2O_7)^{6-}$	monomer/dimer mixtures / dimer	pyrosilicate
1	$(SiO_3)_n^{2n-}$	single chains	metasilicate pyroxene (minerals)
0·75	$(Si_4O_{11})_n^{6n-}$	cross-linked chains	amphibole (minerals)
0·5	$(Si_2O_5)_n^{2n-}$	2-D sheets / cross-linked sheets	layer lattice silicate minerals
0	$(SiO_2)_n$	3-D network	silica

depolymerisation of the 3-D covalent network of silica as a result of the breaking of Si—O—Si bonds:

$$\text{—Si—O—Si—} \xrightarrow{Na_2O} \text{—Si—O}^-Na^+ + Na^+O^-\text{—Si—}$$

If sufficient base is employed, then the depolymerisation may reduce the connectivity progressively through sheet and chain structures to the monomer itself, although the extent to which this occurs depends on the strength of the base employed as well as the B/A ratio.

Hydrothermal methods

In the above method, the acid/base reaction leading to silicate formation made use of dry reactants. Polymeric silicates may also be prepared by an acid/base reaction between silica and a solution of a base in water. This type of reaction occurs only at elevated temperatures (ca. 420°C) and therefore an autoclave suitable for withstanding pressures of 35 MN m^{-2} or more must be used as the reaction vessel. The concentration of base is usually quite small, e.g. 0·025 mole.dm^{-3} and the silica is often used in the form of a glass rod. Reaction occurs at this silica glass surface and if a soluble silicate is formed as the reaction product then the interface remains 'clean' for further reaction; if on the other hand an insoluble silicate is formed as the reaction product, the latter may build up on the surface and thereby decrease the rate of subsequent reaction. In addition to the B/A ratio and the precise nature of the reactants, the pH of the solution is an important variable in this type of reaction and may vary during the course of the reaction. The interaction of pH with the other variables has been studied in detail in the case of the synthesis of α-quartz and α-cristobalite and a mechanism has been proposed involving the copolymerisation of $H_3(SiO_4)^-$ ions and undissociated molecules in the range of pH from 7 to 9:

$$\underset{\underset{OH}{|}}{\overset{\overset{OH}{|}}{HO-Si-O^-}} + \left[\underset{\underset{OH}{|}}{\overset{\overset{OH}{|}}{HO-Si-O}}-H\right]_n \longrightarrow \left[\underset{\underset{OH}{|}}{\overset{\overset{OH}{|}}{HO-Si-O}}-H\right]_{n+1} + OH^-$$

At very high and at very low pH values, depolymerisation occurs.

The mechanism of formation of ionic polymeric silicates by hydrothermal synthesis is less clearly understood, but the method has nevertheless been used successfully to prepare a large number of different polymers ranging from the 2-D sheets of the clay minerals and mica to the 3-D framework structures of the feldspars and zeolites. Unfortunately the method cannot be used for the synthesis of polymeric phosphates owing to the rapid hydrolytic degradation of these compounds in the presence of water. The use of non-aqueous solvents suggests itself as a possibility here but the main methods used so far are those involving condensation polymerisation as described earlier or crystallisation from the melt as mentioned below.

Crystallisation from the melt

This method involves heating up the acidic and basic reactants to a temperature above the melting point, and then allowing the melt to cool under controlled conditions. The factors which control the formation of polymeric phosphates by crystallisation from the melt are in fact very similar to the ones which govern the condensation polymerisation process. The B/A ratio is very important in that it determines the proportion of branched, middle, end and ortho units present in the melt which have sufficient mobility to 'reorganise' themselves into the most stable arrangement. It will be observed from Table 6 that these units have connectivities of 3, 2, 1 and 0 respectively, and that the composition of the melt will depend on the proportion of each type of unit present. The proportion of each type of unit present will be determined by the B/A ratio as set out in Table 7; at any given temperature an equilibrium is set up between the various possible species in the melt and if the temperature is changed then the steady-state concentration of each species changes to the value appropriate to equilibrium at the new temperature. Studies carried out on phosphate melts having $B/A = 1$ indicate that the mean chain length of species in a polyphosphate melt increases with increasing temperature. The time for which the system is held at a particular temperature also has an important influence on the chain length distribution presumably because the high viscosity of the melt allows only a slow approach to the equi-

TABLE 6

SUGGESTED GROUPS PRESENT IN 'REORGANISING' PHOSPHATE AND SILICATE MELTS

Unit	Connectivity	Structure of unit present in Phosphate melt	Structure of unit present in Silicate melt
ortho	0	MO—P(=O)(OM)—OM	MO—Si(OM)(OM)—OM
end	1	—O—P(=O)(OM)—OM	—O—Si(OM)(OM)—OM
middle	2	—O—P(=O)(OM)—O—	—O—Si(OM)(O—)—O—
branched	3	—O—P(=O)(O—)—O—	—O—Si(O—)(O—)—O—
branched	4	none	—O—Si(O—)(O—)—O—

librium composition. As in the condensation polymerisation reactions referred to earlier, so also here the presence of water vapour influences the distribution of reaction products, the mean chain length of the polymeric species formed at any given temperature falling off rapidly with increasing humidity. It has been suggested that the water may exercise its influence either (a) in the pre-natal stage of polymer formation by acting as a chain terminating agent and thereby inhibiting the growth of the polymeric chains or (b) in the post-natal stage of polymer formation by causing hydrolytic degradation of the polymeric chains once they have been formed. If the former mechanism is important, then one might expect a similar sensitivity to water in the silicate melts and since silicate polymers themselves are not as sensitive to hydrolytic degradation as the phosphates, it

TABLE 7

PHOSPHATE UNITS IN THE MELT
AT VARIOUS B/A RATIOS

B/A	Phosphate units in the melt
3	ortho units only
	middle, end and ortho units
1	middle and end groups
	branched, middle and end groups
0	

should be possible to differentiate between the two mechanisms. So far, however, insufficient data are available on this aspect of silicate melts to allow a suitable comparison of the two systems to be made, although of course other properties of silicate melts have been the subject of extensive investigation.[1,2]

Despite the vast amount of work that has been carried out on silicate melts from the geochemical[9] and industrial (slags)[2] point of view, comparatively little is known of the detailed principles which govern the formation of silicates by crystallisation from the melt. It has been assumed by some workers[3] that silicate melts behave as 'reorganising' systems like the phosphates and if this assumption is correct then similar branched, middle, end and ortho units are likely to be involved; these are included in Table 6 for comparison with the corresponding phosphates. Viscosity studies[26] carried out on silicate melts of B/A ratio less than unity have shown that the valency of the cations present has a strong influence on the structure of the melt. Within the range $0 < B/A < 1$, the viscosity of systems incorporating univalent (M^+) cations decreases much more rapidly than the viscosity of systems containing alkaline earth (M^{2+}) cations and this has been attributed to ionic cross-linking of the polymeric chains:

$$\left[\begin{array}{c} O \\ | \\ -O-Si-O^- \\ | \\ O \end{array} \right]_n \cdots Ca^{2+} \cdots \left[\begin{array}{c} O \\ | \\ {}^-O-Si-O- \\ | \\ O \end{array} \right]_n$$

This type of cross-linking may also explain the higher crystallisation temperatures observed for melts containing divalent cations compared with those containing alkali metal ions. Very few investigations appear to have

CRYSTALLINE SILICATES AND PHOSPHATES AS IONIC POLYMERS 321

been made on systems for which B/A is greater than unity, but there are some indications that in such systems the mean chain length decreases with increasing B/A ratio until when B/A reaches the value 3 only the monomeric orthosilicates are formed.

An important industrial application of the melt crystallisation process is in the manufacture of synthetic micas.[14] The most widely used mica is fluorophlogopite ($K_2Mg_6(Al_2Si_6O_{20})F_4$) and this can be made in a state of high purity by melting a finely ground mixture of potash feldspar, potassium fluorosilicate, alumina, magnesia and quartz sand at about 1420°C using either an internal resistance or an arc resistance heating process. The result is a solid mass or 'pig' from which the mica may be extracted after comminution.

Although most of the work on the synthesis of silicates by crystallisation from the melt has been carried out at atmospheric pressure, some studies have also been made at elevated pressures ca. 75 to 150 MN m^{-2} in an attempt to simulate the conditions of formation of deep seated igneous rocks in the earth's crust and a number of small crystals of polymeric silicates have been made in this way.[25]

Precipitation from aqueous solution
Since the phosphate polymers suffer hydrolytic degradation in an aqueous environment this method is inapplicable to the preparation of phosphates. It has however been applied to silicate polymer formation and considerable scope exists for the extension of this work to copolymer formation under similar conditions. So far most of the detailed kinetic work appears to have been carried out on the precipitation of silica, and the important variables have been shown to be pH, temperature and electrolyte concentration. These variables are very important both in the initial stages of the reaction when polymeric molecules are being formed and in the later stages of the reaction when cross-linking (gelation) takes place. Control of the pH between 8 and 9 is important in order to achieve a maximum rate of polymerisation, and at any given pH value, the rate of polymerisation is also increased by an increase in temperature and by an increase in electrolyte concentration. The mechanism proposed for the reaction in alkaline solution involves the attack of an $H_3SiO_4^-$ ion (formed from silicic acid) on a neutral molecule:

$$\underset{\underset{OH}{|}}{\overset{\overset{OH}{|}}{HO-Si-O^-}} + HO\left[\underset{\underset{OH}{|}}{\overset{\overset{OH}{|}}{Si-O}}\right]_n H \longrightarrow HO\left[\underset{\underset{OH}{|}}{\overset{\overset{OH}{|}}{Si-O}}\right]_{n+1} H + OH^-$$

and in acid solution the ionic species which takes part in a similar reaction with an undissociated molecule of silicic acid is the protonated species $(H_5SiO_4)^+$.

Copolymers of silica and alumina, *i.e.* aluminosilicate gels, have also been prepared by precipitation from solution by a slight modification of the above process. These copolymers have been extensively used as petroleum cracking catalysts because of their acidic surface sites which promote carbonium ion reactions[27,28] and the two general methods of preparation involve (a) addition of aluminium sulphate in acid solution to a solution of sodium silicate[29] and (b) hydrolysis of a solution of aluminium isopropoxide in tetra-ethoxysilane.

In recent years, the aluminosilicate cracking catalysts have been almost completely replaced by synthetic zeolite catalysts which have the advantages of higher activity, greater resistance to poisoning and smaller coking tendency. Like the silica gels and the aluminosilicate gels, these zeolites are prepared by precipitation from solution, in this case by the reaction between sodium silicate and sodium aluminate in alkaline solution, and begin life as gels: these gels are then allowed to crystallise slowly at temperatures around 100°C to give the cage-like structure of this industrially important class of materials. The crystallites formed by heat treatment are normally from 1 to 5 μm in diameter; they are mixed with clay, which acts as a binder and plasticiser, and then extruded to the required pellet size for the particular industrial application.

In addition to the synthetic zeolites, a number of other insoluble silicates are prepared synthetically by precipitation from aqueous solution. These include the industrially important calcium, aluminium and magnesium salts[29] which find a wide variety of applications.

7.3.3 Characterisation

The characterisation of naturally occurring and synthetic raw materials is important in industry as is the quality control of the product. Chemical analysis still remains of basic importance in the characterisation of both silicates[30] and phosphates[3] despite the number of more rapid instrumental techniques which are currently available. Differential thermal analysis[31] and X-ray diffraction[32] are widely used to supplement the optical mineralogical methods of characterisation of silicates and X-ray diffraction has also proved successful in the analysis of crystalline phosphate mixtures used for detergent builders since it allows an identification to be made of the crystalline modifications as well as the molecular species in mixtures of crystalline ortho-, pyro-, tripoly- and metaphosphates. Similar analyses have also been made by infra-red spectroscopy.

Various solution methods have also been used for the characterisation of phosphates and soluble silicates. pH titration has been used unequivocally in the analysis of ortho-, pyro- and tripolyphosphate mixtures and NMR[33] at low resolution has also been used to determine the amount of end, middle and branch points in fluid samples where the rate of exchange of a given PO_4 group from one to another of these differentiated positions in the molecule[4] is slow; further information may be obtained at higher resolutions. Other techniques which have been used to characterise phosphate mixtures in terms of their molecular composition include paper chromatography, paper electrophoresis and ion-exchange chromatography, the latter technique being of great value in separating the various sized phosphate ions.

The structures of the colloidal micelles formed by sodium silicate in solution have been studied by light scattering and by gas-chromatographic analysis of the trimethylsilyl derivatives; the polymeric species identified include $(Si_2O_7)^{6-}$, $(Si_3O_{10})^{8-}$, and $(SiO_3)_4^{4-}$ together with polysilicate structures of greater complexity.

The results of the structural characterisation of ionic phosphate and silicate polymers are discussed in the next section.

7.4 STRUCTURES OF CRYSTALLINE IONIC SILICATE AND PHOSPHATE POLYMERS

Many of the naturally occurring silicates and all of the synthetic silicates and phosphates have important industrial and commercial applications. In some cases the reason for the selection of a particular polymer is related more to economics or to chemical composition rather than to its particular properties. Nevertheless a good many ionic silicate and phosphate polymers are used on account of specific properties which are closely related to the structures of the materials concerned and in particular to the combined polymeric/ionic structures which are characteristic of this group of chemical compounds. In the present section, therefore, these structures are reviewed using the unifying concept of connectivity as the basis for classification. The same concept is also useful in discussing the structure and properties of glasses, which have been described elsewhere in this book.

Whilst most of the interest in the following sections centres around the high molecular weight polymers, the oligomers have also been included, partly for the sake of completeness and partly because some of these oligomers, for example $(P_3O_{10})^{5-}$ have considerable commercial importance.

7.4.1 Structures of oligomers

Under the heading of oligomers are grouped together polymeric structures having a connectivity of 1, those having an overall connectivity between 1 and 2 and the low molecular weight ring systems having a connectivity of 2. The various species which are known to exist in the crystalline state are set out in Table 8. The 2-connective rings are dealt with here rather than included with the 2-connective high molecular weight polymers because those which have been isolated contain relatively few phosphorus or silicon atoms in the ring. From the preparative point of view of course, it will be remembered that when B/A = 1, infinite chain and/or ring compounds may be formed since both are 2-connective. We shall now review the structures of each of the three groups of oligomers namely (a) dimers and co-dimers, (b) linear chain polymers and (c) ring polymers.

Dimers and co-dimers (1-connective)

Naturally occurring dimers containing the $(Si_2O_7)^{6-}$ ion are rare. They include the mineral thortveitite $Sc_2(Si_2O_7)$ and hemimorphite $Zn_4(Si_2O_7)(OH)_2 \cdot H_2O$ in which the Sc^{3+} ion (octahedrally co-ordinated by oxygen), and the Zn^{2+} ion respectively, link the $(Si_2O_7)^{6-}$ dimeric units together in the crystal lattice.

Some dimeric silicates are formed as a result of slag attack on silicate refractories. They include the mineral akermanite of ideal formula $Ca_2Mg(Si_2O_7)$ in which the Ca^{2+} and Mg^{2+} ions may be replaced by Na^+ and Fe^{2+} ions. Furthermore one silicon atom in the dimeric ion itself may be isomorphously substituted by Al^{3+} to form a copolymer ion $((Si, Al)_2O_7)^{7-}$. The associated cations may be Ca^{2+}, Na^+, Mg^{2+} or Al^{3+} in various proportions and a general formula $(Ca, Na)_2(Mg, Al)((Si, Al)_2O_7)$ may be written down to express the solid solution range. At one end of the range is the mineral akermanite, $Ca_2Mg(Si_2O_7)$ in which no isomorphous substitution of silicon by aluminium has taken place, and at the other end of the range is the mineral gehlenite, $Ca_2Al(Si, Al)O_7)$ in which half of the silicon atoms in the dimeric ions have been replaced by aluminium ions in tetrahedral co-ordination.

This complexity of isomorphous substitution does not arise in the pyrophosphates containing the ion $(P_2O_7)^{4-}$ which are made synthetically, but of course the condensed phosphate–silicates and phosphate–arsenates

TABLE 8
CRYSTALLINE SILICATE AND PHOSPHATE OLIGOMERS

Connectivity	Silicates			Phosphates	
	Polymeric species	Polymeric anion	Commonly Associated cations	Polymeric anion	Commonly associated cations
1	dimers	$(Si_2O_7)^{6-}$	Sc^{3+}, Ca^{2+}	$(P_2O_7)^{4-}$	H^+, Na^+ K^+, NH_4^+
	codimers	$((Si, Al)_2O_7)^{7-}$	Mg^{2+}, Zn^{2+} $2Ca^{2+}$ Al^{3+}		
1 and 2	short chains	none		$(P_3O_{10})^{5-}$ $(P_4O_{13})^{6-}$ $(P_5O_{16})^{7-}$	as above
2	rings	$(Si_3O_9)^{6-}$ $(Si_6O_{18})^{12-}$	$1Ba^{2+} + 1Ti^{4+}$ $3Be^{2+} + 2Al^{3+}$	$(P_3O_9)^{3-}$ $(P_4O_{12})^{4-}$	as above Al^{3+}, Cu^{2+}
	ring copolymer	$((Si_5Al)O_{18})^{13-}$	$2Mg^{2+} + 3Al^{3+}$		

Note
(a) $(Si_3O_{10})^{8-}$ reported in the literature has been found to consist of $(Si_2O_7)^{6-}$ and $(SiO_4)^{4-}$ ions (e.g. epidote $Ca_2FeAl_2(SiO_4)$-$(Si_2O_7)OH$) in crystalline silicates. However, both $(Si_3O_{10})^{8-}$ and $(Si_4O_{12})^{8-}$ have been identified in colloidal solutions of soluble silicates.
(b) $(P_4O_{13})^{6-}$ which has been reported as a tetraphosphate is in fact a mixture of $(P_3O_{10})^{5-}$ and $(PO_3)_n^{-}$; some tetraphosphates may, however have limited stability at high temperatures.

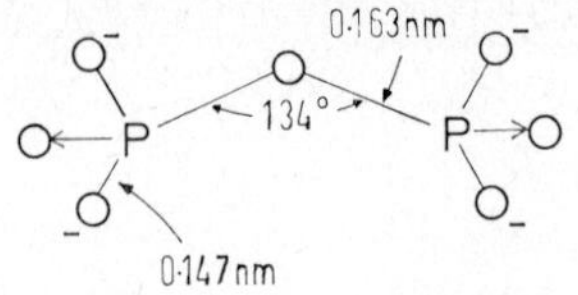

Fig. 13 Structure of the $(P_2O_7)^{4-}$ ion[16]

which can be prepared synthetically[34] are 1:1 copolymers and could be regarded in this way. The structure of the dimeric ion has been studied in detail by X-ray crystallography using a number of salts of trivalent metals and the bond lengths and bond angle are as shown in Fig. 13. Only the normal salts of univalent, divalent and trivalent metals and the alkali metal dihydrogen salts appear to be known in the anhydrous state; the other acid salts of the alkali metals containing 1 and 3 hydrogen atoms exist only in the hydrated form. Once the anhydrous form of the salt has been hydrated, the water cannot be removed without breaking down the anion. There are three crystalline modifications of anhydrous $Na_4(P_2O_7)$ which are stable within the temperature ranges indicated in the following scheme:

$$Na_4(P_2O_7)\text{-III} \underset{}{\overset{410°C}{\rightleftharpoons}} Na_4(P_2O_7)\text{-II} \underset{}{\overset{520°C}{\rightleftharpoons}} Na_4(P_2O_7)\text{-I} \underset{}{\overset{985°C}{\rightleftharpoons}} \text{melt}$$

The transition between the three forms is so rapid that the high temperature forms cannot be obtained by quenching and in fact only $Na_4(P_2O_7)$-III has been obtained at room temperature. The calcium pyrophosphates also exist in two crystalline modifications designated α and β. Several crystalline ammonium phosphates are also known.

Linear chain polymers (1- and 2-connective)
There are no known crystalline silicates which consist of short chains of linked SiO_4 tetrahedra and the only polymeric phosphates which appear to have been obtained in a crystalline state are the tri-, tetra- and pentapolyphosphates containing the anions $(P_3O_{10})^{5-}$, $(P_4O_{13})^{6-}$ and $(P_5O_{16})^{7-}$ respectively. The sodium salt, hexasodium tetrapolyphosphate $(Na_6P_4O_{13})$ is apparently not a tetrapolyphosphate at all but rather a mixture of pentasodium tripolyphosphate $(Na_5P_3O_{10})$ and sodium metaphosphate $(NaPO_3)_n$. The pentapolyphosphate anion has been identified in the calcium mineral trömelite but its linear structure has only been determined in solution and not in the solid state. The family of tripolyphosphates on the other hand has been well characterised in the solid state, and an X-ray study of the low temperature form of $Na_5P_3O_{10}$ has revealed the structure of the anion to be as shown in Fig. 14.

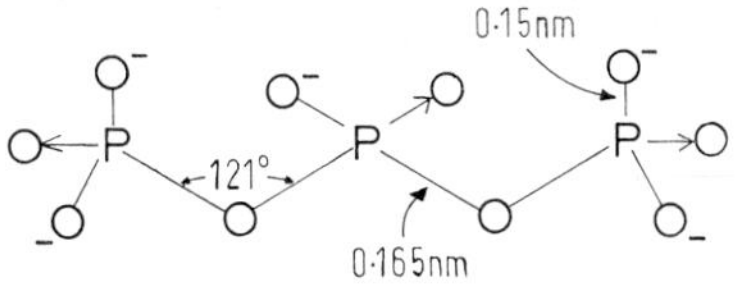

CRYSTALLINE SILICATES AND PHOSPHATES AS IONIC POLYMERS 327

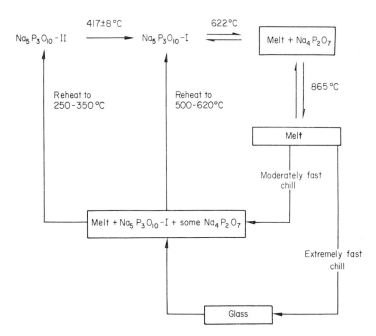

Fig. 14 Structure of the $(P_3O_{10})^{5-}$ ion[16]

Various combinations of cations may be associated with the $(P_3O_{10})^{5-}$ anion. All of the sodium, potassium and ammonium salts from $MH_4(P_3O_{10})$ to $M_5(P_3O_{10})$ are known, but only the $M_5(P_3O_{10})$ salts are of technical interest. Of these, the pentasodium salt is the most widely studied and used. It exists in a high and a low temperature form (Fig. 15) like the corresponding pyrophosphate, but the transition between the two forms is so unpredictable that it is difficult to obtain either form pure at room temperature. By contrast the potassium salt $K_5(P_3O_{10})$ appears to have only one crystalline form.

Fig. 15 Effect of heat treatment on pentasodium tripolyphosphate-II.

Ring silicates and phosphates

The best known example of an $(Si_3O_9)^{6-}$ ring anion is found in the mineral benitoite, in which the negative charge is balanced by the presence of Ba^{2+} and Ti^{4+} cations octahedrally co-ordinated by oxygen. $(Si_6O_{18})^{12-}$ rings are present in the mineral beryl, $Be_3Al_2(Si_6O_{18})$ and are linked together through bridging Al^{3+} and Be^{2+} cations which are octahedrally and tetrahedrally co-ordinated respectively with peripheral oxygen atoms of the rings. The mineral tourmaline, $Na(Mg, Fe, Mn, Li, Al)_3Al_6(Si_6O_{18})(BO_3)_3$-$(OH, F)_4$ which occurs in granite and pegmatites and is often associated with clay deposits, also has a structure composed of hexameric rings; these rings are held together normally by Al^{3+} and B^{3+} cations but in some tourmalines these ions have been isomorphously substituted by other cations.

Isomorphous substitution of silicon by aluminium may also occur in the rings, giving rise to copolymer formation. The substitution of one sixth of the silicon atoms in the ring by aluminium leads to a copolymer anion of formula $((Si_5Al)O_{18})^{13-}$, which is present in the fibrous mineral cordierite- the charge balancing cations being Mg^{2+} and Al^{3+}. In addition to substitution in the ring, the isomorphous substitution of magnesium by iron may occur also, giving rise to an iron cordierite which occasionally occurs in imperfectly fired fireclay products in the ceramics industry.

The trimeric $(P_3O_9)^{3-}$ ring has been shown to exist in the water-soluble sodium salt $Na_3P_3O_9$, which is the compound described as $NaPO_3$-I in the literature (see Fig. 12). The only other known cyclic ion is the tetrameric anion $(P_4O_{12})^{4-}$ which occurs in the sodium salt $Na_4(P_4O_{12})$; its trimeric ring structure has been established in the aluminium $(Al_4(P_4O_{12})_3)$ and ammonium $((NH_4)_4(P_4O_{12}))$ derivatives. Divalent copper also forms a tetramer of composition $Cu_2(P_4O_{12})$.

7.4.2 Structures of infinite chain polymers

Neglecting the two end groups (which are 1-connective), the infinite chain silicates and phosphates are all 2-connective. This group includes (a) the metal silicates of composition $(SiO_3)_n^{2n-}$ in which there are two ionic bonds per monomer unit in the structure and (b) the metaphosphates of composition $(PO_3)_n^{n-}$ in which only one ionic bond per monomer unit is present. The cations normally associated with these infinite chain silicates and phosphates are shown in Table 9.

Infinite chain silicates

The silicates containing the anion $(SiO_3)_n^{2n-}$ comprise (i) the synthetic sodium and lithium salts (Na_2SiO_3 and Li_2SiO_3) in which the structure of

CRYSTALLINE SILICATES AND PHOSPHATES AS IONIC POLYMERS   329

TABLE 9

INFINITE CHAIN SILICATE AND PHOSPHATE POLYMERS

Species	Polymeric anion	Commonly associated cations
Silicates	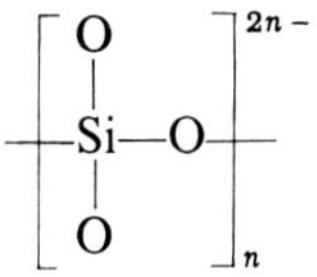 $\left[\begin{array}{c} O \\ \mid \\ -Si-O- \\ \mid \\ O \end{array}\right]_n^{2n-}$	Na^+, Ca^{2+} Mg^{2+}, Fe^{2+}, Li^+, Mn^{2+} Al^{3+}, Fe^{3+}, Ti^{4+}
Phosphates	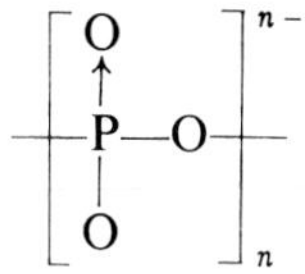 $\left[\begin{array}{c} O \\ \uparrow \\ -P-O- \\ \mid \\ O \end{array}\right]_n^{n-}$	Na^+, K^+, NH_4^+, (Rb^+), (Ag^+) Cu^{2+}, Al^{3+}

the anion is reported to be [16] as shown in Fig. 16(a) and (ii) the naturally occurring pyroxene minerals in which the arrangement is as depicted in Fig. 16(b). The pyroxenes, of general formula:

$$A_{1-x}(BC)_{1+x}((Si, Al)_2O_6)$$

where $A = Na^+$, Ca^{2+}; $B = Mg^{2+}$, Fe^{2+}, Li^+, Mn^{2+}; $C = Al^{3+}$, Fe^{3+}, Ti^{4+} and the aluminium substitutes for silicon in the anion only in small amounts, include the well-known orthorhombic minerals enstatite

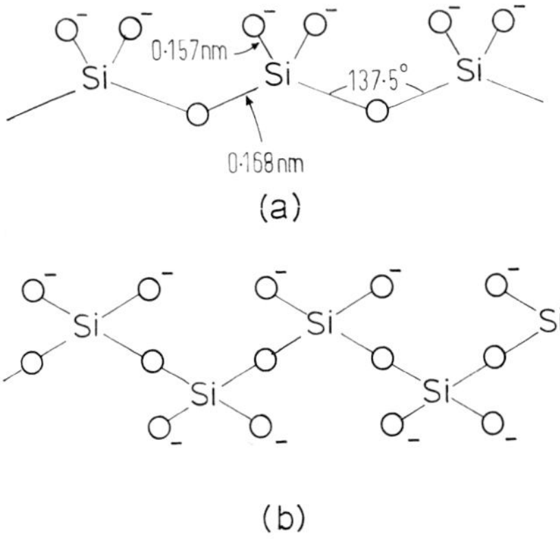

Fig. 16 Infinite chain anions in polymeric silicates: (a) part of the chain in sodium metasilicate; (b) part of the chain in the pyroxene minerals.

($MgSiO_3$) and hypersthene ((Mg, Fe)SiO_3) together with the monoclinic pyroxenes diopside ($Ca, Mg(SiO_3)_2$), spodumene ($LiAl(SiO_3)_2$) and the rare mineral jadeite ($NaAl(SiO_3)_2$). Spodumene and jadeite may be regarded as being derived from diopside by the isomorphous substitution of ($Li^+ + Al^{3+}$) and ($Na^+ + Al^{3+}$) respectively for ($Ca^{2+} + Mg^{2+}$).

The prismatic habit of these and the many other pyroxenes reflect the internal structure of the crystals in which the polymeric chains lie parallel to the vertical crystallographic axes and are held together by cations between them. The difference in ionic size of the metal ions causes slight angular adjustments in the relative positions of one $(SiO_3)_n^{2n-}$ chain to another and this results in the two different symmetry classes, monoclinic and orthorhombic, mentioned above; the pyroxenes in these two classes are usually referred to as clinopyroxenes and orthopyroxenes respectively. For example in diopside, 6 co-ordinated Mg^{2+} cations and the larger 8 co-ordinated Ca^{2+} cations act as ionic cross-links between the parallel anionic chains, whereas in enstatite which contains only Mg^{2+} as cations the $(SiO_3)_n^{2n-}$ chains are arranged rather differently to allow all of the Mg^{2+} ions to occupy sites of six-fold co-ordination.

The distribution of cations amongst the available sites which lie between the $(SiO_3)_n^{2n-}$ chains has recently been studied.[35] In the bands of cations which stretch along the chains in the c-direction, there are two positions of 6-fold co-ordination, which are designated M_1 and M_2. Cations in the M_1 position are co-ordinated to six oxygen atoms each of which is linked to one silicon atom; the M_1 sites approximate octahedral symmetry and the cations lie almost at the centre of symmetry of the site. Cations in the M_2 position on the other hand are co-ordinated by four oxygen ions each linked to one silicon atom and two bridging oxygen atoms which are shared by two silicon atoms; this M_2 site is considerably distorted from octahedral symmetry. It has been found that Fe^{2+} ions show a preference for the larger M_2 sites and an explanation for this is given in terms of crystal field theory;[11] Mn^{2+} ions also show a preference for the larger M_2 sites but in this case, since the ion acquires zero crystal field stabilisation energy in the silicates, an explanation has to be sought in terms of ionic size and the balance of electrical charge.

In addition to the pyroxene minerals, there are a number of other compounds in which infinite chain $(SiO_3)_n^{2n-}$ anions occur. We have already mentioned that the tetrahedra are disposed similarly in the sodium and lithium salts and in the pyroxenes; the pattern repeating every two units. Other arrangements are also possible; β-wollastonite has 3 tetrahedral units in the repeat pattern (Fig. 17(b)), rhodonite ($CaMn_4(SiO_3)_5$) has five and pyroxymanganite ((Ca, Mg)(Mn, Fe)$_6(SiO_3)_7$) seven.

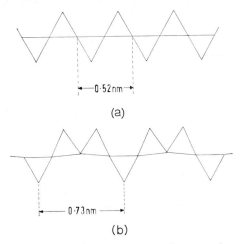

Fig. 17 Arrangement of tetrahedral units in polymeric silicates: (a) pyroxene chain; (b) β-wollastonite (β-CaSiO₃) chain.

Infinite chain phosphates

The most important long chain polyphosphates containing the anion $(PO_3)_n^{n-}$ are the sodium, potassium and ammonium salts. The sodium salt $(NaPO_3)_n$ exists in three crystalline modifications II, III and IV (Fig. 12). So far the structure of modification III has not been determined but the structures of forms II and IV are known; they differ essentially in the way the tetrahedra are linked together in the chain (Fig. 18).

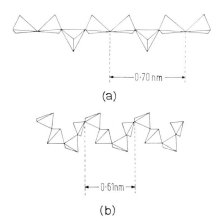

Fig. 18 Arrangement of tetrahedral units in polymeric phosphates: (a) straight chain in (NaPO₃)-II; (b) spiral chain in (NaPO₃)-IV.

NaPO$_3$-II, the high temperature form of Maddrell's salt, consists of long straight chains of linked PO$_4$ tetrahedra, the pattern within the chain repeating itself every three tetrahedra over a distance of 0·70 nm. NaPO$_3$-IV (Kurrol's salt) differs from this in having spiral rather than straight chains of interconnected tetrahedra with a repeat unit of four tetrahedra along the chain and a repeat distance of 0·61 nm.

The silver salt (AgPO$_3$)$_n$ has been shown to have a structure similar to NaPO$_3$-IV and the rubidium salt also forms a spiral chain but in this case only two tetrahedral units form the repeat pattern and the repeat distance is 0·42 nm. The crystalline potassium salt, of which there only appears to be one modification, is also thought to be a long chain polymer; ultracentrifuge and viscosity studies indicate a relative molecular mass varying from 5×10^4 to 3×10^6 depending on the B/A ratio and heat treatment during preparation, which corresponds to a mean chain length ranging from 400 to 20 000 P atoms/chain, and supplementary gel permeation chromatography results taken in conjunction with sedimentation curves suggest a sharp distribution of chain lengths about the mean. The ammonium salt ((NH$_4$)PO$_3$)$_n$ appears to exist in several crystalline forms and in this respect resembles the sodium salt; however, the X-ray diffraction patterns are very much like those for the potassium salt, and light scattering and viscometric data show a relative molecular mass as high as 2×10^6 for one of the crystalline modifications isolated so far.

7.4.3 Structures of cross-linked and branched chain polymers
In this group of polymers, some units are 2-connective and others 3-connective; the overall connectivity is therefore less than 3 but greater than 2. The most important silicate polymers in this group are the amphiboles containing the (Si$_4$O$_{11}$)$_n^{6n-}$ anion which result from the cross-linking of pyroxene chains. Very little is known about the structures of the phosphates in this group which are known as ultraphosphates. The various species to be considered are listed in Table 10.

Double chain silicate polymers
The polymers containing the (Si$_4$O$_{11}$)$_n^{6n-}$ anion are very similar in many respects to the infinite single chain silicates containing the (SiO$_3$)$_n^{2n-}$ anion which were discussed in the previous section. The compositional difference between the two is very small (Si$_4$O$_{11}$ compared with Si$_4$O$_{12}$) and it is not surprising therefore that criteria other than chemical analysis have to be used to distinguish between them. It will be clear from Fig. 19 that the amphibole structure will be formed if two pyroxene chains in juxtaposition

TABLE 10
BRANCHED-CHAIN AND CROSS-LINKED SILICATE AND PHOSPHATE POLYMERS

Species	Polymeric anion	Commonly associated cations	Commonly associated anions
Silicates	$(Si_4O_{11})_n^{6n-}$	Ca^{2+}, Na^+, Mg^{2+} Fe^{2+}, Ti^{4+}, Al^{3+}	OH^-, F^-
	$((Si_3Al)O_{11})_n^{7n-}$	Fe^{3+}	
	$(Si_6O_{17})_n^{10n-}$	Ca^{2+}	OH^-
	$((SiAl)O_5)_n^{2n-}$	Al^{3+}	
Phosphates	$(P_4O_{11})_n^{2n-}$	Ca^{2+}	none
	$(P_6O_{17})_n^{4n-}$	Ca^{2+}	none

form covalent cross-links through oxygen atoms of every alternate tetrahedron. The number of tetrahedra in the repeat unit is therefore two.

The structures of the other polymeric anions in this group (Fig. 20) are also related to infinite single chain ions of different spatial orientation. $(Si_6O_{17})_n^{10n-}$ is formed by the cross-linking of β-wollastonite (β-$CaSiO_3$) chains having 3 tetrahedra in the repeat unit and an $(Si_2O_5)_n^{2n-}$ anion would be obtained by the cross-linking of the type of $(SiO_3)_n^{2n-}$ anion reported present in Na_2SiO_3, if such a structure were stable. In fact this structure does not appear to exist, but a copolymer $(AlSiO_5)_n^{3n-}$ is known in which chains of topologically equivalent $(SiO_3)_n^{2n-}$ tetrahedra cross-link with similar chains of $(AlO_3)_n^{3n-}$ tetrahedra (Fig. 20(c)); this copolymer anion exists in sillimanite. Another type of copolymer is formed if the isomorphous substitution of silicon atoms in the chains by aluminium occurs,

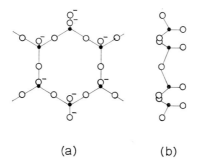

(a) (b)

Fig. 19 Cross-linked polymeric silicates: (a) part of the chain in the amphibole minerals; (b) end-on view of the chain. ● = silicon; ○ = oxygen.

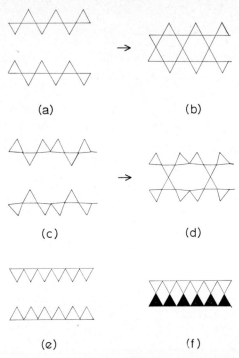

Fig. 20 Relationship between chain configurations in the infinite single and double chain silicate polymers: (a) pyroxene chains; (b) amphibole chain; (c) β-wollastonite chains; (d) xonotlite chain; (e) hypothetical isotactic silicate chains; (f) sillimanite chain. △ = SiO_4 tetrahedra; ▼ = AlO_4 tetrahedra.

giving an anion of composition $((Si_3Al)O_{11})_n^{7n-}$. The structure of mullite, which is related to that of sillimanite, is referred to later.

In sillimanite, which is an important ceramic raw material, octahedrally co-ordinated Al^{3+} cations lie between the double chains and link them together by means of ionic bonds. In the amphiboles, the double chains can be linked together by a whole variety of cations and OH^- ions (sometimes isomorphously substituted by F^-) which can be conveniently accommodated within the chains are an essential part of the structure. A general formula for the amphiboles is:

$$(ABC)_{7-8}((Si, Al)_4O_{11})_2(OH, F)_2$$

in which A = Ca^{2+}, Na^+ (K^+); B = Mg^{2+}, Fe^{2+} (sometimes Mn^{2+}); C = Ti^{4+}, Al^{3+}, Fe^{3+}. It is interesting to note that (a) aluminium may replace up to one quarter of the silicon atoms in the chains (high temperature formation may lead to a greater extent of substitution); (b) Fe^{2+} and Mg^{2+} are completely interchangeable; (c) total (Ca, Na, K) may be zero or near zero or may vary from 2–3; however the total Ca^{2+} can never exceed 2 and potassium is present only in minor amounts; (d) OH^- and F^- are completely interchangeable. With all these possibilities, it is clear that the composition of the amphiboles may be very complex; however, on the basis of chemical composition and crystal structure two distinct groups of series are recognised:

orthorhombic
1. Anthophyllite series $(Mg, Fe)_7(Si_4O_{11})_2(OH)_2$ $(Mg \gg Fe)$

monoclinic
2. Cummingtonite series $(Fe, Mg)_7(Si_4O_{11})_2(OH)_2$ $(Fe > Mg)$
3. Tremolite series $Ca_2(Mg, Fe)_5(Si_4O_{11})_2(OH)_2$
4. Hornblende series $Ca_2Na_{0-1}(Mg, Fe, Al)_5((Al, Si)_4O_{11})_2(OH)_2$

5. Alkali amphibole series
   Glaucophane $Na_2Mg_3Al_2(Si_4O_{11})_2(OH)_2$
   Riebeckite $Na_2Fe_3^{2+}Fe_2^{3+}(Si_4O_{11})_2(OH)_2$
   Arfvedsonite $Na_3Fe_4^{2+}Fe^{3+}(Si_4O_{11})_2(OH)_2$

In the anthophyllite series, the percentage of Fe^{2+} in the cationic bands ranges from 0 to 50% and Al^{3+} sometimes replaces silicon in the double chains up to a composition $((Si_3Al)O_{11})_n^{7n-}$ with corresponding cation substitutions to balance the extra negative charge. This series overlaps the iron-rich cummingtonite series in which the percentage of Fe^{2+} in the cationic bands varies from 33·3% up to 100%.

In the tremolite series, Mg^{2+} is replaceable by Fe^{2+} and also in part by Al^{3+} and Fe^{3+}; silicon in the chains is partly replaceable by Al^{3+} as in the anthophyllite series and furthermore Ti^{4+} and F^- may be present as well as an additional Na^+ cation for every two $(Si_4O_{11})_n^{6n-}$ groups of the structure. The product of all these substitutions is the hornblende series which has therefore a very wide range of composition. Finally the alkali amphiboles may be considered to be derived from the hornblende series by the partial or complete substitution of Na^+ for Ca^{2+}. The best known of these soda amphiboles are glaucophane, arfvedsonite and riebeckite. The composition of riebeckite corresponds to that of crocidolite or blue asbestos $(Na_2Fe_3^{2+}Fe_2^{3+}(Si_4O_{11})(OH)_2)$ and the other commercially important amphibole is amosite, which has a composition $Fe_{5.5}^{2+}Mg_{1.5}^{2+}(Si_4O_{11})_2(OH)_2$. The other three forms of amphibole asbestos, namely tremolite

($Ca_2Mg_5(Si_4O_{11})_2(OH)_2$), anthophyllite ($Mg_7(Si_4O_{11})_2(OH)_2$) and actinolite ($Ca_2Mg_4Fe^{2+}(Si_4O_{11})_2(OH_2)$) complete the range of fibrous amphiboles, but these are found in nature only in relatively small amounts and their commercial importance in the asbestos industry is limited on account of their low tensile strengths. Rules governing the distribution of cations on the seven sites ($2M_1$, $2M_2$, M_3, $2M_4$) of the unit cell have been given by Whittaker[36] and the results of spectroscopic studies on cation distribution amongst these sites has been presented by Burns.[11]

A somewhat different chain from those found in the amphiboles occurs in the mineral xonotlite. This mineral contains the anion of composition $(Si_6O_{17})_n^{10n-}$ which has the structure shown in Fig. 21. (CaO_n) groups are interlinked between the double chains giving the mineral an overall composition of $3CaSiO_3 \cdot xH_2O$. The mineral tobermorite, a synthetic silicate of composition $Ca_5(Si_6O_{17})5H_2O$ formed by autoclaving mixtures of lime and silica probably also has this structure. Gimblett[22] has drawn attention to an interesting correlation first suggested by Belov that Mg, Fe and Al silicates form preferably the simpler chain types of the pyroxenes and amphiboles while the Ca silicates prefer the wollastonite or xonotlite type of chain development; the reason for this lies in the different co-ordination preferences of the interchain cations which hold the structures together.

Ultraphosphates

According to Van Wazer[3] only two crystalline ultraphosphates are known with certainty. These are the calcium salts CaP_4O_{11} and $Ca_2P_6O_{17}$ which are found in the phase diagram of the CaO/P_2O_5 system. Apparently no structural investigations have yet been carried out on these materials, but if the structure of $(P_4O_{11})_n^{2n-}$ is analogous to that of the amphiboles it will

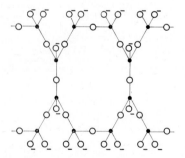

Fig. 21   Double chain in xonotlite. ● = silicon; ○ = oxygen.

have the double chain configuration shown in Fig. 21 although of course the anion charge will be different; similarly the anion $(P_6O_{17})_n^{4n-}$ may have a similar type of structure to that of xonotlite.

7.4.4 Structures of 2-D sheet polymers

The tetrahedral phosphate group has a maximum connectivity of three. If all three labile oxygen atoms of the tetrahedron are used in forming covalent network links, then the completely covalent structure of $(P_2O_5)_n$ results, with no possibility of covalent/ionic bonding characteristic of an ionic polymer. The tetrahedral SiO_4 group on the other hand has a maximum connectivity of 4 and is capable of forming an ionic polymer since three of the four oxygen atoms can take part in the formation of a network of composition $(Si_2O_5)_n^{2n-}$ leaving the fourth oxygen atom to form an ionic bond with suitable cations. In practice it is found that the cations associated with these 2-D sheet anions are of two kinds: (a) complex cations which themselves have a sheet structure, e.g. $(Al_2(OH)_4)_n^{2n+}$ and $(Mg_3(OH)_4)_n^{2n+}$ and form with the anion a composite layer and (b) simple (often hydrated) cations which in some structures (e.g. the micas) hold the composite layers together. Sheet silicates consisting of single layers in the form of a 4:8 network [16] are extremely rare and will not be considered here. Double layer silicates only exist as copolymers of composition (Si, Al)O_2 and are of no particular interest. The structures of the layer lattice minerals are reviewed below.

Layer lattice silicates

Various structures may be built up from polymeric layers of composition $(Si_2O_5)_n^{2n-}$ and suitable complex cations. The simplest of these composite structures arises when one $(Si_2O_5)_n^{2n-}$ layer is linked to one aluminium hydroxide layer $(Al_2(OH)_4)_n^{2n+}$ in which the aluminium is in octahedral co-ordination. The resulting composite layer structure is shown in Fig. 22. It is electrically neutral and forms the basic unit of structure of the kaolin group of clay minerals. In these minerals, the neutral composite layers are held together by hydrogen bonding between the hydrogen atoms of the free OH groups on aluminium and the oxygen atoms of the next Si_2O_5 layer. The differences between the individual members of this group, which includes nacrite, dickite, kaolinite and halloysite, result from the way in which the many composite layers which go to form a single particle of the clay mineral are stacked together one on top of another.

An interesting variation on this structure arises when $(Mg_3(OH)_4)_n^{2n+}$ takes the place of $(Al_2(OH)_4)_n^{2n+}$ in the kaolinite structure, for this gives

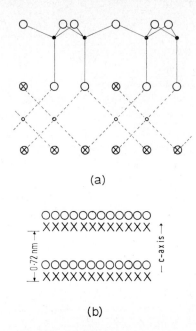

Fig. 22 Structure in kaolinite (a) schematic drawing ○ = oxygen; ● = silicon; ⊗ = hydroxy ; ○ = aluminium: (b) simplified representation of the same structure; ○ = $(Si_2O_5)_n$ layer; × = $(Al_2(OH)_4)_n$ layer.

rise to the serpentine group of minerals of which chrysotile (white asbestos) is probably the most important member. The sheet structure of chrysotile is of particular interest because it does not form planar sheets as in the kaolin group or montmorillonite group of minerals but rather curved sheets in the form of scrolls (Fig. 23). The reason for the curved sheet structure compared with the planar sheet structure of kaolinite is that whereas in the latter the dimensions of the hexagonal $(Si_2O_5)_n^{2n-}$ and the $(Al_2(OH)_4)_n^{2n+}$ (gibbsite) layers are closely similar, in chrysotile there is a considerable mismatch between the $(Si_2O_5)_n^{2n-}$ and the $(Mg_3(OH)_4)_n^{2n+}$ (brucite) layers; better matching of the layers can be achieved by curvature of the composite sheet with its tetrahedral component on the inside of the curve.

A second type of layer structure involves the $(Al_2(OH)_2)_n^{4n+}$ or the $(Mg_3(OH)_2)_n^{4n+}$ cation, bonding together two $(Si_2O_5)_n^{2n-}$ layers in the form of a 'sandwich' (Fig. 24) giving a structure of overall composition $Al_2Si_4O_{10}(OH)_2$ or $Mg_3Si_4O_{10}(OH)_2$ respectively. This sandwich structure, which is electrically neutral, occurs in the montmorillonite group of minerals

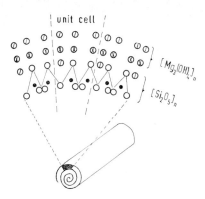

Fig. 23   Curved sheet structure of chrysotile. ① $=$ *hydroxyl;* ⊗ $=$ *magnesium;* ● $=$ *silicon;* ○ $=$ *oxygen.*

and in talc. As in the previous group of minerals, particles are formed by the stacking together of these sandwich layers one on another, but the forces which hold them together are no longer hydrogen bond forces since the hydrogen atoms of the hydroxyl groups which were responsible for the bonding in kaolinite have been removed by condensation with the second silica layer. In the montmorillonites and in talc, therefore, the only forces available for holding the neutral sandwich layers together are the weak Van der Waals' forces and this, as we shall see later, has important consequences for the properties of this group of minerals.

Another structural feature which has a marked effect on properties is the widespread occurrence of isomorphous substitution in the $(Al_2(OH)_2)_n$ $(Mg_3(OH)_2)_n$ and $(Si_2O_5)_n$ layers of the 'sandwich' minerals. Aluminium is the most ubiquitous of the substituting cations on account of its small

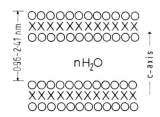

Fig. 24   Simplified representation of montmorillonite structure; ○ $= (Si_2O_5)_n$ *layer;* × $= (Al_2(OH)_2)_n$ *layer.*

size, which allows it to enter sites of tetrahedral as well as those of octahedral co-ordination. Other cations such as Mg^{2+}, Fe^{2+}, Fe^{3+} and Li^+ which are too large to substitute for silicon in the tetrahedral sites can nevertheless replace aluminium in the $(Al_2(OH)_2)_n$ layer. Two examples may be given by way of illustration of substitution in pyrophyllite, which may be regarded as the parent mineral of the montmorillonite group:

(1) *Substitution in the $(Al_2(OH)_2)_n$ layer.* Since the formula for pyrophyllite is $Al_2Si_4O_{10}(OH)_2$, the substitution of one sixth of the aluminium atoms by magnesium leads to a lattice of composition $((Al_{1.67}Mg_{0.33})Si_4O_{10}(OH_2))^{0.33-}$. Since magnesium, a divalent ion, is replacing trivalent aluminium, the composite layer will have an overall fractional negative charge of $0.33-$. This negative charge is balanced by the presence of alkali metal ions between the stacked layers and at crystal edges giving an overall composition of $Na_{0.33}(Al_{1.67}Mg_{0.33})Si_4O_{10}(OH)_2$, which is the formula for montmorillonite, the mineral from which the group as a whole takes its name.

(2) *Substitution in the $(Si_2O_5)_n$ layers.* Starting once again with the formula for pyrophyllite $Al_2Si_4O_{10}(OH)_2$, substitution of one sixth of the silicon atoms by aluminium gives rise to a lattice of composition $(Al_2(Si_{3.67}Al_{0.33})O_{10}(OH))^{0.33-}$ having a net negative charge of $0.33-$ due to the substitution of tetravalent silicon by trivalent aluminium. Once more the net negative charge on the layer is balanced by the presence of alkali metal ions situated between the stacked layers and at crystal edges. This leads to the mineral of composition $Na_{0.33}(Al_2(Si_{3.67}Al_{0.33})O_{10}(OH)_2$ which is known as beidellite.

There are of course many other possible substitutions in addition to the ones given as examples above, but one which is of particular interest arises from the more extensive substitution of silicon in the $(Si_2O_5)_n$ layers by aluminium. If, for example, one quarter of the silicon atoms in pyrophyllite are replaced by aluminium, a composite layer having unit negative charge and a composition $(Al_2(Si_3Al)O_{10}(OH)_2)_n^{n-}$ is formed. The negative charge is balanced by the presence of a potassium ion K^+ and the resulting mineral is potash mica, $K(Al_2(Si_3Al)O_{10}(OH)_2)$, having the structure shown in Fig. 25.

If further isomorphous substitution occurs such that half the silicon atoms in the $(Si_2O_5)_n$ layers are replaced by aluminium, then a lattice of composition $(Al_2(Si_2Al_2)O_{10}(OH)_2)_n^{2n-}$ is formed having $2n$ negative charges; in the mineral margerite (a calcium mica) this net negative charge is balanced by the presence of calcium (Ca^{2+}) ions. In both the potash and

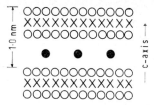

Fig. 25 *Simplified representation of the structure of potash mica* $\bigcirc = (Si_2O_5)_n$ *layer;* $\times = (Al_2(OH)_2)_n$ *layer;* ● $= K^+$ *ions.*

calcium micas the sandwich layers are held together by ionic bonding resulting from the K^+ and Ca^{2+} cations between the layers.

In addition to K^+ and Ca^{2+} as charge balancing cations in the micas, it appears that the hydroxonium ion H_3O^+ may also be able to function in this capacity, and the species of minerals called the illites (or hydrous micas), if they can be regarded as a separate species, probably contain this ion together with K^+, Na^+ and Ca^{2+}.

Finally it should be mentioned that complex cations may also act as charge balancing cations. An important example of such a cation is $((Mg_2Al)(OH)_6)^+$ which is known to be present in the chlorite group of minerals of composition $((Mg_2Al)(OH)_6)(Mg_3(Si_3Al)O_{10}(OH)_2)$. This formula represents the parent substance and other members of the group are formed from this by isomorphous substitution as in the montmorillonite series.

The kaolins, montmorillonites and illites comprise the three main groups of clay minerals. The fourth group of layer lattice minerals is the vermiculite group of which the mineral vermiculite is probably the most important member. Structurally, vermiculite bears the same relationship to the parent substance talc as mica bears to its parent substance pyrophyllite. Talc, as we have seen earlier, consists of a layer of composition $(Mg_3(OH)_2)_n$ sandwiched between two $(Si_2O_5)_n$ sheets; substitution of silicon by aluminium and of magnesium by ferric iron results in a complex anion of composition $((Mg_{4\cdot84}Fe_{0\cdot50})(Si_{5\cdot5}Al_{2\cdot5})O_{18}(OH)_6)_n^{1\cdot32n-}$, and if the charge-balancing cations present are magnesium, then the formula for vermiculite results, *i.e.* $Mg_{0\cdot66}(Mg_{4\cdot84}Fe_{0\cdot50})(Si_{5\cdot5}Al_{2\cdot5})O_{18}(OH)_6 7H_2O$. The cations, as in hydrous forms of mica but unlike those in talc, are hydrated and occupy definite positions with respect to the oxygen atoms of neighbouring talc-like layers. The water molecules may be regarded as forming layers held by hydrogen bonding on the faces of the $(Si_2O_5)_n$ layers of neighbouring sandwiches (Fig. 26); they are held together within

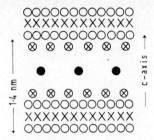

Fig. 26 Simplified representation of the structure of vermiculite $\bigcirc = ((Si_{1.5}Al_{0.5})O_5)_n$ *layer;* $\times = (Mg_3(OH)_2)_n$ *layer;* $\otimes = H_2O$ *molecules;* $\bullet = Mg^{2+}$ *ions.*

the same plane by hydrogen bonding and the hydrous surfaces so formed are then linked together by the cations which lie midway between them.

7.4.5 Structure of 3-D framework polymers

Strictly speaking, 4-connective groups are unable to form ionic polymers since (a) the SiO_4 tetrahedron, which has a maximum connectivity of four, is able to form only a completely covalent network when it shares all of its four corners with neighbouring tetrahedra and (b) the PO_4 group is incapable of forming even a 4-connective covalent network by sharing its corners with other PO_4 tetrahedra on account of the 'inert' (electronically stable) fourth oxygen atom present in the group.

We have seen earlier that, despite its electronic stability, this fourth oxygen atom can act as an electron pair donor (Lewis base) in the presence of a Lewis acid such as BO_3 or AlO_3 and therefore gives rise to co-ordinate bond formation and the production of 4-connective network copolymers of formula $(BPO_4)_n$ and $(AlPO_4)_n$ respectively, both of which exist as 3-D framework structures. The P—O—B and P—O—Al bonds in these framework structures have a large amount of ionic character but there are no —O^-M^+ bonds present such as would be required for the structures to be included under the heading of ionic polymers.

Ionic bonds of the type —O^-M^+ do not occur either in a 4-connective $(SiO_2)_n$ network. However, isomorphous substitution of aluminium for silicon in such networks does lead to 3-D framework structures having ionic bonds of this kind. This arises of course because Al^{3+} is replacing Si^{4+} in the structure and therefore an additional M^+ cation must be present in the lattice to maintain electrical neutrality. Since the net negative charge resulting from substitution resides on the oxygen ions of the framework,

—O^-M^+ linkages of the type found in polymers of lower connectivity are present. These copolymers may therefore be classed as ionic polymers and are discussed below.

3-D Framework silicate polymers
There are three main types of framework silicate of general formula $M_x((Al_xSi_{y-x})O_{2y})_n$; these are: (a) the feldspars, (b) the zeolites and (c) the ultramarines.

The feldspars may be regarded as being formed from four units of an $(SiO_2)_n$ framework, *i.e.* Si_4O_8 by replacement of either one quarter or one half of the silicon atoms by aluminium. Replacement of one quarter of the silicon atoms by aluminium leads to the composition $(AlSi_3O_8)^-$ for this four unit part of the structure, and an overall composition of $(AlSi_3O_8)_n^{n-}$ for the framework as a whole. The net negative charge is balanced by the presence of univalent cations, *e.g.* Na^+ and K^+, and feldspars of formula $KAlSi_3O_8$ (orthoclase) and $NaAlSi_3O_8$ (albite) are very common in nature. Further substitution of aluminium for silicon such that half the silicon atoms have been replaced leads to a framework having a composition $(Al_2Si_2O_8)_n^{2n-}$; the divalent cations Ca^{2+} and Ba^{2+}, are present in the structures of $Ca(Al_2Si_2O_8)$ (anorthite) and $Ba(Al_2Si_2O_8)$ (celsian) respectively as charge balancing ions.

Crystallographers normally classify the feldspars according to the symmetries of their structures, which are related to the sizes of the cations present:

Group 1: large cations present, *e.g.* K^+ (0·133 nm); Ba^{2+} (0·135 nm)
        orthoclase $KAlSi_3O_8$
        celsian $BaAl_2Si_2O_8$
Group 2: smaller cations present, *e.g.* Na^+ (0·095 nm); Ca^{2+} (0·099 nm)
        albite $NaAlSi_3O_8$
        anorthite $CaAlSi_2O_8$

It is interesting to note that the small transition metal ions which commonly occur in the chain and layer lattice silicates are not found as charge balancing cations in the feldspars probably because the framework which has contracted somewhat in the formation of the group 2 feldspars is unable to contract further around the much smaller transition metal ions in order to form a stable structure. The framework of the feldspars may be considered to be built up from SiO_4 and AlO_4 tetrahedra situated at the points of intersection of a 4:8 network and the tetrahedra at these points may point up or down giving rise to a number of possible arrangements and therefore a number of different 3-D frameworks.[16]

A second important group of 3-D framework silicates consists of the naturally occurring and synthetic zeolites. A good example of a natural zeolite is the mineral analcite $Na(AlSi_2O_6) \cdot H_2O$; the framework may be thought of as formed from $(SiO_2)_n$ by replacing one third of the silicon atoms with aluminium, and the cation Na^+ serves as the charge balancing cation; the water molecules are accommodated in the large channels which occur in the structure. In analcite the 16 Na^+ ions per unit cell are distributed at random amongst the 24 possible positions and each one is octahedrally co-ordinated by four oxygen atoms of the framework and two oxygen atoms of neighbouring water molecules. Other important naturally occurring zeolites include chabazite $((Ca, Na_2)(Al_2Si_4O_{12})6H_2O)$ and faujasite $(NaCa_{0.5}(Al_2Si_5O_{14})10H_2O)$ which, like analcite, may be thought of as formed from $(SiO_2)_n$ by isomorphous substitution and contain water as an essential constituent.

The presence of water in the structure is characteristic not only of naturally occurring zeolites but also of synthetic zeolites as well. One example of such a synthetic zeolite is the compound of formula $Na_{12}(Al_{12}Si_{12}O_{48})27H_2O$. The structure of this and of the naturally occurring zeolites may be represented most simply by the linking together of truncated octahedra, the apices of which denote the positions of each Si and Al atom of the network; oxygen atoms are omitted for simplicity. In three-dimensional models of the structures where oxygen atoms are included within these octahedral outlines, it can be seen that there is ample space for the accommodation of water molecules; these spaces within the framework are sometimes called type A sites. When the truncated octahedra are joined together in groups to form the three-dimensional framework structure, a second set of spaces (type B sites) is created; these spaces are rather larger than the ones within the truncated octahedra. The structures of the zeolites consist therefore of these small (site A) and large (site B) holes distributed throughout the 3-D network; it is these holes which are responsible for the high internal specific surface areas of the zeolites and which are responsible for many of their important properties.

All of the zeolites discussed above have 3-D framework structures but there are some other zeolites which have lamellar or fibrous structures. The lamellar zeolites, of which heulandite $(Ca_2(Al_4Si_{14}O_{36})12H_2O)$ is one example have been comparatively little studied, but much more is known of the fibrous zeolites, which resemble the feldspars quite closely. They may be regarded as comprised of chains of tetrahedra, cross-linked to other chains by only a few covalent Si—O—Si linkages giving the crystals a fibrous rather than a massive nature; an example is the mineral natrolite $(Na_2(Al_2Si_3O_{10}) \cdot 2H_2O)$. Viewed end-on the chains in these fibrous zeolites have a square appearance and the structures of these compounds can be most easily represented on paper in terms of patterns made up from these squares, the points of intersection of one square with another indicating the positions of the cross-links.

The third main group of framework silicate polymers are the ultramarines; they are of interest not only because of their colour, which is the basis of their commercial use as pigments but also because, from the structural point of view they are closely related to the zeolites referred to above. They are based on $((Si, Al)O_2)_n$ frameworks and like the zeolites contain positive ions e.g. Na^+ in the internal cavities of the cage structure. They differ from the zeolites, however, in two respects: (a) they contain negative ions e.g. S^{2-}, Cl^- and SO_4^{2-} as well as positive ions within the cavities and (b) they are anhydrous (like the feldspars). Examples of this group of 3-D framework silicates, which differ only in the nature of the anion present within the cage structure are:

ultramarine	$Na_8((Al_6Si_6)O_{24})S_2$
sodalite	$Na_8((Al_6Si_6)O_{24})Cl_2$
noselite	$Na_8((Al_6Si_6)O_{24})SO_4$

All three may be regarded as formed from $(SiO_2)_n$ by replacement of half the silicon with aluminium as in other framework silicates. A further kind of substitution takes place in the mineral helvite, $(MnFe)_8((Be_6Si_6)O_{24})S_2$ where Be^{2+} substitutes for Si^{4+} in tetrahedral sites and the charge balancing cations are Mn^{2+} and Fe^{2+}. There are many other known examples of ultramarines containing cations such as Li^+, Tl^+, Cu^{2+} or Ag^+ and others in which Se^{2-} and Te^{2-} are present in the cages instead of S^{2-}.

7.5 STRUCTURE/PROPERTY RELATIONSHIPS OF IONIC SILICATE AND PHOSPHATE POLYMERS

Ionic polymers are by definition polymers which contain both covalent and ionic bonds. In the previous section it has been shown that as the connectivity of the polymeric structure increases, the proportion of covalent bonds and hence the relative molecular mass of the polymer increases and the proportion of ionic bonds decreases until the maximum connectivity value for each unit is reached and the structure then consists ideally of only covalent bonds as are present in $(SiO_2)_n$ and $(P_2O_5)_n$.

The presence of both covalent and ionic bonds in the structures of the polymeric silicates and phosphates leads to interesting properties which are not exhibited either by the fully covalent polymeric structures on the one hand or by the largely ionic monomeric structures of the orthosilicates and orthophosphates on the other hand. The distinctive properties of this group of ionic polymers arises therefore from the simultaneous presence of both (a) the polymeric anion and (b) the cations present in the structure. In the present section the relationships between the structural features of these polymers and their properties will be reviewed.

7.5.1 Cleavage of infinite chain and 2-D sheet polymers

The way in which the polymeric ion and the cation of the structure contribute to mechanical properties may be seen in the cleavage properties of the infinite chain silicates—the pyroxene and amphibole minerals. In these minerals, as we have seen earlier, $(SiO_3)_n^{2n-}$ and $(Si_4O_{11})_n^{6n-}$ chains respectively lie along the crystallographic axes and the chains are held together by bands of cations lying between them. Since the bonding within the chains is covalent and strong whilst that between the chains is ionic and rather weaker, cleavage in these crystals will take place preferentially along the cation bands; furthermore, since these cation bands lie along the direction of the chains, it is in fact the orientation of these chains which gives the minerals their characteristic cleavage planes and which determines the cleavage direction.

In the ortho-pyroxenes this direction is (210) and for the clino-pyroxenes (110), giving angles between cleavage planes of 87° and 93°. Figure 27(a) shows typical cleavage pathways through a pyroxene crystal; in this diagram the crystallographic axis is perpendicular to the plane of the paper with the chains appearing trapezoidal in shape when viewed end-on (Fig. 27(b)).

The structures of the amphiboles may be similarly represented, as shown in Fig. 28. These minerals show planes of well-developed prismatic cleavage which intersect at angles of about 124° and 56° compared with the approximately 90° angle in the pyroxenes. It is in fact this difference in the angle of cleavage which allows the pyroxene and amphibole minerals to be distinguished one from another. Figure 28(b) shows the structure of the trapezoidal units in the amphiboles and Fig. 28(a) indicates how they are packed together. The cation bands are shown in this diagram together with two possible cleavage routes (1) and (2) through the structure, both

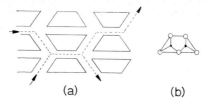

Fig. 27 Cleavage in the pyroxene minerals: (a) packing of chains (seen in end-on view) showing cleavage directions; (b) structure of the trapezoidal units; ● = silicon; ○ = oxygen.

of which give rise to the same cleavage angle. In route (1) suggested by Warren[37] the cleavage passes along cation bands and involves the breaking of seven bonds between oxygen and cations in sites of 6-fold co-ordination. The alternative route (2) suggested by Taylor et al.[38] requires the breaking of fewer and weaker bonds between oxygen and cations in sites of 8-fold co-ordination and therefore requires the least expenditure of energy.

A similar type of correlation between cleavage and structure is found in the layer lattice minerals. Here the structure most closely resembling the pyroxenes and amphiboles is that of the micas where covalently bound composite layers of the structure are held together by ionic bonding resulting from the cations (*e.g.* K$^+$) which lie between them. The perfect

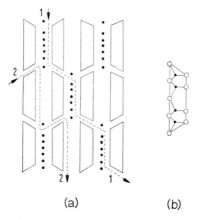

Fig. 28 Cleavage in the amphibole minerals: (a) packing of chains (seen in end-on view) showing two possible routes for cleavage; ● = cations: (b) structure of the trapezoidal units; ○ = oxygen; ● = silicon.

laminar cleavage shown by mica is a matter of common experience, and results from the breaking of the weaker ionic bonds between layers in preference to the strong covalent bonds within the polymeric sheets.

7.5.2 The strength of asbestos

In the above examples, the characteristic cleavage properties of the infinite chain and 2-D sheets silicates depend on the combination of the weakness of the ionic bonding between chains and sheets and the strength of the covalent bonds within the chains and sheets. The strength of this covalent bonding can be illustrated by reference to the asbestos minerals which include the amphibole minerals amosite and crocidolite (blue asbestos) as two of the most important commercial forms and the mineral chrysotile (white asbestos) as the third. In these fibrous amphiboles the covalently bound double chains lie along the fibre axis whilst in chrysotile, the form of asbestos having the 'scroll' structure (Fig. 23), curved covalently bound planes of atoms are aligned with the axis of the scroll pointing along the fibre axis. The axial strength of all three forms of asbestos would therefore be expected to be determined largely by the strength of the polymeric silicate anion and therefore to be large. Orowan[39] has calculated a theoretical strength for asbestos of about 10 000 MN m^{-2} and practical strengths approaching 60% of this value have been observed in very short crocidolite fibres. The longer fibres which are of commercial interest have rather lower strengths, and typical values are given in Table 11 for standard 4 mm length samples.[40] Crocidolite has the highest room temperature strength of any known mineral, and it is interesting to compare the higher strengths found in the well-orientated crystalline structures with the rather lower strengths quoted for typical 'E-glass' fibres in which the silicate chains are arranged at random; values for carbon fibres are also included for comparison. The asbestos fibres, like E-glass and carbon fibres, are used on account of their high strength as reinforcement for composites. The largest outlet for asbestos fibres is for use in asbestos–cement products for the building industry. These products normally contain from 10 to 16% of asbestos fibres; chrysotile asbestos forms the main bulk of the reinforcement in this application, but up to 40% of the fibres added may consist of amphibole fibres since these help to increase the strength, assist in the dispersion of the fibres and enhance the drainage properties of the asbestos–cement mix. Both chrysotile and the amphibole fibres are used for the reinforcement of thermosetting resins, crocidolite being selected on account of its superior strength as the reinforcement for structural plastics.

TABLE 11

TYPICAL PROPERTIES OF HIGH STRENGTH NATURAL AND SYNTHETIC FIBRES[40]

Property	Amosite	Crocidolite	Chrysotile	E-glass fibres[41]	High strength carbon fibres[42]
Tensile strength ($MN\ m^{-2}$)	2 450	3 450	3 050	1 730	2 600
Young's Modulus ($GN\ m^{-2}$)	160	185	160	70	210–260
Fibre diameter	0·1–1	0·1–1	0·01–1	3·7	8·0
Specific gravity	3·2–3·45	3·37	2·55	2·55	1·75

Other advantages of the amphibole group of minerals as reinforcement are high Young's modulus, high surface area, offering an extended fibre/resin bond, and high thermal stability. The high thermal stability of asbestos is well known from its use in protective clothing. However, if asbestos is exposed to elevated temperatures for any length of time the strength is adversely affected as shown in Table 12. The drop in strength of amosite and crocidolite appears to be associated with the loss of hydroxyl (OH) groups from the structure and the effect may be understood if these hydroxyl groups assist the cations at low temperature in holding together series of overlapping crystallites of various lengths by means of H—O—H bridges. Loss of such combined water at elevated temperatures would then lead to the weakening of the longitudinal adhesion between crys-

TABLE 12

EFFECT OF HEAT ON THE TENSILE STRENGTH OF ASBESTOS FIBRES[43]

Temperature °C	Percentage of original strength retained		
	Amosite	Crocidolite	Chrysotile
320	58	70	91·6
430	32	32	73·3
550	15	20	59·3
Time of heat soak (ks)	0·18	14·4	14·4

tallites and thus the fibres as a whole would be weakened in the axial direction.[43]

7.5.3 Thermal stability

Dehydration/dehydroxylation

The dehydroxylation process which occurs in the case of asbestos minerals at elevated temperatures is also characteristic of other naturally occurring silicate polymers which contain chemically or physically combined water. The layer lattice minerals which all contain water in one form or another suffer from dehydration and dehydroxylation at elevated temperatures and the changes involved can be followed by differential thermal analysis (DTA).[31] Endothermic troughs in $\Delta T/T$ plots appear at 20–130°C for most air-dried naturally occurring clays due to the evolution of physically adsorbed water and further peaks arise at temperatures between 520–720°C due to dehydroxylation of the structures and the formation of new mineral species. Kaolinite, for example, on loss of its hydroxyl groups is converted to metakaolin ($Al_2O_3 \cdot 2SiO_2$); this is thought to be a layer lattice structure in which the two internal hydroxyl groups (Fig. 22) and two external hydroxyl groups in the unit cell are converted to oxide ions and four of the external hydroxyl groups are lost. The basal spacing decreases from 0·72 to 0·63 nm as a result of this process. Other clay minerals behave in an analogous way on dehydroxylation.

Vermiculite, a non-clay mineral, also loses water on heating. As we have seen earlier (Fig. 26) this mineral has cushions of inter-layer water on each of its internal surfaces. Loss of water occurs in three stages; up to 10°C, 9·7% of inter-layer water is lost reversibly and the basal spacing contracts from 1·4 to 1·28 nm. Between 200 and 450°C another 5·1% of inter-layer water is lost (again reversibly) with a further decrease in basal spacing to 0·93 nm; the X-ray pattern of the mineral at this stage resembles that of talc. At 550–750°C a further 5·4% this time of chemically combined water in the form of OH groups is evolved and the product resembles enstatite, a mineral which is stable up to the melting point. If, instead of being gradually heated, the raw vermiculite flakes are subjected to temperatures of 800–1100°C for periods of 2 seconds to 4 minutes, the water between the layers is rapidly converted to steam and exerts a disruptive effect on the structure; the vermiculite expands by as much as 12 to 26 times its original volume into a concertina-like mass and the colour changes from brown-black or yellow to a silvery or golden tint. Vermiculite is always used in this exfoliated condition and since the grains are very porous, the material finds application as a thermal and acoustic insulant often in the form of a 'loose-fill' but sometimes also incorporated in plasters. It is also used as an aggregate in lightweight concrete and as a moisture retaining soil conditioner. As a thermal insulant ($k = 0.05$ W m^{-1} k^{-1} for particles of about 10 to 25 mm diameter) for both radiant and convective heat, vermiculite may be used up to about 1100°C but prolonged exposure at such temperatures leads to a brittle product due to further structural changes involving the loss of chemically combined water.

High temperature changes
At temperatures within the range 930–1000°C the DTA curves for the layer lattice minerals show exothermic peaks due to the formation of new mineral species. The changes involved have been studied in detail in the case of kaolinite; on prolonged heating at 950°C metakaolin splits off silica and is converted to an aluminium/silicon defect spinel having a lattice parameter of about 0.7886 nm. At higher temperatures still, in the range 1100–1200°C, the aluminium/silicon spinel is further converted to mullite ($3Al_2O_3 \cdot 2SiO_2$) which is the stable phase of the mineral of this composition up to the melting point as the phase diagram[44] for the SiO_2/Al_2O_3 system indicates. In this high temperature range, the silica split out is present as cristobalite and amorphous silica which has been formed at lower temperatures during the conversion of metakaolin to aluminium/silicon spinel also crystallises as cristobalite. The overall scheme may be written:

$$6(Al_2O_3 \cdot SiO_2 \cdot 2H_2O) \qquad \text{kaolinite}$$
$$\downarrow 450\text{–}550°C$$
$$6(Al_2O_3 \cdot 2SiO_2) \qquad \text{metakaolin}$$
$$\downarrow 950\text{–}1000°C$$
$$3SiO_2 + 3(2Al_2O_3 \cdot 3SiO_2) \qquad \text{Al/Si spinel}$$
$$\text{(amorphous)} \quad \downarrow 1100\text{–}1200°C$$
$$8SiO_2 + 2(3Al_2O_3 \cdot 2SiO_2) \qquad \text{mullite}$$

and the overall reaction is:

$$3(Al_2Si_2O_5(OH)_4) \longrightarrow 3Al_2O_3 \cdot 2SiO_2 + 4SiO_2$$
$$\text{kaolinite} \qquad\qquad \text{mullite}$$

The structure of mullite is not known with certainty on account of (a) its variable composition (from $2Al_2O_3 \cdot SiO_2$ to $3Al_2O_3 \cdot 2SiO_2$) and (b) the difficulty of obtaining single crystals suitable for X-ray work. Recently the structure of a mineral of composition $2Al_2O_3 \cdot SiO_2$ has been elucidated.[45] The structure is closely related to that of sillimanite (Fig. 20(f)) but having the following differences: (a) a greater extent of isomorphous substitution of aluminium for silicon, (b) one fifth of the oxygen atoms of sillimanite removed to balance the loss in positive charge resulting from (a), and finally (c) silicon and aluminium atoms more randomly positioned.

The breakdown of the magnesium analogue to kaolinite, *i.e.* serpentine, proceeds according to the equation:

$$2(Mg_3Si_2O_5(OH)_4) \longrightarrow 3(Mg_2SiO_4) + SiO_2 + 4H_2O$$

the monomeric orthosilicate being produced at all temperatures below about 1000°C. At temperatures above this, the reaction product consists of an equimolar mixture of the monomer and the infinite chain silicate enstatite $(MgSiO_3)_n$, the overall reaction then being:

$$Mg_3Si_2O_5(OH)_4 \longrightarrow Mg_2SiO_4 + MgSiO_3 + 2H_2O$$

The corresponding overall reaction for pyrophyllite, the parent mineral of the montmorillonites is:

$$3(Al_2Si_4O_{10}(OH)_2)) \longrightarrow 3Al_2O_3 \cdot 2SiO_2 + 10SiO_2 + 3H_2O$$

and for its magnesium analogue, talc:

$$Mg_3Si_4O_{10}(OH)_2 \longrightarrow 3(MgSiO_3) + 2SiO_2 + 2H_2O$$

Enstatite is the sole product of decomposition for material of the talc composition (B/A ratio) as indicated by the phase diagram for the MgO/SiO_2 system.[44] Thus we see that the long chain silicate structure is thermally stable up to the melting point provided that there is insufficient base present to cause depolymerisation. If Lewis bases are present in the system then reaction with the formation of monomer occurs.

Depolymerisation also occurs in polyphosphate systems at elevated temperatures in the presence of Lewis bases such as O^{2-} and F^-, which are able to combine with one half of the depolymerised structure, *e.g.*

$$(P_3O_{10})^{5-} \xrightarrow{O^{2-}} (PO_4)^{3-} + (P_2O_7)^{4-}$$

The analogy between this type of reaction and the pH dependent depolymerisation of silicates in alkaline solution has been discussed by Gimblett.[22]

In the absence of excess base the polyphosphate anions have high thermal stability.

7.5.4 Hydrolytic degradation of polyphosphates

Although the polyphosphates are thermally stable in the absence of Lewis bases, they are unstable in aqueous environments. In the presence of water all polyphosphates undergo slow or rapid hydrolysis to the monomeric form, the rate of degradation depending on (a) the type of polyphosphate involved and (b) the environmental conditions.

Ultraphosphates containing 3-connective tetrahedral units are the most susceptible to hydrolysis (the anti-branching rule[3]), rupture of one of the three P—O—P linkages occurring rapidly on dissolution. The products of this hydrolysis, which may be 2-connective rings or chains, degrade, like other 1 to 2-connective polyphosphates at a much slower rate giving the monomeric $(PO_4)^{3-}$ ion as a final product.

The rate of hydrolysis of the 1- to 2-connective polyphosphates depends on the environmental conditions. Of these, (a) temperature and (b) pH are the most important but (c) catalytic effects of a number of different kinds (e.g. the presence of enzymes, colloidal gels, complexing cations, etc.) may also exert a considerable effect. At room temperature and neutral pH, the rate of hydrolysis is extremely slow. For example tetramethylammonium tripolyphosphate has a half-life of the order of years at 25°C and pH = 7. But the rate of hydrolysis increases as the temperature increases and as the pH decreases so that at 100°C and pH = 1 (strongly acid solution) the corresponding half-life is just under 1 minute. Hydrolysis itself, particularly in the case of the very long chain polyphosphates, leads to a decrease in pH and therefore to an increase in the rate of hydrolysis, since the rupture of P—O—P linkages by neutral water molecules leads to the conversion of the hydrogen of water to the hydrogen of a phosphoric acid.

The hydrolysis of pyrophosphates $((P_2O_7)^{4-})$ and tripolyphosphates $((P_3O_{10})^{5-})$ at constant pH and temperature has been shown to be first order and to involve the reactions:

$$(P_3O_{10})^{5-} \xrightarrow{H_2O} (P_2O_7)^{4-} + H_2PO_4^-$$
$$\downarrow H_2O$$
$$2HPO_4^{2-}$$

A very large increase in the first order rate constant is observed above about pH = 8 for pyrophosphates and above pH = 6 for tripolyphosphates, and temperature dependence experiments show an average value for the activation energy of 92–108 kJ mol^{-1}.

In the case of tetrapolyphosphates and longer chain structures, ring formation leading to predominantly $(PO_3)_3^{3-}$ proceeds concurrently with hydrolysis but independently of it. The hydrolysis of rings and of the long chain polyphosphates is found to be catalysed by both acids and alkalis. In neutral and alkaline solution, hydrolytic degradation occurs predominantly by the removal of end PO_4 groups, so that the concentration of orthophosphate builds up rapidly in solution. In acid solutions, some random scission also occurs along the chain so that shorter chain polyphosphates are also produced in addition to orthophosphate. The presence of Ca^{2+} and other cations which readily form complexes with polyphosphates have a marked effect on the rate of hydrolysis but do not affect the product distribution.

The branching points in ultraphosphates are, as mentioned earlier, very susceptible to rapid hydrolysis, so that essentially all branching points are destroyed by the time dissolution is complete. The reasons for this appear to be twofold: (a) a lower activation energy (79 kJ mol^{-1}) compared with that observed for 1 to 2-connective structures and (b) the zero overall charge on the PO_4 group (due to the absence of ionic —O$^-$ groups) which allows easier access of the negatively charged oxygen atoms of the water molecules to the phosphorus atoms during nucleophilic attack.

7.5.5 Hydrolytic stability of silicate polymers

By contrast with the polyphosphates, the soluble crystalline polysilicates hydrolyse in water giving alkaline solutions. The mechanism of hydrolysis involves two stages: (a) exchange of alkali metal ions in the silicate surface with hydrogen ions from dissociated water molecules,

$$\begin{bmatrix} & ONa & ONa & \\ & \diagdown & \diagup & \\ & Si & & \\ & \diagup & \diagdown & \\ O & & O & \end{bmatrix}_n + 2n\ H_2O \longrightarrow \begin{bmatrix} & HO & OH & \\ & \diagdown & \diagup & \\ & Si & & \\ & \diagup & \diagdown & \\ O & & O & \end{bmatrix}_n + 2n\ NaOH$$

followed by (b) depolymerisation of the polysilicic acid formed as the pH of the solution increases:

$$\begin{bmatrix} & HO & OH & \\ & \diagdown & \diagup & \\ & Si & & \\ & \diagup & \diagdown & \\ O & & O & \end{bmatrix}_n \xrightarrow{H_2O/OH^-} n \begin{bmatrix} & HO & OH & \\ & \diagdown & \diagup & \\ & Si & & \\ & \diagup & \diagdown & \\ & HO & OH & \end{bmatrix}$$

The insoluble crystalline silicate polymers are much more resistant to chemical attack but over long periods of time and in a suitable environment depolymerisation will take place. Such depolymerisation accounts for example for the formation of layer lattice minerals from the parent feldspars during the course of weathering (Fig. 11) and for the existence of other species in sedimentary deposits of neutrally occurring minerals. There are indications[9] that during initial attack, the silicate minerals dissolve, even the silica and alumina being at least for a short time in true ionic solution. The first stage involves hydration and hydrolysis whereby alkali and alkaline earth metal cations are leached out of the mineral and oxide ions are partly replaced by hydroxyl groups; both Al^{3+} and Si^{4+} form stable octahedral and tetrahedral arrangements respectively with hydroxyl and when set free from the lattice form such complex ions in solution. In a second stage, these co-ordinated ions unite to form colloidal aggregates, as outlined in an earlier section, initially probably in an amorphous form, which develop crystal lattices such as those of the clay minerals on ageing.

Behaviour of the kind just described appears to apply in the case of the feldspars, 3-D framework structures in which the basic cations are held within a densely packed polymeric framework and whose leaching out can only occur with a concurrent depolymerisation of the 3-D network as a whole. Some silicate minerals may not undergo complete breakdown of the lattice during weathering; for example the micas present in igneous rocks only lose the basic cations present between the 2-D sheets on weathering and the resulting fragments of the 2-D sheet structure may then be directly incorporated into the new (clay) mineral. Following Goldich,[46] the order of stability of the minerals in igneous rocks to weathering may be written:

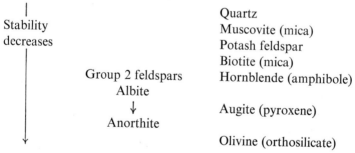

The rate of weathering is dependent on a number of factors including temperature and pH (carbonic acid content of the water) as two main variables. The altered rock crumbles under the mechanical effects of erosion and its constituents may (a) remain *in situ* or (b) be transported by wind, water and ice and redeposited or (c) remain in solution. Since hydrated alkali metal ions are very stable in solution, they usually remain as such, but silicon, aluminium and iron are normally soon deposited as insoluble compounds leading to new minerals at an early stage of weathering.

7.5.6 Hydration/dehydration properties of zeolites and bentonites

Some polymeric silicates show interesting reversible hydration/dehydration properties which are closely related to their structures. These include the industrially important zeolites (3-D copolymers) and the bentonite clays which contain montmorillonite (2-D sheet polymer) as a major constituent. Alkali or alkaline earth metals exist in both structures as charge balancing cations in association with water.

In the case of the zeolites, which contain up to 50% of large cavities with interconnecting windows within the structure, this associated water moves in and out of the network reversibly without collapse or rearrangement of the crystal structure; the phenomenon occurs in both the naturally occurring zeolites (faujasite, chabazite, Na-mordenite and K-mordenite) and in the synthetic zeolites (*e.g.* the 'Union Carbide' molecular sieves having window sizes varying from 0·3 nm (type 3A) to 1·0 nm (type 13X)). If the zeolites are activated by heating in air, the water is driven off and the crystals on cooling show a strong tendency to recapture water molecules or in dry atmospheres to capture other molecules that happen to be present, provided these are not too large to be able to penetrate the windows and enter the pore spaces. The strong preferential adsorption of water and other polar molecules makes the synthetic zeolites particularly suitable for use as industrial desiccants and the high adsorption capacity for other molecules is used to good effect in separation and purification processes, in adsorption gas chromatography and in their employment as pore selective catalysts.[47,48]

The reversible loss and uptake of water which occurs with the zeolites is also observed in the montmorillonite group of minerals. The structure of the parent mineral of the montmorillonite series, pyrophyllite, is depicted in Fig. 24, but of course the montmorillonite minerals derived from this incorporate also charge balancing cations which are located at random positions between the layers and on the edges of crystals; water may also be present in both pyrophyllite (as indicated) and in the montmorillonites. In the dry condition, the basal spacing in the montmorillonites is 0·96 nm but this spacing may increase to as much as 2·41 nm as a result of the uptake of inter-layer water, the actual extent of the expansion depending on the nature of the charge balancing cation present. Sodium montmorillonites which contain Na^+ as the charge balancing cation may absorb up to about five times their own weight of water and increase in volume up to fifteen times their dry bulk volume; they are classified commercially as 'swelling' bentonites. Calcium montmorillonites, on the other hand, in which the charge balancing cation is Ca^{2+} do not swell to the same extent and are classified as 'non-swelling' bentonites. The reasons for the marked difference in swelling behaviour of the sodium and calcium montmorillonites are not as yet very clearly understood but several suggestions have been made[8] of the way in which the cation may exert its influence on water uptake: (a) if the cation serves as a bond holding the layers in the particles together (as in the micas for example) the strength of the bonding may limit the distance to which they can be separated; (b) the size of the adsorbed cation and its tendency to hydrate will influence the overall nature of the arrangement of water molecules and the thickness to which the orientation can develop and (c) the geometry of the adsorbed ions may also be of importance in relation to their possible fit into the water network with the disruption or weakening of the hydrogen bonded water configuration. There are indications that in the sodium montmorillonites the Na^+ ions are not hydrated whereas in the calcium montmorillonites the Ca^{2+} ions are octahedrally co-ordinated by water as a first stage during water uptake and that this is followed by completion of a water layer having a hexagonal type of structure.

7.5.7 Ion exchange

Analogous to the reversible uptake and loss of water in zeolites and the montmorillonites, is the reversible exchange of charge balancing cations in the crystal lattices with cations in solution. Since this process (a) is reversible and (b) involves no permanent change in the structure of the solid it is one of true ion exchange.[49]

As long ago as 1870, Lemberg[50] observed that the natural zeolite leucite $(K_2(Si_4Al_2)O_{12})$ could be converted to analcite $(Na_2(Si_4Al_2)O_{12} \cdot 2H_2O)$ by leaching with sodium chloride solution and that the analcite could be transformed back to leucite quantitatively by rinsing with a strong solution of potassium chloride in water; this clearly established the stoichiometry and the reversibility of the exchange reaction. Synthetic zeolites, which have similar but more closely controlled pore sizes, show similar effects. Recently catalytically active ions have been exchanged with the indigenous sodium ions of the 'as prepared' synthetic zeolites to give catalysts of high intrinsic activity and high internal specific surface area which may also be used as pore selective catalysts by choice of a zeolite having the appropriate window size; such catalysts are now widely used in hydrocarbon conversion reactions. Natural and (later) synthetic zeolites have also been used since about 1910 for water softening,† this application involving an ion-exchange reaction of the type:

$$2(Na^+(zeolite)^-) + Ca^{2+}_{aq} \longrightarrow Ca^{2+}(zeolite)^-_2 + 2Na^+_{aq}$$

regeneration
strong NaCl solution

The cation exchange capacities of zeolites may be as high as 500 to 600 milliequivalents per 100 grammes (me/100 g).

In the montmorillonite clays, charge balancing cations such as Na^+ and Ca^{2+} can migrate readily along the water filled planes between the composite layers of the crystals to exchange with ions in solution; such clays therefore show relatively high cation exchange capacities of the order of 80 to 150 me/100g. Vermiculite, which has a similar layer structure exhibits a cation exchange capacity of about the same order of magnitude. Other layer lattice minerals like the micas and the kaolin group of minerals show much lower cation exchange capacities. In the mica group, as we have seen earlier, the sheets are strongly bound together by inter-layer cations and the basal spacing is about 1·0 nm; consequently the potassium ions in potash mica are unable to migrate freely and exchange is confined largely to the edges and faces of the mineral particles. Similarly in kaolinite, the only exchangeable cations are those held at the edges of the crystallites by 'broken bonds' and at the external surfaces; in this case too, the cation exchange capacity is low, typical figures being 10–40 me/100g for illite (a finely divided micaceous clay) and 3–15 me/100g for kaolinite. If it were not for the very small sizes of these mineral particles when dispersed in water (the majority of particles are of colloidal dimensions) the cation exchange capacities would be almost zero.

7.5.8 The stability of colloidal suspensions

As we have seen, the naturally occurring layer lattice silicate minerals are largely composed of very small particles of the polymeric anion (say 90% less than 1 μm in diameter) supporting charge balancing cations at the edges and faces of crystals and, in the case of the expanding layer lattice

---

† The zeolites have now been largely superseded by synthetic ion-exchange resins for this purpose.

minerals, within the crystal lattice as well. When these minerals are suspended in water, an equilibrium is set up between the cations in the solid and the cations in the water of the type:

$$A^+ (clay)^- + B^+_{aq} \rightleftharpoons B^+ (clay)^- + A^+_{aq}$$

the polymeric anion (clay)$^-$ being of colloidal dimensions (diameter within the range 0.001 to 1 μm). Suspensions of clays in water therefore exhibit phenomena typical of polyelectrolyte solutions containing species of high relative molecular mass e.g. Brownian motion, light scattering and electrophoresis. The zeta potential of such a system is defined by the equation:

$$\zeta = \frac{\sigma d}{\varepsilon \varepsilon_0}$$

where d is the thickness of the double layer surrounding the negatively charged particle, σ is the surface charge, ε is the permittivity of the medium and ε_0 is the permittivity of a vacuum. The nature of the cation has a strong influence on d in this equation and therefore on the resulting zeta potential. The zeta potential is low when the associated cations are H^+, Ca^{2+}, Mg^{2+} or polyvalent cations and high when the counterions are Li^+, Na^+ or K^+. Since a high zeta potential signifies strong particle–particle repulsion, the particles present in aqueous suspensions of lithium, sodium or potassium clays will repel each other and will therefore remain as discrete particles. Conversely, since a low zeta potential signifies a weak particle–particle repulsion, particles of hydrogen, calcium and magnesium clays in aqueous suspension are able to approach each other sufficiently closely to allow the Van der Waals' attractive forces to become operative and therefore for agglomeration or flocculation of the particles to occur. Sodium and similar clays therefore form stable suspensions in water whereas calcium and similar clays form unstable suspensions. Naturally occurring clays usually exist in the calcium form and often contain soluble calcium salts as associated impurities (e.g. gypsum, $CaSO_4$); in order to form a stable suspension of such a clay (e.g. for use in slip-casting) it is necessary to get rid of the associated impurities and to replace the indigenous calcium ions of the clay by sodium ions. This can be done by ion exchange of the type mentioned previously but since the firmness with which cations are held to the polyanion decreases in the order:

$$H^+ > Al^{3+} > Ba^{2+} > Sr^{2+} > Ca^{2+} > Mg^{2+} > NH_4^+ > K^+ > Na^+ > Li^+$$

(the Hofmeister series) either (a) excess of the required cation must be present in solution for the equilibrium:

$$Ca^{2+}(clay)_2^- + 2Na_{aq}^+ \rightleftharpoons 2Na^+(clay)^- + Ca_{aq}^{2+}$$

to be pushed to the right or (b) the unwanted ion must be precipitated as an insoluble salt. In ceramic practice, deflocculation is normally carried out by addition of a solution of sodium carbonate in water, which not only exchanges Na^+ for Ca^{2+} but also removes Ca^{2+} ions from solution as insoluble calcium carbonate. Such deflocculation has a marked effect on the rheological properties of clay suspensions and an analogous effect is observed with highly concentrated suspensions or pastes; sodium clays flow readily, have a low 'water tolerance' (*i.e.* very little excess water is needed to render them fluid) and a low yield value, whereas calcium clays exhibit just the opposite properties.

A second way in which the zeta potential of a suspension may be increased and deflocculation enhanced is by increasing the surface charge (σ) on the particles. This can be done by adding to the suspension polyvalent anions of high surface charge which are strongly adsorbed on to the surface. The long chain polyanions $(SiO_3)_n^{2n-}$ and $(P_nO_{3n+1})^{(2n+1)-}$, where n in the latter case is between 5 and 15, are particularly effective and are widely used as deflocculants. In ceramic practice, solutions of sodium carbonate and sodium silicate are normally used synergistically not only for deflocculation but also to control the rheological properties (fluidity and thixotropy) of the clay suspensions as well. It has been suggested in the case of the polyphosphates that the strong adsorption and deflocculating power is related to the complexing ability of these anions and a related effect, namely the ability of the polyphosphates to suspend inorganic materials such as calcite ($CaCO_3$) colloidally in aqueous media, is attributable to the same cause.

The complexing ability of the polyphosphates has been discussed by Van Wazer.[3] Cyclic phosphates do not form complexes to an appreciable extent whereas the chains form very strong soluble complexes. The complexing is non-specific in the sense that polyphosphates form complexes with all metallic cations and the chain length appears to have little effect on complexing ability, the complexes formed by $(P_2O_7)^{4-}$ being almost as stable as those formed by long chain polyanions. Sodium tripolyphosphate has a high sequestering action and this combined with its crystallinity, good deflocculating properties, mildly alkaline nature (pH = 9·8) and good buffering capacity makes it particularly suitable as a builder for synthetic detergents. Although it is the newest member of the family, sodium tripolyphosphate is commercially the most important polyphosphate produced today.

REFERENCES

1. Eitel, W. (1964–6). *Silicate Science* (5 Vols.), Academic Press, New York.
2. Eitel, W. (1954). *The Physical Chemistry of Silicates*, University of Chicago Press.
3. Van Wazer, J. R. (1958). *Phosphorus and its Compounds*, Vol. 1, Interscience, New York.
4. Turner, F. J. and Verhoogen, J. (1960). *Igneous and Metamorphic Petrology*, 2nd edn, McGraw-Hill, New York.
5. Hatch, F. H. and Rastall, R. H. (1971). *Petrology of the Sedimentary Rocks*, 5th edn, Thomas Murby, London.
6. Deer, W. A., Howie, R. A. and Zussman, J. (1966). *An Introduction to Rock-Forming Minerals*, Longman, London.
7. Dana, E. S. (1932). *A Textbook of Mineralogy*, 4th edn, Wiley, New York.
8. Grim, R. E. (1953). *Clay Mineralogy*, McGraw-Hill, New York.
9. Mason, B. (1952). *Principles in Geochemistry*, 3rd edn, Wiley, New York.
10. Grimshaw, R. W. (1971). *The Chemistry and Physics of Clay and Other Ceramic Materials*, 4th edn, Benn, London.
11. Burns, R. G. (1970). *Mineralogical Aspects of Crystal Field Theory*, Cambridge University Press.
12. Clews, F. H. (1969). *Heavy Clay Technology*, 2nd edn, Academic Press, London and New York.
13. Norton, F. H. (1949). *Refractories*, 3rd edn, McGraw-Hill, New York.
14. Johnstone, S. J. and Johnstone, M. G. (1961). *Minerals for the Chemical and Allied Industries*, 2nd edn, Chapman and Hall, London.
15. Bragg, W. L. (1937). *Atomic Structure of Minerals*, Cornell University Press.
16. Wells, A. F. (1962). *Structural Inorganic Chemistry*, 3rd edn, Oxford University Press.
17. Sidgwick, N. V. (1927). *The Electronic Theory of Valency*, Oxford University Press.
18. Pauling, L. (1940). *The Nature of the Chemical Bond*, 2nd edn, Oxford University Press, p. 69.
19. Van Vlack, L. H. (1964). *Elements of Materials Science*, 2nd edn, Addison-Wesley, New York, p. 41.
20. Holliday, L. (1968). *J. Appl. Polym. Sci.*, **12**, 333.
21. Van Wazer, J. R. and Callis, C. F. (1962). 'Phosphorus-Based Macromolecules,' in *Inorganic Polymers*, Ed. F. G. A. Stone and W. A. G. Graham, Academic Press, New York and London.
22. Gimblett, F. G. R. (1963). *Inorganic Polymer Chemistry*, Butterworths, London.
23. Vail, J. G. (1952). *Soluble Silicates*, Vol. 1, ACS Monograph No 116, Reinhold, New York.
24. Welch, A. J. E. (1955). 'Solid–Solid Reactions,' in *Chemistry of the Solid State*, Ed. W. E. Garner, Butterworths, London.
25. Barrer, R. M. (1949). *Disc. Faraday Soc.*, **5**, 326.
26. Bockris, J. O'M., Mackenzie, J. D. and Kitchener, J. A. (1955). *Trans. Faraday Soc.*, **51**, 1734.
27. Greensfelder, B. A., Voge, H. H. and Good, G. M. (1945). *Ind. Eng. Chem.*, **37**, 1038.
28. Ciapetta, F. G. and Plank, C. J. (1954). 'Catalyst Preparation,' in *Catalysis*, Ed. P. H. Emmett, Reinhold, New York, Vol. 1.

29. Wills, J. H. (1969). 'Silicon Compounds (Silicates)' in *Encyclopedia of Chemical Technology*, 2nd edn, Ed. R. E. Kirk and D. F. Othmer, Interscience, New York.
30. Bennett, H. and Reed, R. A. (1971). *Chemical Methods of Silicate Analysis*, Academic Press, London and New York.
31. Mackenzie, R. C. (Ed.) (1957). *The Differential Thermal Investigation of Clays*, Mineralogical Society (Clay Minerals Group), London.
32. Brindley, G. W. (Ed.), (1951). *X-ray Identification and Crystal Structures of Clay Minerals*, The Mineralogical Society (Clay Minerals Group), London.
33. Corbridge, D. E. C. (1969). 'The Infra-red Spectra of Phosphorus Compounds,' in *Topics in Phosphorus Chemistry*, Ed. M. Grayson and E. J. Griffith, Interscience, New York, Vol. 6, 235.
34. Crutchfield, M. M., Dungan, C. H., Letcher, J. H., Mark, V. and van Wazer, J. R. (1967). P^{31} Nuclear Magnetic Resonance,' in *Topics in Phosphorus Chemistry*, Ed. M. Grayson and E. J. Griffith, Interscience, New York, Vol. 5.
35. Ghose, S. (1965). *Zeit. Krist.*, **122**, 81.
36. Whittaker, E. J. W. (1960). *Acta Crystallogr.*, **13**, 291.
37. Warren, B. E. (1929). *Z. Kristallogr. Kristallgeom.*, **72**, 42.
38. Hodgson, A. A., Freeman, A. G. and Taylor, H. F. W. (1965). *Mineralog. Mag.*, **35**, 5.
39. Orowan, E. (1948–9). *Rep. Prog. Phys.*, **12**, 186.
40. Hodgson, A. A. (1965). *Fibrous Silicates*, London: Royal Institute of Chemistry Lecture Series No. 4, 24.
41. Benjamin, B. S. (1969). *Structural Design with Plastics*, Van Nostrand Reinhold, New York, 21.
42. Courtaulds, Ltd., Carbon Fibres Unit, Graphil Data Sheets FSL8, August, 1972.
43. Hodgson, A. A. (1965). *Fibrous Silicates*, London: Royal Institute of Chemistry Lecture Series No. 4, 26.
44. Levin, E. M., Robbins, C. R. and McMurdie, H. E. (1964). *Phase Diagrams for Ceramists*, The American Ceramic Society, Columbus, Ohio.
45. Sadanaga, R., Tokonami, N. and Takeuchi, Y. (1962). *Acta Crystallogr.*, **15**, 65.
46. Goldich, S. S. (1938). *J. Geol.*, **46**, 17.
47. B.D.H. Ltd., 'Union Carbide' *Molecular Sieves for Selective Adsorption*, 3rd Edn, 46 pp.
48. Chen, N. Y. and Weisz, P. B. (1967). *Kinetics and Catalysis*, American Institute of Chemical Engineers, Symposium Series 73, Vol. 63, 86.
49. Kirk, R. E. and Othmer, D. F. (Eds.) (1966). *Encyclopedia of Chemical Technology*, 'Ion Exchange,' Interscience, New York, Vol. 11, 871.
50. Lemberg, E. (1870). *Z. Deut. Geol. Ges.*, **22**, 335.

CHAPTER 8

INORGANIC GLASSES AS IONIC POLYMERS

N. H. RAY

8.1 INTRODUCTION

There are several excellent textbooks on inorganic glasses [7,8,9] and recent developments in glass science have been the subject of a number of reviews.[14,15] No attempt is made to cover the same ground in this chapter; instead it is the intention to present a different approach to the understanding of inorganic glasses by regarding them as polymers: oxide glasses are considered as polymers of oxygen, and chalcogenide glasses as polymers of sulphur, selenium or tellurium. Inorganic glasses were not generally regarded as polymers until quite recently, and even now there is some reluctance to consider glass as a polymer; for example there is no entry under 'Glass' in the *Encyclopedia of Polymer Science and Technology*. Zachariasen[1] was the first to recognise that oxide glasses have a random network structure, and he proposed a number of rules to determine the glass-forming ability of different oxides. According to Zachariasen's rules, simple oxides of the type A_2O and AO should not form glasses, and with the possible exception of water, none does; on the other hand, oxides of the type A_2O_3, AO_2 and A_2O_5 can form glasses which have random network structures composed of AO_3 triangles or AO_4 tetrahedra with shared vertices. More complex glasses, that is glasses containing several different oxides, necessarily contain appreciable amounts of at least one oxide that is glass-forming; other oxides which are not glass formers can also be present and are regarded as modifiers. Zachariasen suggested that the modifier cations fit into the interstices of the network. Warren's[2] investigations of the structure of simple glasses by X-ray diffraction methods substantially confirmed Zachariasen's views, and in spite of criticisms by

Morey[3] and Hägg[4] the random network theory has been widely accepted. More recently it has been recognised that many oxide glasses exist which do not obey the Zachariasen rules and a number of other theories of glass formation have been proposed. These are of two types: those which attempt to relate glass-forming ability to bond character or bond strength, (for example Stanworth's correlation[5] and Sun's bond strength criterion[6]), and those which depend upon a consideration of the processes that occur during solidification of a melt. Theories of glass formation and structure that are not restricted to oxide glasses alone have obvious merit and fall into the latter category.

8.1.1 Polymeric nature of glasses

Factors concerned in glass formation
Whether or not a glass is formed when a melt solidifies depends upon the extent of reorganisation required for crystallisation to occur, and the thermal energy available for the necessary structural changes to take place. There are in general two types of crystallisation process; in the first the units of crystal structure are the same as those present in the liquid. No structural reorganisation is needed for crystallisation to occur and the rate of crystallisation is ultimately dependent on the rate of diffusion of the units through the liquid. Unless diffusion is hampered by molecular interactions or intermolecular forces such as hydrogen bonds, glass formation is only observed under special conditions of extremely rapid cooling and such glasses are very unstable.

In the second type of crystallisation process the units of crystal structure are different from those present in the liquid, and a reorganisation involving the rupture of interatomic bonds must take place before crystals can be formed. The rate of crystallisation in this case depends not only upon the rate of diffusion in the liquid but also upon the rate at which bonds can be broken and reformed. The lower the melting temperature of the solid in relation to the bond energies concerned in this process, the smaller is the thermal energy available at the melting point for reorganisation into a crystalline form. Consequently glasses are most likely to be formed by substances of relatively low melting point which have a chain or network structure in the molten state. For this reason, the majority of stable glasses are polymers. Furthermore, since linear polymers such as the alkali metal polyphosphates can assume a crystalline configuration without structural reorganisation, it is to be expected that the most stable glasses are formed by highly branched or cross-linked polymers. However, not all substances that form glasses are polymeric; phosphoric acid, zinc chloride and beryllium fluoride all form stable glasses on cooling; glycerol, ethylene glycol, and many monohydric alcohols form glasses readily; stable glasses are also formed by mixtures of simple salts such as potassium and calcium nitrates. In most of these cases it is evident that the liquid has a quasi-polymeric structure from its relatively high viscosity and low vapour pressure; such liquids consist of chains or networks composed of bridging halide ions or hydrogen bonds. Nevertheless there are a few instances of glass formation for which it is difficult to postulate any sort of polymeric structure in the liquid, for example, mixtures of alkali and alkaline earth metal nitrates.

Types of inorganic glasses
Although most glasses are polymers, using this term in its broadest sense, it is convenient to classify inorganic glasses into those which have chain or network structures composed principally of covalent bonds, and those which are either not polymeric, or which are associated in the liquid state only through the formation of hydrogen bonds, bridging halide ions, or other interatomic forces that are about an order of magnitude weaker than a covalent bond. Here we shall only be concerned with polymeric inorganic glasses, of which there are two main types, the oxide glasses and the chalcogenide glasses.

Oxide glasses consist of hetero-atom chains or networks composed of oxygen atoms linked together by covalent bonds to elements such as boron, silicon, germanium, phosphorus, and arsenic whose oxides are polymeric in the molten state, and are referred to as network formers. They may also contain metal cations randomly distributed through the network and linked to oxygen atoms by ionic bonds. Such cations interrupt the mainly covalent network and can be regraded as chain terminators, but in addition they contribute to inter-chain bonding through co-ordination to more than one oxygen atom. Some elements, such as aluminium, lead, zinc and cadmium can occupy both types of position in the structure, sometimes appearing as network forming elements and sometimes as metallic cations. The essential difference between network formers and cations is clearly shown by the process of ion exchange: glass containing, say sodium ions immersed in molten potassium nitrate at 350°C undergoes an exchange of cations, so that potassium ions enter the glass and sodium ions are transferred into the melt. This process can occur at temperatures below the transformation range even though the potassium ion is larger than the sodium ion and its introduction into the surface imposes compressive stresses of the order of 10^4 psi. By contrast, there is no process apart from chemical disruption of the network by which network former atoms can be removed from or substituted in a glass except at much higher temperatures of the order of 1000°C.

The chalcogenide glasses consist of homopolymeric chains of sulphur, selenium or tellurium cross-linked through atoms of silicon, germanium, phosphorus, arsenic, antimony or bismuth. They may also contain halogens which act as chain terminators.

Because all the network-forming elements have valencies of three or more, most inorganic glasses are cross-linked and consequently insoluble except as a result of chemical reactions such as hydrolysis. For this reason the methods available for characterising inorganic glasses are more limited than they are for organic polymers, and the study of glasses as polymers has not kept pace with the tremendous progress that has been made over the last thirty years in the field of organic polymers. Much valuable information about polymeric organic systems has been obtained from studies of their behaviour in solution and in the molten state, but cross-linked organic polymers usually decompose before melting, and consequently the study of highly cross-linked polymers has received comparatively little attention from polymer science. Inorganic glasses constitute a unique class of cross-linked polymers which cannot be studied in solution but which melt without decomposition; they have to be characterised by measurements of properties in the molten or solid state and by observations of their behaviour on melting. Our knowledge of their constitution is limited to what can be derived from this type of information.

8.2 OXIDE GLASSES

Structure

Oxide glasses constitute the largest group of inorganic glasses and include the oldest of all synthetic polymers, for glass was made in Egypt in 1500 B.C. The compositions and properties of the many different kinds of oxide glasses have been described in considerable detail in a number of textbooks[7,8,9] but there have been few attempts at a unified correlation of structure and properties. The customary classification of oxide glasses into silicates, borates, borosilicates, phosphates, etc., is useful for descriptive purposes but draws the attention away from the one constituent that is common to them all, oxygen. If a typical oxide network, such as vitreous silica, is drawn to scale (Fig. 1) it can be seen that the structure is an

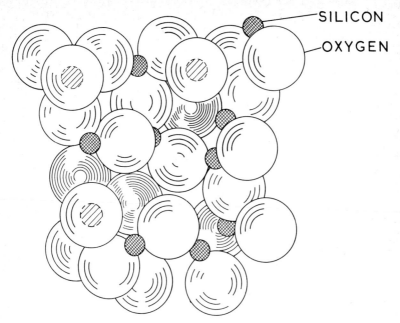

Fig. 1 *Silica network approximately to scale.*

assembly of oxygen atoms connected by mainly covalent bonds to the much smaller network-forming atoms which occupy the interstices of the assembly. The packing of this assembly would not be greatly affected if the silicon atoms were replaced by phosphorus (about the same size) or boron (about 50% smaller); only the number of network bonds per network atom would be altered. Calculation of the volume per gram-atom of oxygen for some simple oxide glasses shows that similar results are obtained for a number of quite different networks (Table 1). It is only when metal cations are either too large to fit into the spaces available or form sufficiently strong ionic bonds to two or more oxygen atoms to cause a contraction of the assembly, that the packing of the oxygen atoms in an oxide glass is significantly altered. Oxide glasses can therefore be regarded as polymers of oxygen; but in contrast to organic polymers which consist of chains of carbon atoms linked together directly, oxygen polymers consist of chains and networks of oxygen atoms linked by mainly covalent bonds through intermediate, multivalent atoms of network-forming elements such as boron, silicon and phosphorus. In addition to the network-forming elements, metal cations and hydrogen may be present. They modify the

TABLE 1

SPECIFIC VOLUMES OF SOME
NETWORK OXIDES

Oxide	Density of glass g cm^{-3}	Volume per g-atom of oxygen cm^3
As_2O_3	3·701	17·8
B_2O_3	1·812	12·8
GeO_2	3·628	14·4
P_2O_5	2·235	12·7
SiO_2	2·203	13·6

network in two ways; they act as chain terminators, interrupting the covalent backbone and reducing the cross-link density, and they can also modify the density of packing of the oxygen atoms. These are therefore two of the factors which are important in relating the properties of an oxide glass to its composition and structure; namely the density of cross-linking in the network and the density of packing of the oxygen atoms. The cross-link density in organic polymers has been defined as the fraction of monomer units which are cross-linked;[10] it is not easy to extend this definition directly to complex oxide glasses because it is not generally obvious what constitutes a 'monomer unit'. However, a corresponding and mathematically equivalent definition can be formulated in terms of network connectivity. In any oxide network there can be two kinds of oxygen atom; those which link together two other atoms that are part of the network (*i.e.* network-forming atoms such as B, Si, P) and those which are linked to only one network atom; the latter either carry a negative charge or constitute part of a hydroxyl group. Oxygen atoms of the former kind are referred to as 'bridging' oxygens. If there are not more than two 'bridging' oxygens per network-former atom, the structure is equivalent to that of a linear polymer; each additional 'bridging' oxygen atom constitutes a cross-link. The cross-link density may therefore be defined as the average number of 'bridging' oxygen atoms in excess of two per network-forming atom. In other words the cross-link density is equal to the average connectivity of the network atoms minus two. The connectivity and hence the cross-link density of most oxide glasses can be calculated from their composition if the system is anhydrous; for example the connectivity and cross-link density of pure silica are 4 and 2 respectively. The cross-link density of silicate glasses can vary between 2 and zero according to the proportions of cations. The cross-link densities of anhydrous phosphoric oxide and boric oxide are equal to unity and are reduced by combined water in the form of hydroxyl groups, as well as by cations. Glasses can also be obtained in which the ratio of metal ions to network-former atoms is greater than the valency of the network former minus 2. Such glasses cannot be cross-linked and must be of comparatively low molecular weight; they are mostly water-soluble and correspond formally to the oligomers of organic chemistry.

The oxygen packing density can be calculated from the empirical composition of a glass and its actual density; it is conveniently expressed as the number of gram-atoms of oxygen per litre of glass. The relationship between these parameters and the properties of glasses will be discussed in greater detail below.

Transformation temperature

The glass transition in organic polymers is a second order transition at which many physical properties undergo a fairly abrupt change. Many polymers change from being hard brittle solids to rubbery materials above the transition temperature, and the tensile modulus decreases by two to three orders of magnitude. The transition is normally accompanied by an absorption of heat and an increase in the coefficient of expansion.[11]

With inorganic glasses there is a corresponding second order transition which is observable at temperatures in the transformation range, which is a temperature range over which the thermal expansion coefficient increases considerably and the rate of change of viscosity with temperature shows a marked increase. The transformation range begins at a temperature (the transformation temperature) which approximates to the 'strain point', which is the highest temperature from which the glass can be cooled rapidly without introducing permanent stress. (*Note*: Glass technologists invariably use the word strain in this context, but stress is what is meant.) The transformation range extends up to the annealing point, which is the highest temperature at which a piece of glass can retain its shape under stress for a significant time, and is usually defined as the temperature at which stresses relax within a few minutes. There is generally a much more gradual change in the properties of glass over the transformation range than is the case with an organic polymer near its glass transition temperature, and except with very simple one- or two-component glasses it is not usually possible to locate a single transition temperature. The transformation temperature of an inorganic glass can be defined as the point at which a significant change in the slope of the expansion–temperature curve occurs; this corresponds to the beginning of the transformation range and to a viscosity of 10^{14} poise. As the temperature is raised further the viscosity decreases at an increasing rate as far as the annealing point, where the viscosity is 10^{13} poise, and which is usually from 50 to 100°C above the transformation temperature.

As with organic polymers, the specific heat of a glass undergoes significant changes in the transformation range, and with fairly simple glasses a well-defined dip in a differential thermogram can usually be observed. The magnitude of this endotherm varies with the previous thermal history of the glass, and when the sample has been rapidly cooled there may be an exotherm before the endotherm. This results from the rearrangement of the metastable high temperature structure, which had been frozen in by quenching, towards an equilibrium configuration. Observations of the transformation temperature may also be affected by the rate of heating, which should not be too great.

With complex glasses of five or more components it may not be possible to observe a transformation temperature by differential thermal analysis, because the change in heat capacity is too gradual. In such cases measurements of the change in thermal expansion coefficient with temperature provide the only satisfactory way to determine the start of the transformation range.

At low temperatures, oxide glasses are rigid elastic solids, with a highly cross-linked network structure. In the transformation range the network begins to acquire a degree of mobility that increases as the temperature rises. This can only occur by a process of bond-switching, or transfer of covalent bonds between oxygen atoms, whereby segments of the network can become detached and able to move relative to the rest of the network. At the transformation temperature the internal energy of the glass becomes large enough for a sufficient proportion of network bonds to be loosened in this way. The minimum number of network bonds per atom of a typical segment that must be relocated to allow the segment to become mobile is equal to the average cross-link density of the network; consequently the transformation temperature of an oxide glass depends upon the cross-link density of the network as well as the strength of the bonds of which it is composed. For example, the bond strengths in arsenious oxide, boric oxide, and phosphoric oxide glasses are similar and all three networks have a cross-link density of one; accordingly their transformation temperatures are nearly the same (Table 2). On the other hand, although the bond strengths in germania and silica glasses are not very much greater, the

TABLE 2

PROPERTIES OF SIMPLE OXIDE GLASSES

Glass	As_2O_3	B_2O_3	P_2O_5	GeO_2	SiO_2
M—O bond energy kcal/mole	70	110	88	104	106
Density g/cm^3	3·701	1·812	2·235	3·628	2·203
Oxygen density g-atoms per litre	56·2	78·0	78·7	69·5	73·5
Network connectivity	3	3	3	4	4
Cross-link density	1	1	1	2	2
Transformation temperature °C	160	260	270	800	1200
Viscosity at crystalline m.p. poise	10^6	10^5	5×10^6	7×10^5	10^7
Activation energy of viscous flow kcal/mole	23	40	41·5	75	180

cross-link density of these 4-connective networks is twice as great and their transformation temperatures are correspondingly much higher.

The transformation temperature also depends upon the tightness of packing in the network; the more open the network, the smaller is the increase in internal energy needed to attain the degree of mobility required for the transition, and hence the transformation temperature decreases when the network becomes less tightly packed. The closeness of packing of an oxide network is reflected in the oxygen density, and at a constant cross-link density the transformation temperature therefore increases with oxygen density. This is evident from the data in Table 2; of the 3-connected oxide glasses, arsenious oxide has the lowest oxygen density, even though its actual density is the highest, and accordingly it has the lowest transformation temperature. Similarly, germania glass has a lower oxygen density than silica and a correspondingly lower transformation temperature.

Calculations of oxygen density also reveal another characteristic feature of polymeric oxide glasses. The maximum possible packing density of oxygen atoms is to be found in crystalline oxides such as beryllium oxide in which the cations are so small that the oxygen atoms are in hexagonal close packing. The oxygen density of crystalline beryllium oxide is approximately 120 g.-atom per litre, while the oxygen densities of the commoner oxide glasses are in the region of 70–75 g.-atom per litre. It follows that a typical oxide glass only contains about 60% of the number of oxygen atoms that could be packed into the same volume.

Effects of cations

The cross-link density of an oxide network is generally reduced by the addition of cations (boric oxide is exceptional in that its cross-link density is increased by alkali metal cations) and the transformation temperature would be expected to decrease. If the only effect of cations were to reduce the cross-link density, then equivalent proportions of all cations would have the same effect on transformation temperature, but this is not the case. Cations introduce additional, non-directional forces between segments of the network, the number and strength of which depend on the nature of the ions involved. All cations, even those of the alkali metals, form co-ordinate links to several oxygen atoms in the network, the co-ordination number increasing with ionic radius. These inter-segmental forces are weaker than the covalent cross-links they replace, but they have the effect of raising the transformation temperature in comparison with the value that would result from the reduction in cross-link density alone.

The relative effects of different cations on the transformation temperature of an oxide glass therefore depend on two factors: the number and strength of the co-ordinate links formed and the change in network packing, or oxygen density that is produced; these factors tend to oppose one another. The average co-ordination number and single-bond strength to

TABLE 3

CO-ORDINATION NUMBER AND OXYGEN-BOND STRENGTH OF SOME CATIONS IN OXIDE GLASSES

Cation	Average co-ordination number	Single-bond strength to oxygen kcal/mole
Li	4	36
Na	6	20
K	9	13
Rb	10	12
Cs	12	10
Mg	6	37
Ca	8	32
Sr	8	32
Ba	8	33

oxygen of a number of typical cations are listed in Table 3. For cations of the same valency, the larger ions form more co-ordinate links and these tend to offset the reduction in transformation temperature due to the lower cross-link density; but this effect is opposed by the reduced packing density of the network because a more open structure is needed to accommodate them. This is reflected in a lower oxygen density. On the other hand, strong inter-segmental forces produced by small multivalent cations are reflected in an increased oxygen density, the result of a tightening of the network. The higher the oxygen density of the network the higher the temperature to which it must be heated to obtain the same degree of mobility, so that for a given cross-link density the transformation temperature increases with oxygen density. This is illustrated in Fig. 2, in which the transformation temperature of some metaphosphate glasses, which are essentially linear polymers, are plotted against oxygen density. The observed temperatures lie close to two approximately parallel curves, that for the monovalent cations being almost 200°C lower than that for the divalent cations. In this system of glasses the effect of network packing is dominant, so that the transformation temperature is higher, the smaller the cation. In other systems, however, such as the alkali silicate glasses, the co-ordination number of the cation has the greater influence and for monovalent ions the transformation temperature increases with the size of the cation.

Relationship between transformation temperature and composition
Since the internal energy of a glass at the transformation temperature must be just sufficient to loosen a proportion of network bonds that is equal to the average cross-link density, an empirical relationship between transformation temperature and composition can be derived for oxide glasses generally.[12] The fraction of all bonds of a certain type which have energies equal to or greater than some chosen value E is given by an equation of the form

$$f = \exp(-E/kT)$$

where k is a constant and T is the absolute temperature. In an oxide network containing one or more network formers and various cations there will in general be several types of bond between oxygen and other atoms. If $n_1, n_2, n_3, \cdots$ are the average numbers of each type of bond per unit segment of network (*i.e.* per network atom in a simple network with only one kind of network-forming atom) and $E_1, E_2, E_3, \cdots$ are the internal energy levels at which the respective bonds are sufficiently loosened for the

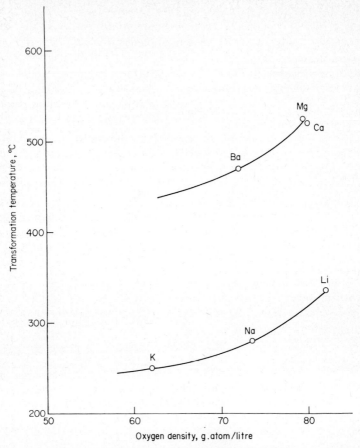

Fig. 2   Transformation temperatures of metaphosphate glasses: upper curve, alkaline earth phosphates; lower curve, alkali metal phosphates.

segment to become mobile, the condition that T = transformation temperature can be expressed:

$$n_1 \exp(-E_1/kT) + n_2 \exp(-E_2/kT) + \cdots = D$$

where D is the cross-link density of the network. In forming this equation, bonds between cations and oxygen are to be regarded as one kind of cross-link just as much as bonds between oxygen and network-forming atoms since appropriate energy terms are inserted for each type of bond. The value of cross-link density to be used is therefore the cross-link density of the fully cross-linked network, *i.e.* two less than the maximum connectivity.

The values of $E_1, E_2, \cdots$ in this equation are not the same as the bond dissociation energies because the process by which the network transforms from a rigid structure to one possessing mobility cannot involve actual dissociation of covalent bonds, but rather a rapid transfer of bonds from one location to another. Consequently the values of $E_1, E_2, \cdots$ must be derived empirically from experimental measurements of transformation temperature. By fitting equations of this form to experimental results on glasses of known composition, values for these constants can be obtained which give reasonably accurate predictions of the effects of compositional changes on the transformation temperature of borate, silicate and phosphate glasses, provided that the glasses are anhydrous.

Melt viscosity

Molecular solids, salts and metals melt to liquids of low viscosity, usually less than 1 poise at the melting temperature. By contrast, polymeric solids such as polyethylene, nylon, and polymethylmethacrylate melt to liquids of very high viscosity, often of the order of 10^6 poise at the crystalline melting point. Network-forming oxides such as silica, boric oxide and phosphoric oxide, whether in the crystalline or glassy state, being polymers of oxygen, also give extremely viscous melts with viscosities in the range 10^5–10^7 poise at the crystalline melting point. The case of phosphoric oxide is particularly interesting. When the hexagonal form, which is a molecular solid made up of P_4O_{10} units, is melted, a comparatively fluid liquid of high vapour pressure is obtained; but if either of the ortho-rhombic forms or the glass is melted, very viscous liquids of low vapour pressure are obtained and these give a glass on cooling.

The flow of a viscous liquid may be regarded as the relaxation of applied stress by the successive displacement of small units of the liquid relative to the bulk. In simple liquids such as water, the units are H_2O molecules; in the case of long-chain polymers such as polyethylene and nylon the mobile unit is a relatively short segment of the chain. Viscous flow in cross-linked polymers must involve at least the temporary rupture of chemical bonds and the activation energy for this process will therefore depend both on the bond energies involved and on the cross-link density, just as the transformation temperature does. In Table 2 the activation energies for viscous flow of some network oxides are compared. The activation energies for the 3-connective networks As_2O_3 and B_2O_3 are similar and much lower than the values for GeO_2 and SiO_2 which have 4-connective networks. Like the transformation temperature, the activation energy of viscous flow also depends on the packing of the network, a more open network undergoing flow more readily than a tightly bound network. This is also shown in Table 2; for a given cross-link density, the higher the oxygen density, the higher the activation energy of flow.

The introduction of cations into an oxide network changes the activation energy of viscous flow in the direction to be expected from the resulting change in cross-link density. Addition of cations to silica invariably reduces the activation energy, monovalent cations causing a larger reduction than divalent cations. On the other hand, addition of cations to boric oxide, which has the effect of increasing the cross-link density (*see below*), causes an increase in the activation energy of viscous flow.

Mechanical properties

The Young's modulus of nearly all inorganic oxide glasses lies in the range 100–1000 kbar, which is about an order of magnitude higher than that of most organic polymers. The modulus of many multi-component glasses is not greatly different from that of the corresponding parent oxide network; for example the modulus of most silicate glasses is close to that of vitreous silica (700 kbar). This suggests that the stress required to distort their structure is mainly determined by the magnitude of the forces between the oxygen atoms. Where differences occur they probably reflect differences in the closeness of packing—that is, differences in oxygen density. For example, the presence of large cations such as potassium which cause the network to expand but which form comparatively weak bonds to oxygen brings about a lowering of modulus; small cations which form strong bonds to oxygen such as aluminium and magnesium increase the packing density and raise the modulus.

It is noteworthy that the three commonest glass-forming oxides have closely similar oxygen densities (Table 1). Unfortunately the value of Young's modulus has only been measured for vitreous silica so it is not possible to compare the moduli of the pure oxide glasses; however, the moduli of binary phosphate and borate glasses that have similar oxygen densities to the parent oxides fall into the same range as silicate glasses.

Plots of Young's modulus versus oxygen density for phosphate and silicate glasses containing a variety of cations show an approximately linear correlation for both networks (Fig. 3).

The tensile strength of oxide glasses is practically independent of composition, being almost entirely determined by surface condition. In the form of thin fibres, tensile strengths as high as 30–35 kbar can be attained, whereas the practical strength of oxide glasses in bulk form rarely exceeds

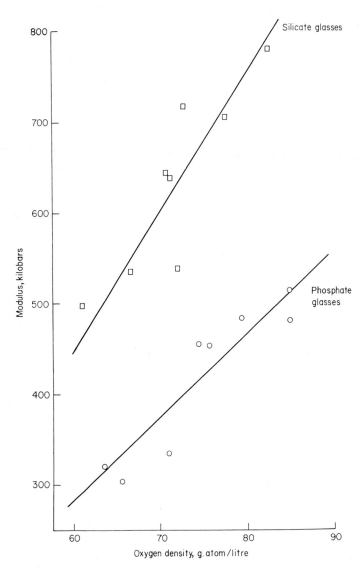

Fig. 3 Correlation of Young's modulus with oxygen density for silicate and phosphate glasses: upper curve, silicate glasses; lower curve, phosphate glasses.

1 kbar. The reasons for this are firstly the absence of any process permitting plastic flow in the structure at ordinary temperature, so that sharp-ended surface cracks remain sharp under stress; and secondly, because of the relatively high modulus of an oxide glass, the work required to extend a crack is less than the strain energy released in the process. Because of the large stress-magnifying factor at the tip of a sharp crack the ultimate strength of the network can be locally exceeded even though the overall stress is still far below the value required to disrupt the oxide structure. The surface crack therefore opens up and deepens, but the strain energy released in this process greatly exceeds the work done in opening up the crack; consequently the crack propagates rapidly, resulting in catastrophic fracture. With materials that behave in this manner the actual stress level at which fracture occurs is a matter of chance, and is determined more by the frequency and size of surface defects than composition or structure.[13] Only in the case of specially prepared specimens with a pristine (defect-free) surface, and with very thin fibres which by their flexibility can retain the perfection of surface that is produced in the drawing process, does the fracture stress approach a significant fraction of the modulus. Even then, most measured values of strength fall far short of the theoretically attainable value for a material having a continuous covalent network, which is about one-fifth of the modulus. Marsh[56] has suggested an explanation for this, which is that stresses in excess of 35 kbar can produce plastic flow in an oxide glass and the observed fracture stress is in fact a yield stress. Plastic yield is not observed in ordinary circumstances because the fracture stress is too low. If Marsh's explanation is correct, it is to be expected that pristine glass strengths would be composition dependent, but there are insufficient data to confirm this.

Mobility of cations

A common feature of all oxide glasses is the easy mobility of simple cations in the structure. The use of a glass electrode to measure hydrogen ion concentration depends upon this property and demonstrates the ease of transport of hydrogen ions from aqueous solution through silicate glass at room temperature. Alkali metal cations are readily transferred from molten salts into glass, and this is the basis of one way of increasing the useful strength of glass; when a glass containing sodium ions, for example, is immersed in a molten potassium salt such as potassium nitrate at a temperature below the strain point, potassium ions enter the glass network in place of sodium ions. The reduced density of the exchanged layer produces a compressive stress in the surface which materially increases the overall stress the glass can support without fracture.[45] Alkali metal cations in glass can be exchanged for silver ions, and alkaline earth ions will also undergo exchange, although much more slowly. The relative rates of diffusion of cations through silicate glasses have been measured by Schulze,[46] Halberstadt,[47] Johnson,[48,49] LeClerc,[50] and others. At temperatures below the transformation range the diffusion coefficients of the univalent ions Li, Na, K and Ag change with temperature according to the equation

$$D = D_0 \exp(-E/RT)$$

where D_0 (the pre-exponential factor) and E (the activation energy) are independent of temperature but vary both with the size and charge of the diffusing ion, and with the composition of the glass. The diffusion coefficients increase markedly in the transformation range, and above the softening point the activation energies for diffusion decrease with temperature. The diffusion coeffiicent of calcium ions in a soda-lime glass is about 100 times smaller than the diffusion coefficient of sodium ions in the same glass at the same temperature. The diffusion coefficients for all the alkali cations increase as the alkali content of the glass increases, and this is attributable to a decrease in activation energy. The self-diffusion coefficient of sodium in a binary sodium silicate glass is about ten times the diffusion coefficient of potassium in a binary potassium silicate with the same molar proportion of alkali, but sodium ions diffuse about five times faster in the potassium glass which has a more open network. Thus the activation energy for diffusion increases with increasing oxygen density as well as with increasing size of the diffusing ion.

8.3 PARTICULAR OXIDE GLASSES

8.3.1 3-Connective networks—phosphate glasses

Background
Although the phosphate network is built up of tetrahedral PO_4 units, only three of the four oxygen atoms in each unit are capable of forming bonds to other atoms so that the maximum connectivity of this network is 3, and the cross-link density can vary between 0 (for linear polyphosphates) and 1 (for phosphoric oxide glass itself). Of all oxide glasses, the phosphates most closely resemble organic polymers and for this reason have often attracted the attention of polymer scientists. The chemistry of the

vitreous polyphosphates has been reviewed by Van Wazer[14] and by Westman.[15]

The techniques of polymer chemistry have been applied with considerable success to the study of chain phosphates; for example measurements of molecular weight distribution have been found to agree closely with the reorganisation theory developed by Van Wazer, and the structure of some soluble polyphosphate glasses is known in great detail. By contrast, highly cross-linked, insoluble polyphosphate glasses, and the multi-component phosphate glasses of importance in glass technology have received very little detailed study.

The fully cross-linked network structure from which phosphate glasses are derived is vitreous phosphoric oxide with a cross-link density of unity. Two series of glasses can be derived from phosphoric oxide by introduction of increasing proportions of metal cations in the one case, and of hydroxyl groups in the other. Both types of substituent reduce the cross-link density, and the limiting compositions corresponding to an infinite linear chain polymer are the metal 'metaphosphate' and metaphosphoric acid respectively. Any composition containing a higher proportion of P_2O_5 than the metaphosphate is classified as an ultraphosphate.

Formation

Phosphate glasses are obtained by polycondensation reactions from mixtures of metal hydrogen phosphates and phosphoric acid, or metal oxides and ammonium phosphate; for example, by heating sodium dihydrogen phosphate at temperatures in the range 400–700°C, linear polymers with the structure HO—[—P(O)(ONa)—O—]$_n$—H are obtained. Similar products are obtained by heating mixtures of alkali carbonates and ammonium dihydrogen phosphate, and more complex glasses can be prepared by melting together mixtures of various metal oxides and/or carbonates with phosphoric acid or ammonium phosphate.[16] When there is stoichiometric equivalence of cations and phosphate units or a surplus of cations, only linear polymers are obtained; but with a stoichiometric deficiency of cations, polymers having various degrees of cross-linking can be obtained, the cross-link density increasing as the condensation proceeds. The reaction

$$2 \begin{array}{c} \mathrm{O} \\ \| \\ -\mathrm{P}-\mathrm{OH} \\ | \\ \mathrm{O}^- \end{array} \rightleftharpoons \begin{array}{c} \mathrm{O} \quad\;\; \mathrm{O} \\ \| \quad\;\; \| \\ -\mathrm{P}-\mathrm{O}-\mathrm{P}- \\ | \quad\;\; | \\ \mathrm{O}^- \quad \mathrm{O}^- \end{array} + \mathrm{H_2O}$$

is reversible and the water formed is only eliminated slowly at temperatures in the range 700–800°C. Since phosphorus pentoxide forms an azeotrope with water boiling at 850°C,[14] attempts to complete the reaction by heating at higher temperatures alter the composition towards that of the metaphosphate. Consequently most phosphate glasses contain combined water in the form of residual hydroxyl groups and this has an important effect on their properties.[16]

The effect of residual hydroxyl groups on the transformation temperature of barium phosphate glasses containing from 30 to 55 mole % barium oxide was studied by Namikawa and Munakata[17] who found that the transformation temperature was very dependent on the conditions of preparation and the variations could be attributed to changes in combined water content in the range 1–3%. They also concluded that the preparation of anhydrous barium phosphate glass is impossible.

Properties

(i) *Transformation temperatures.* The transformation temperature of phosphate glasses increases with continued heating towards a maximum value corresponding to the minimum attainable water content. The residual water content is a measure of the extent to which cross-linking has occurred, because the number of hydroxyl groups per phosphate unit corresponds to the number of potential cross-linking sites that have not yet reacted. In a glass containing a mole fraction y of P_2O_5, the proportion of phosphorus atoms linked to metal cations is $(1 - y)/y$, so that the proportion of phosphorus atoms available for cross-linking is $(2y - 1)/y$. If the cross-link density defined as the fraction of P atoms linked to three other P atoms is x, then the proportion of P atoms carrying a hydroxyl group is

$$z = \frac{2y - 1}{y} - x$$

and this must be equal to the proportion of combined water expressed as moles H_2O per mole P_2O_5. Hence if M is the weight of (anhydrous) glass containing one mole of P_2O_5, the weight fraction of combined water is

$$W = 18z/(M + 18z)$$

whence

$$x = \frac{2y - 1}{y} - \frac{WM}{18(1 - W)}$$

Measurements of the transformation temperatures of a number of multi-component phosphate glasses after refining to different hydroxyl contents have shown that the relationship between transformation temperature and cross-link density in this system is the same as that observed in organic polymers.[18] The relationship can be expressed by an equation derived on theoretical grounds by Di Marzio:[19]

$$\frac{T(X) - T(0)}{T(0)} = \frac{Kx}{1 - Kx}$$

where $T(X)$, $T(0)$ are the transformation temperatures (in °K) of a polymer with a cross-link density of x, and of the corresponding linear polymer respectively; K is a constant for the system, which when x is expressed as the fraction of cross-linked phosphorus atoms has the value 0·6 approximately. Typical curves for some multi-component phosphate glasses are shown in Fig. 4.

The value of $T(0)$, the transformation temperature of the linear polymer, depends both on the nature and on the proportions of the cations present. In the absence of any cation the only inter-chain forces are hydrogen bonds and for this reason polyphosphoric acid, $(HPO_3)_n$, has the very low transformation temperature of $-10°C$. The successive replacement of hydrogen by cations introduces ionic forces between the chains and the transformation temperature rises in direct proportion to the mole fraction of cations towards the value for the appropriate metaphosphate. With monovalent cations the transformation temperature of the linear metaphosphate lies in the range 250–350°C, and with divalent cations which introduce stronger inter-chain forces it falls in the range 450–550°C. The smaller the cation the higher is the oxygen density of the glass and therefore the higher the transformation temperature for the same ionic charge (Fig. 2).

The effects of different cations on the transformation temperature of metaphosphate glasses have also been studied by Eisenberg,[20] who found that

$$T_g = K(q/r)$$

where q is the charge on the cation and r is the sum of the cation and oxygen radii. These results, together with some additional measurements made by the author are given in Table 4.

(ii) *Durability*. All oxide glasses are attacked by water because the equilibrium in reactions such as

$$P_2O_5 + 3H_2O \rightleftharpoons 2H_3PO_4$$
$$B_2O_3 + 3H_2O \rightleftharpoons 2H_3BO_3$$
$$Si{-}O{-}Si + H_2O \rightleftharpoons 2Si{-}OH$$

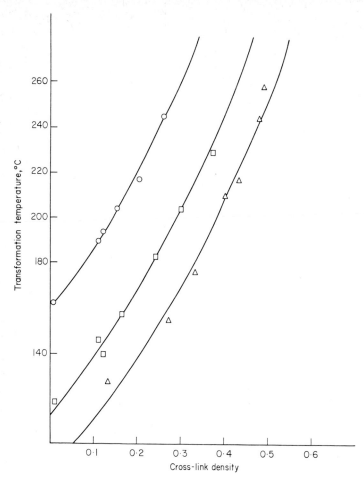

Fig. 4 Change of transformation temperature with cross-link density of phosphate glasses. ⊙ = 60 mole % P_2O_5; ▫ 65 mole % P_2O_5; △ = 68·3 mole % P_2O_5.

lies well to the right at ordinary temperatures. Nevertheless the rate of attack by water and aqueous solutions on most silicate glasses is sufficiently small for them to be useful household materials; although the domestic dishwasher, operating as it does at elevated temperatures and usually with alkaline detergents, readily demonstrates that even apparently durable glass dissolves at a measurable rate.

TABLE 4

TRANSFORMATION TEMPERATURES OF BINARY PHOSPHATE GLASSES

Cation	Ion radius Å	Charge ÷ sum of cation + oxygen radii	Mole % P_2O_5 50	60	70
Li	0·60	0·5	335		
Na	0·95	0·42	280		
K	1·3	0·366	243	252	260
Ca	0·99	0·84	520		
Sr	1·13	0·79	485		
Ba	1·35	0·73	470	400	343
Zn	0·74	0·93	520		
Cd	0·97	0·84	450		
Pb	1·2	0·76	320	305	290
(none, *i.e.* H)			−10		

Simple chain phosphate glasses such as sodium or potassium polyphosphates dissolve in water with very little hydrolysis at room temperature; lithium polyphosphates dissolve less readily. Alkaline earth phosphates, and cross-linked phosphate glasses have only a limited solubility in water and dissolve by hydrolysis of P—O—P bonds, which may be a relatively slow process in some cases. In the case of the alkaline earth phosphate glasses $RO \cdot (P_2O_5)_x$, Kamazawa and co-workers[21] found that the rate of solution in water is a maximum for $x = 1$, that is, at the metaphosphate composition. In acidic conditions the rate of solution decreases as x increases, while in alkaline media this trend is reversed. The durability of ultraphosphate glasses has been shown to increase with cross-link density;[22] furthermore glasses containing small cations such as lithium and magnesium are more resistant to hydrolysis than those containing large cations such as potassium and barium. These observations suggest that the more tightly bound the phosphate network, the more resistant it is to hydrolysis. If this were true generally there should be a correlation between oxygen density and durability, but zinc phosphate glasses which have oxygen densities in the range 80–84 g-atoms per litre are less durable than lead phosphate glasses with oxygen densities around 70 g-atoms per litre.[23] It is evident that certain cations can exert at least as much influence on the rate of hydrolysis of the phosphate network by chemical effects as by purely structural effects.

The introduction of other network formers, especially boric oxide, silica, and alumina, can produce dramatic improvements in the durability of phosphate glasses. Takahashi[24] found that the durability of sodium phosphate glasses was increased by addition of silica, which formed a mixed network; addition of boric oxide produced little or no change in durability unless the P_2O_5 content was 50 mole % or more, when a large increase in hydrolytic stability was observed over the range 10–30 mole % B_2O_3. A similar result was obtained in the case of alumina which produced a very great increase in durability with increasing concentrations in the range 0–10 mole %, provided that the $P_2O_5:Na_2O$ ratio was greater than one. Both boron phosphate and aluminium phosphate have 4-connective networks, and the effects of boric oxide and alumina in phosphate networks can be attributed to the formation of the more highly cross-linked structures associated with BPO_4 and $AlPO_4$ groups. These are more stable in acidic conditions and therefore more likely to be formed in glasses containing a stoichiometric excess of phosphoric oxide.

(iii) Melt viscosity. In common with most oxide glasses, and in contrast to organic polymers, the viscosity of molten phosphate glasses is practically independent of shear rate, so that their flow is Newtonian, and the temperature dependence of viscosity is given by an equation of the form

$$\eta = A \exp(E/RT)$$

For simple binary phosphate glasses such as the sodium polyphosphates, Van Wazer[14] found that the activation energy E increased as the $Na_2O:P_2O_5$ ratio decreased towards the metaphosphate composition; that is, the activation energy increased with molecular weight. In a series of multi-component ultraphosphate glasses containing from 60–70 mole % P_2O_5, the activation energy of viscous flow was found to be approximately independent of cross-link density[23] and not greatly different from the value for phosphoric oxide (41 kcal mole^{-1}). However, the viscosity of a particular composition at a constant number of degrees above the transformation temperature was found to increase markedly with cross-link density; unless, therefore, there is an abrupt change in activation energy in the transformation range, the transformation temperature of these glasses is not an isoviscous temperature.

(iv) Modulus. The Young's modulus of most oxide glasses lies in the range 200–800 kbar, and phosphate glasses have moduli that fall towards the lower part of this range as is to be expected from the lower connectivity of the phosphate network compared to either the borate or silicate networks. As the cross-link density of a phosphate glass is increased by continued dehydration the modulus hardly changes until a cross-link density of 0·25 is reached, after which the modulus increases rapidly with

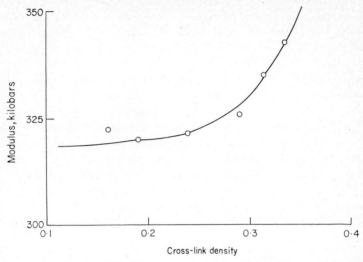

Fig. 5 Effect of cross-link density on Young's modulus of a typical phosphate glass.

continued cross-linking (Fig. 5). As with other oxide glasses, the modulus increases markedly with increasing oxygen density (Fig. 3), and hence bulky cations like potassium that form relatively weak bonds to oxygen tend to reduce the modulus, while small cations of high charge density like lithium and magnesium tend to increase the modulus.

8.3.2 3 to 4-Connective networks—borate glasses

Background

Glasses are formed by boric oxide itself and by mixtures of boric oxide with a very wide range of metal oxides including all the alkali and alkaline earth metal oxides, ZnO, CdO, PbO, Bi_2O_3, La_2O_3, and Tl_2O. The comparatively low softening points of borate glasses make them useful as solder glasses, which are used to join together other glasses such as the face and cone of TV tubes, and as sealing glasses that are used to seal metal and ceramic parts into glass. Borate glasses are of particular interest to the polymer scientist because of the variable connectivity of the boric oxide network that results from the possibility of boron exhibiting valencies of either 3 or 4.

Formation

Boric oxide glass is obtained by dehydration of boric acid; this is a polycondensation reaction resulting in a gradual increase in cross-link density and consequently the viscosity of the melt is continually rising. For this reason the last traces of water are very difficult to remove, even at temperatures as high as 1200°C. Borate glasses containing various cations can be obtained either by dissolving the appropriate metal oxides in molten boric oxide or by condensing boric acid with mixtures of metal oxides or carbonates. The composition ranges over which stable glasses are formed from boric oxide and various metal oxides have been determined by Imaoka,[25] and it is interesting to observe that with zinc, cadmium, lead and bismuth oxides, stable glasses are formed with considerably less than 50 mole % B_2O_3. Stanworth[22] has suggested that these oxides can be incorporated into the glass as network formers as well as modifying cations.

Properties

(i) *Transformation temperature.* The transformation temperature of alkali borate glasses increases with alkali content, steeply at first up to about 20 mole % alkali, then levelling out. Nuclear magnetic resonance measurements on alkali borate glasses[27] show that with increasing alkali content an equivalent proportion of boron atoms become tetravalent. If x is the mole fraction of alkali oxide the fraction of tetravalent boron atoms is equal to $x/(1-x)$. The cross-link density, defined as the fraction of BO_3 or BO_4 units linked to more than two other such units, is therefore equal to $1/(1-x)$. Up to about 20 mole % alkali the transformation temperature increases linearly with cross-link density (Fig. 6). At higher alkali contents, the rise in transformation temperature is much smaller and it seems as if further addition of alkali beyond about 20 mole % no longer causes an equivalent number of boron atoms to become four co-ordinate, but instead results in the formation of non-bridging oxygen atoms so that the cross-link density increases much more slowly and ultimately remains constant. It has been suggested[28] that this happens because adjacent BO_4 groups cannot share a common oxygen atom, so that each tetravalent boron atom must be surrounded by four trivalent borons. On this basis the change in slope of the curve relating transformation temperature and alkali content should occur at $16\frac{2}{3}$ mole % alkali and a cross-link density of 1·2. The nuclear magnetic resonance studies of Bray and O'Keefe[29] contradict this explanation, however, for their results clearly show that the fraction of boron atoms in tetrahedral co-ordination increases linearly with the mole fraction of alkali at least as far as 30 mole % in each of the alkali borates. In this composition range, the number of tetravalent boron atoms is in fact equal to the number of cations present, so that there can be no non-bridging oxygen atoms, and the cross-link density must increase with the

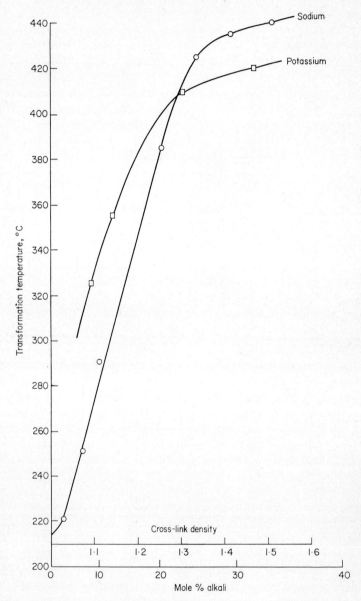

Fig. 6 Change of transformation temperature with cross-link density for alkali borate glasses. ⊙ *sodium borate;* ⊡ *potassium borate.*

mole fraction of alkali well past the composition at which the transformation temperature begins to level out. Another explanation must therefore be sought for the way the transformation temperature changes with composition. It is significant that up to about 20 mole % alkali, potassium borate glasses have higher transformation temperatures than sodium borate glasses with the same mole fraction of alkali, whereas above 23 mole % alkali this order is reversed (Fig. 6). Since both cations produce the same increase in cross-linking, the difference in their effects on transformation temperature must be due to differences in the packing of the network. Calculations of the changes in oxygen density with alkali content (Fig. 7) indicate that below 20 mole % alkali sodium borate glasses have a more open network, but at higher alkali contents potassium borate glasses have the more open structure. It is possible that these structural changes are the result of there being a limit to the number of structural sites where a particular cation can be accommodated without distorting the network. When all such sites are filled, any further increase in alkali content causes a structural change that tends to counteract the effect of the increased cross-link density.

(ii) Expansion coefficient. In oxide glasses such as silicates and phosphates, introduction of alkali cations other than lithium produces an increase in thermal expansion coefficient, as expected from the reduction in both cross-link density and oxygen density, resulting in a more open network. Lithium ions raise the oxygen density of phosphate glasses and accordingly decrease the expansion coefficient. When alkali metal oxides are introduced into boric oxide glass, however, an unusual change in coefficient of expansion occurs. Up to about 15 mole % alkali, the expansion coefficient decreases with alkali content irrespective of which cation is present. At about 17 mole % alkali, the expansion coefficient reaches a minimum value and then increases again in the composition range over which the transformation temperature levels out (Fig. 8).[30] The initial decrease in expansion coefficient can be explained by the increase in cross-link density due to the increase in co-ordination number of some boron atoms, but as already noted the cross-link density continues to increase linearly with alkali content well beyond the composition at which the expansion coefficient passes its minimum value and begins to increase again. So far there is no completely satisfactory theory to account for these observations. A suggestion by Abe that BO_4 groups are unstable at higher temperatures could explain why these changes in expansion coefficient become less pronounced as the temperature increases, but does not help

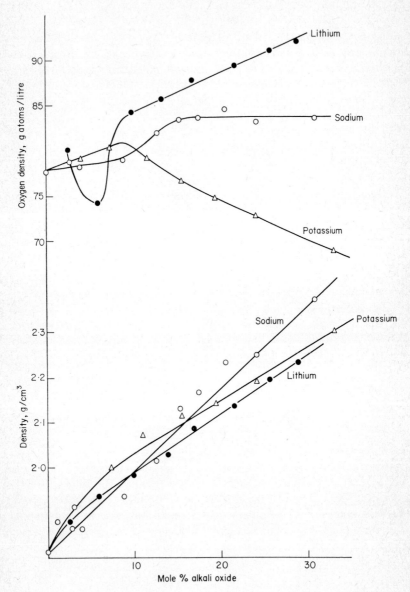

Fig. 7 Densities and oxygen densities of alkali borate glasses. ● *lithium;* ⊙ *sodium;* △ *potassium.*

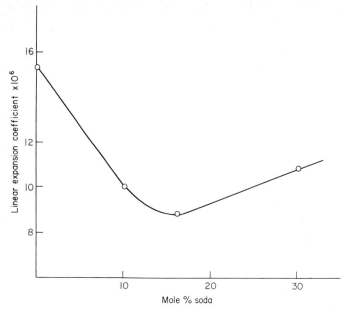

Fig. 8 Thermal expansion coefficient of sodium borate glasses.

to account for the occurrence of a minimum expansion coefficient at room temperature. The changes in oxygen density previously noted offer a partial explanation.

As shown in Fig. 7, the oxygen densities of both sodium and potassium borate glasses rise to a maximum value in the region of 10–15 mole % alkali. At higher alkali contents, the oxygen density either falls, as in the case of potassium borates, or remains almost constant as with sodium borate glasses. Consequently although the cross-link density continues to increase, the network ceases to contract further and in the case of potassium borate it begins to expand. The coefficient of expansion therefore increases again towards the value appropriate to a more open network.

(*iii*) *Modulus.* There have been few recorded measurements of the Young's modulus of borate glasses. Ciesla[31] found that increasing the proportion of lead in a lead borate glass from 30 mole % to 70 mole % reduced the Young's modulus from 560 to 370 kbar. This result supports the view that lead oxide in high concentrations contributes to the network bonding, because this reduction in modulus is no more than would be expected from the decrease in oxygen density, and is much less than would result from the replacement of almost all the covalent B—O—B bonds by relatively weak ionic forces.

8.3.3 4-Connective networks—silica and silicate glasses

The silicate glasses constitute the largest and most important class of inorganic glasses. Their properties range from those of vitreous silica which is notable for its extreme chemical and thermal stability, through the whole range of common soda-lime glasses with varying degrees of chemical resistance, to the simple alkali silicates that are readily soluble in water. Because it is the parent oxide from which all silicate glasses are derived, vitreous silica has been studied in considerable detail and its structure may be regarded as established with considerable certainty. The structures of the many different multi-component glasses in common everyday use have not been studied in any detail, but at least for those containing a relatively high proportion of silica, a reasonably acceptable structure may be deduced from that of silica itself. The position is quite different, however, in the case of the apparently much simpler alkali silicates that are water-soluble, and those silicate glasses that contain relatively high proportions of other oxides such as lead oxide and alumina. The simplest model is based on Zachariasen's theory and assumes that silica has a random 4-connected network structure with a cross-link density of 2. According to this model, addition of metal oxides causes a progressive reduction in cross-link density at random positions in the network, until at the metasilicate composition, when there are two cations per silicon atom in the case of a monovalent metal, or one cation per silicon atom in the case of a divalent metal, the nominal cross-link density has been reduced to zero. Further addition of metal oxide should result in a gradual reduction in the average molecular weight of the anions until the orthosilicate composition is reached. On this basis there should be a gradual and systematic change in physical properties with composition, but in fact this is not observed and there are several features of the property/composition relationships in even the simplest binary silicates that have not yet been satisfactorily explained; in particular the melt viscosity and other properties undergo changes at SiO_2 contents higher than 50 mole % which show that this simple model is not adequate.

Soluble silicates
Since much of what is known about organic polymers has been learned from studies of their behaviour in solution, it might be thought that

investigating the properties of the water-soluble alkali silicates would throw considerable light on the structure of silicate glasses in general, but this is not the case. Although the two common alkali metal silicates form a continuous series of homogeneous mixtures with water from the solid partially hydrated glasses through highly viscous liquids to dilute aqueous solutions, the physical properties of these systems show such a strong dependence on composition, and in certain composition ranges exhibit such variations with concentration, that it seems likely that more than one kind of polymer is involved. The dissolution of an anhydrous alkali silicate glass in water is a complex process involving both chemical change as well as physical solution; the solution first formed in contact with the solid generally contains alkali and silica in different proportions from the glass, and the apparently simple process of diluting a concentrated alkali silicate solution with water can cause changes in pH, electrical conductivity, viscosity, and other properties which continue to alter towards their equilibrium values during the course of several hours or even days.[32] Consequently it would be unwise to try to infer the structure of solid alkali silicate glasses from the properties of their aqueous solutions. It will suffice to mention two properties of these solutions which are in direct conflict with what can be deduced about the structure of the solid glasses from their properties in the molten state. Firstly, lithium silicate glasses made by fusion of lithium carbonate and silica cannot be dissolved in water to form concentrated, viscous solutions like those that are readily obtained by treating sodium of potassium silicate glasses with hot water or steam. Lithium silicate solutions can, however, be prepared without difficulty by dissolving precipitated silica in concentrated aqueous solutions of lithium hydroxide and they then exhibit properties very similar to solutions of sodium and potassium silicates. Secondly, the molecular weight of sodium metasilicate glass in aqueous solution, as determined either by light-scattering[33] or by viscosity measurements,[34] is about 71 which corresponds to complete dissociation into discrete SiO_3 anions. As the ratio of silica to alkali metal oxide in the glass is increased, the weight average molecular weight increases (Fig. 9) but even at the molar ratio $Na_2O \cdot 3SiO_2$ the average molecular weight in dilute solutions is only 300, corresponding to chains of 3 to 4 SiO_4 units.[35] The molecular weight of potassium silicate at similar concentrations is higher but still only corresponds to chains of 4 to 5 SiO_4 units. In more concentrated solutions containing from 30 to 50% by weight of glass the viscosity at constant solids content passes through a pronounced minimum at the disilicate composition $Na_2O \cdot 2SiO_2$,[36] rising steeply with increasing silica content, and more gradually with increasing

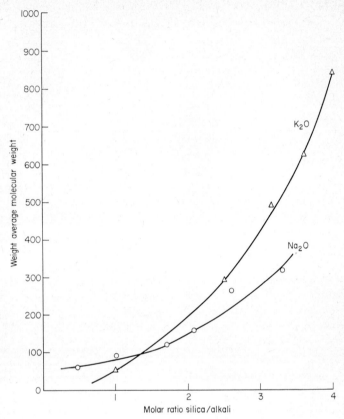

Fig. 9 Molecular weights of alkali silicates in aqueous solution. ⊙ sodium silicate; △ potassium silicate.

alkali. These observations indicate that water can enter into alkali silicate glasses in two ways; the process of dissolution initially involves the hydrolysis of Si—O—Si bonds, and subsequently the hydration of the fragments produced. The extent of hydrolysis must vary not only with concentration but also with alkali/silica ratio, so that the molecular weight of the dissolved glass changes with concentration as well as with composition.

Anhydrous alkali silicates

(i) *Density.* At 25°C, the densities of lithium, sodium and potassium silicates increase almost linearly with alkali content up to about 40 mole %. The oxygen densities of sodium and potassium silicates decrease with alkali content, while the oxygen density of lithium silicate increases with alkali (Fig. 10). At higher temperatures the relationship between density and composition alters; at 1000°C, the density of sodium silicate glasses is nearly independent of composition over the range 20–60 mole % Na_2O, and at 1200°C and above, the density decreases with alkali content in this composition range. Consequently there is a maximum density at about 20 mole % Na_2O at temperatures above 1000°C.[37]

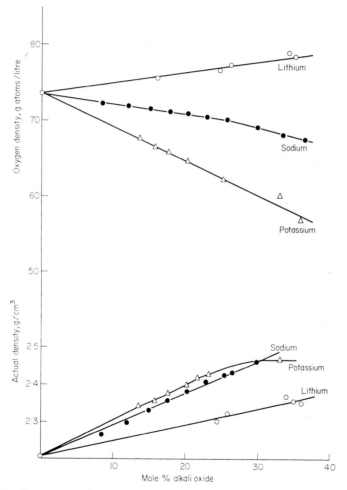

Fig. 10 Densities and oxygen densities of binary alkali silicates. ⊙ lithium silicate; ● sodium silicate; △ potassium silicate.

(*ii*) *Viscosity and transformation temperature.* At temperatures near the melting points of the crystalline metasilicates, the melt viscosities of the anhydrous alkali silicates decrease rapidly with increasing alkali content from values greater than 1000 poise at the composition $R_2O \cdot 3SiO_2$ to less than 10 poise at the metasilicate composition. The low melt viscosity of alkali metasilicate glasses indicates that they are probably not linear polymers of high molecular weight, as might be expected from their stoichiometry; it seems more probable that they consist of rather small though highly branched aggregates with cyclic structures. Glasses are not readily formed by cooling metasilicate compositions except on a small scale and they crystallise readily. Near the metasilicate composition, the melt viscosities of the alkali silicates at 1000°C and above decrease with increasing size of the cation, lithium metasilicate being the most viscous; while at compositions corresponding to the disilicate $R_2O \cdot 2SiO_2$ and at still higher silica contents, the order is reversed.[38] At lower temperatures also, melt viscosity increases in the same order as the radius of the cation, and similarly the transformation temperatures and Littleton softening points (*i.e.* $10^{7.6}$ poise temperatures) increase in the order Li < Na < K, which is the reverse of what would be expected from the oxygen densities (Fig. 11). A possible explanation of this is that while the proportion of Si—O—Si cross-links is reduced to the same extent by any monovalent cation, alkali cations tend to be co-ordinated to more than one oxygen atom in an oxide network, and these weaker links partly counteract the reduction in transformation temperature due to the severance of Si—O—Si linkages. Consequently the transformation temperature of silica is reduced least by the largest alkali cations, which co-ordinate to 9 or more oxygen atoms, and to the greatest extent by lithium which is only 4 co-ordinate at most. At temperatures well above the transformation range, however, these co-ordinate linkages are no longer effective in binding the network and the viscosity depends mainly upon the compactness of the structure; for a fixed proportion of alkali, the viscosity then increases with oxygen density, that is, in the order K < Na < Li.

(*iii*) *Expansion coefficient.* The expansion coefficient of silica is considerably increased by all alkali cations, and for a constant molar proportion of alkali, the coefficient at room temperature increases in the order Rb < K < Na < Li. This is a further manifestation of the effect of the increased number of co-ordinate links that are introduced by the larger cations. At much higher temperatures in the range 900–1200°C, however, the order is reversed and the expansion coefficients increase with increasing

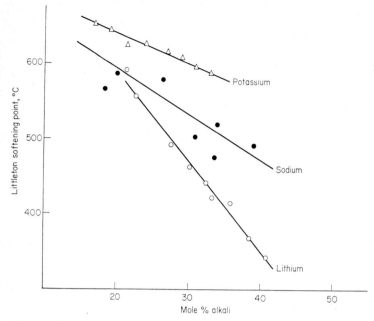

Fig. 11   Littleton softening points of alkali silicate glasses. ⊙ *lithium;* ● *sodium;* △ *potassium.*

size of cation. As was suggested in the case of the viscosity/composition relationship, it is probable that the co-ordinate linkages that are introduced by the cations become much less effective at the higher temperatures and the effects of increased packing density become dominant.

Multi-component silicate glasses

(i) *'Invert' glasses.* Although the majority of industrially important silicate glasses contain from 70 to 85 mole % silica, or a combined content of network formers such as alumina, boric oxide, and silica in this range, Trap and Stevels[39] showed that with a sufficiently complex mixture of Group I and II cations, stable glasses can be formed that contain as little as 40 mole % of silica and no other glass-forming oxide. At this level of network former the covalent bonding only extends to dimer units, *i.e.* to SiO_4 tetrahedra linked in pairs. Trap and Stevels considered that when the silica content is reduced below 50 mole % the role of cations in the glass structure undergoes an inversion, from that of chain terminators to that of chain extenders, and that the glass must be stabilised by the ionic bonds between the isolated groups of SiO_4 tetrahedra. For this reason the name 'invert' glasses was suggested for such compositions. In support of this hypothesis, they showed that the melt viscosity of glasses of this type passes through a distinct minimum at 50 mole % SiO_2. It is significant that all the compositions that give rise to stable glasses in this range contain at least three different cations one of which is divalent. It is possible that the stability of these glasses is due to the combination of a comparatively high degree of co-ordination of cations to oxygen together with relatively strong ionic forces (*cf.* Table 3), so that the essential polymeric structure of the network is maintained.

A similar situation is found with lead and zinc silicates; stable glasses can be formed in the binary system $PbO–SiO_2$ with up to 77 mole % PbO, and lithium–zinc–silicate glasses have been prepared with as little as 35 mole % silica. In these glasses, Stanworth[26] has suggested that lead and zinc atoms can form part of the covalent network structure, linking together pairs of oxygen atoms as in the case of borate glasses of high lead content. It is evident that there is really no clear cut distinction between network-forming elements and modifying cations; both in silicate and borate glasses there is evidence that some divalent elements, including the alkaline earth metals, lead and zinc, can contribute to the network structure.

(*ii*) *Modulus of silicate glasses.* The Young's modulus of vitreous silica is 700 kbars, and the addition of cations may either raise or lower the modulus. Most silicate glasses have Young's moduli in the range 400 to 1000 kbars; but certain constituents, notably alumina and beryllium oxide, give rise to very high values of modulus. There have been numerous attempts to correlate the modulus of glasses with composition; Clarke and Turner[40] found that the modulus of a series of soda-lime–silica glasses of the general formula $(Na_2O + CaO) \cdot 3SiO_2$ could be calculated by means of the equation

$$E \text{ (kbar)} = 13 \cdot 9x + 565 \cdot 6$$

where

$$x = \text{wt. \% CaO}$$

Winkelmann and Schott[41] after extensive measurements on a series of optical glasses obtained a set of factors for calculating Young's modulus from composition by means of an equation of the form

$$E = \sum C_i p_i$$

where p_i is the weight % of metal oxide i and C_i is a 'constant' for the oxide in question but it was found to be necessary to use different values of C according to the type of glass involved, so that many different values had to be used in order to fit measurements to calculations. A more successful correlation of this type is due to Phillips [42] in which p is expressed in mole % and C in kbar per mole %. The values of C for SiO_2, Al_2O_3, and CaO are constant but the values of C for other oxides vary with composition—for example, the coefficient for an alkali oxide varies with the silica content. This approach was extended to alkali-free glasses of unusually high modulus by Williams and Scott,[43] and Bacon and others [44] have shown that certain multi-component glasses containing substantially less than the 50 mole % silica can also have unusually high moduli up to 1500 kbar. The relative effects of different ions on the modulus of silicate glasses is shown in Table 5, from which it can be seen that the modulus contribution per mole of oxide increases with oxygen density (cf. Fig. 3). The highest oxygen densities, and hence the greatest increases in modulus, are provided by the oxides of beryllium, zirconium, aluminium, cerium, yttrium, and lanthanum. Although none of these oxides is a glass former in its own right they are all capable of contributing to the network-bonding providing continuity of the network at low silica contents. The high modulus glasses described by Bacon are therefore distinct from the 'invert' glasses of Trap and Stevels because, unlike the latter, they probably have a continuous covalent oxide network.

TABLE 5

EFFECTS OF DIFFERENT CATIONS ON THE MODULUS OF SILICATE GLASSES

Ion	Relative contribution to modulus, per mole of oxide	Ion	Relative contribution to modulus, per mole of oxide
Li	7	Be	19
Na	4·5	Mg	12
K	3	Ca	13
		Sr	9
Ce	19	Ba	8
Y	24		
La	22	Ti	13
		Zr	18

(*iii*) *Diffusion in silicate networks.* The diffusion of molecules and ions through glass has received much investigation, partly because many industrially important characteristics of glass and glass-making processes depend upon diffusional processes, and also because such studies can yield useful information about the structure of glasses. The activation energy for the diffusion through silica of uncharged molecules such as H_2, N_2 and the inert gases increases as the size of the molecule increases,[51] while the diffusion coefficient at constant temperature decreases with the size of the diffusing molecule. The rate of diffusion of a particular molecule decreases, and the activation energy increases, as increasing amounts of cations are introduced into silica. It is possible that the activation energies of diffusion may be correlated with the oxygen density of the glass, but there are not enough data available to confirm this. By contrast, the rate of diffusion of alkali cations through silicate glasses increases, and the activation energies decrease, with increasing proportions of alkali in the glass. Alkaline earth metals, however, have the reverse effect, for the diffusion coefficient of sodium in a soda-lime glass is lower than in a binary soda–silica glass of the same silica content. The effects of composition on the rates of diffusion of ions have been reviewed by Stevels.[52]

8.4 CHALCOGENIDE GLASSES

General

The chalcogenide glasses are polymers or copolymers of the elements sulphur, selenium or tellurium that are cross-linked through atoms of silicon, germanium, phosphorus, arsenic, antimony or bismuth. Unlike oxygen, the elements sulphur, selenium and tellurium are able to form homopolymers of quite high molecular weight and, with the exception of tellurium, the elements themselves are readily obtained in the solid state as glasses. In oxide polymers, no oxygen atoms are linked directly together; they are connected only through an intermediate (network-former) atom. Since nearly all the intermediate atoms that occur in oxide polymers are multivalent, the cross-link density of oxide glasses tends to be high. In the chalcogenide glasses, on the other hand, chains of varying length of directly linked sulphur, selenium or tellurium atoms are connected by intermediate, multivalent atoms. Consequently the cross-link densities of these glasses is generally much lower than those of oxide glasses. A further important difference between these two systems is to be found in the types of network-limiting atoms or chain terminators that can be incorporated. In oxide polymers, the cross-linking of the network can be reduced, or chains terminated, by cations such as Na^+ or Ca^{++}; in chalcogenide glasses this role is played by the halogens, chlorine, bromine and iodine, together with thallium. Metal cations as such are not found in chalcogenide glasses, although the elements bismuth and cadmium have been introduced into certain chalcogenides.

Constitution

The compositions and properties of the chalcogenide glasses have been reviewed by Pearson[53] and by Rawson.[9] The viscosities of molten selenium and tellurium, and of sulphur above 160°C, are relatively high, indicating that these elements are polymeric in the molten state. Addition of arsenic or antimony to any of these melts causes a marked increase in viscosity up to 40 atom % arsenic; addition of bromine, iodine, or thallium even in quite small amounts causes a marked fall in viscosity. Glasses are formed over wide ranges of composition in both binary and ternary systems containing these elements, with transformation temperatures extending over a very wide range from almost 200°C down to −60°C. Some of these materials have been used for a long time as immersion liquids of unusually high refractive index; others as infra-red transparent materials; and more recently it has been found that many of them are semi-conductors.

Transformation temperature

Tobolsky[54] studied the variation of transformation temperature with composition in the binary systems S—Se and S—As, and also in the ternary system S—Se—As. He found that the transformation temperature of sulphur–selenium copolymers increased linearly with the mole fraction of selenium, which is in accordance with the theory put forward by Di Marzio and Gibbs[55] to predict the glass transition temperatures of copolymers. The transformation temperatures of arsenic–sulphide glasses were also found to increase linearly with the mole fraction of arsenic from −27°C. for pure sulphur up to 160°C. at 40 mole % arsenic. Similar results were obtained by Arai and Saito[61] for the transformation temperatures of arsenic–selenium and arsenic–tellurium glasses (Fig. 12). If each arsenic atom introduced were a cross-linking point, the transformation temperature should increase much more steeply according to Di Marzio's equation[19] which was found to apply to phosphate glasses.[18] Either the number of cross-links formed is less than the number of arsenic atoms introduced, or there is some other factor that changes with arsenic content in such a way as to reduce the transformation temperature. Were the addition of arsenic

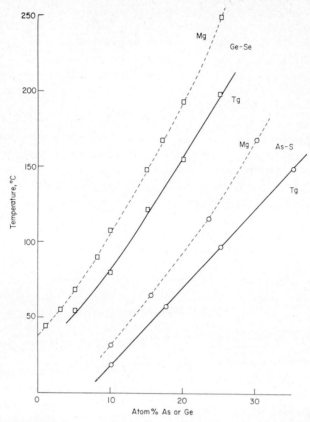

Fig. 12 Deformation temperatures (M_g, dotted lines) and transformation temperatures (T_g, solid lines) of arsenic sulphide and germanium selenide glasses. ⊙ arsenic sulphide; ▫ germanium selenide.

to reduce the packing density of the network *i.e.* to increase the free volume of the polymer, as well as introducing cross-links, this could explain a smaller increase in transformation temperature than would be expected from the increased cross-link density alone; however, Arai and Saito showed that the addition of arsenic reduced the specific volume of arsenic selenide and arsenic telluride glasses, so that this explanation cannot be correct. An alternative hypothesis is that polymeric sulphur contains both rings and chains and that introduction of arsenic alters the ring-chain equilibrium. The rings that are opened provide sites for some arsenic atoms as chain terminators and branch-points, so that only a proportion of the added arsenic actually gives rise to cross-links. The transformation temperatures of arsenic/selenium/tellurium glasses having the stoichiometric compositions $n \cdot As_2Se_3 \cdot As_2Te_3$ were measured by Kolomiets and Pozdnev.[57] They found that the transformation temperature increases with selenium content more steeply than the linear relationship expected for a random copolymer; it is possible that a higher proportion of the arsenic provides cross-linking points in arsenic selenide than in arsenic telluride.

It might be expected that the tetravalent elements germanium and silicon should be twice as effective as arsenic in cross-linking sulphur and selenium chains, but this does not appear to be the case. Nemilov[58] measured the deformation temperature of germanium selenide glasses with increasing germanium contents from 0–25 atom %, and his results show that the softening point rises only slightly faster than corresponds to a linear relationship (Fig. 12). The structure of crystalline silicon sulphide, which consists of chains of four-membered Si_2S_2 rings sharing silicon atoms at opposite corners, suggests that in silicon and germanium chalcogenide glasses also the potential cross-linking effect of Ge and Si atoms is reduced by ring formation. In view of these observations it is not realistic to calculate the cross-link density in chalcogenide glasses from their composition.

Melt viscosity

Because of a rapid increase in the proportion of long chain molecules and a decreasing concentration of S_8 rings, the viscosity of molten sulphur rises steeply just above the melting point to a maximum at 180°C, above which it decreases slowly; there is no corresponding maximum in the viscosity–temperature behaviour of selenium which shows a steady decrease in viscosity as the temperature rises from the transformation temperature upwards. The melt viscosity of tellurium is high near the melting point and falls suddenly at higher temperatures to a value that is indicative of a monomeric liquid. Over the temperature ranges within which each of these elements exhibits the high viscosity characteristics of a long-chain polymer, addition of halogens or thallium causes a considerable reduction in viscosity; for example 0·02% of iodine reduced the viscosity of sulphur from 950 poise to 60 poise.[59] Thus thallium and the halogen elements act as chain terminators, reducing the molecular weight of the polymeric chalcogenides. Phosphorus, arsenic, and antimony have the reverse effect, causing a marked increase in viscosity. The viscosity of arsenic selenide glasses increases steadily with arsenic content up to a maximum value in the region of 45–50 mole % arsenic; at higher arsenic contents the viscosity falls rapidly.[60] Presumably arsenic atoms begin to act as chain terminators when there is more than one per selenium atom; the low viscosity of molten arsenic indicates that As—As bonds would be readily dissociated in the melt.

Mechanical properties
The chalcogenide glasses have much lower tensile strengths and elastic moduli than most oxide glasses; for example arsenic sulphide glasses approximating to As_2S_3 in composition have Young's moduli in the range 70–100 kbars and tensile strengths of about 0·14 kbars,[53] values which are about one-tenth of those typical of silicate glasses. The relatively low moduli of chalcogenide glasses can be partly attributed to the lower cross-link density in these polymers, for some phosphate glasses of very low cross-link density also have moduli in this range. At temperatures in the transformation range, arsenic–sulphur glasses closely resemble organic polymers in viscoelastic properties; with arsenic contents in the range 3–8 atom %, plots of Young's modulus against temperatue show a clearly defined rubbery region.[62] At higher arsenic contents the region of rubbery behaviour becomes smaller, until at 25 atom % arsenic and more, no rubbery region is evident. This behaviour is very similar to that of natural rubber cross-linked with sulphur and demonstrates that similar structural relationships probably apply in both polymer systems.

8.5 CONCLUSIONS

In this survey an attempt has been made to show how the application of some of the concepts of polymer science to inorganic glasses can lead to a rationalisation of many of the observed relationships between their composition, structure, and properties. The properties of phosphate glasses in particular seem to be adequately explained, at least in general terms, by the changes in oxygen density and cross-link density that result from variations in composition. In other systems, however, especially in the case of silicate glasses, there remain a number of features that still lack a satisfactory explanation in structural terms. A serious limitation to the further development of this approach is the lack of detailed and reliable quantitative information about the viscoelastic properties of glasses of simple composition, and this appears to be a fruitful field for research. The possibility of investigating the rheology of cross-linked polymers over a wide range of temperature does not exist in the field of organic materials and should make the study of inorganic glasses particularly attractive to polymer scientists.

REFERENCES

1. Zachariasen, W. H. (1932). *J. Am. Chem. Soc.*, **54**, 3841.
2. Warren, B. E. (1938). *J. Am. Ceram. Soc.*, **21**, 259.
3. Morey, G. W. (1934). *J. Am. Ceram. Soc.*, **17**, 315.
4. Hägg, G. (1935). *J. Chem. Phys.*, **3**, 42.
5. Stanworth, J. E. (1946). *J. Soc. Glass Tech.*, **30**, 54T.
6. Sun, K. H. (1947). *J. Am. Ceram. Soc.*, **30**, 277.
7. Morey, G. W. (1954). *The Properties of Glass*, A.C.S. Monograph no. 124, Reinhold, New York.
8. Stanworth, J. E. (1950). *Physical Properties of Glass*, Clarendon Press, Oxford.
9. Rawson, H. (1967). *Inorganic Glass-forming Systems*, Academic Press, London.
10. Flory, P. J. (1953). *Principles of Polymer Chemistry*, Cornell University Press, New York.
11. Neilsen, L. E. (1962). *Mechanical Properties of Polymers*, Reinhold, New York.
12. Ray, N. H. IXth International Congress on Glass, Versailles (1971), A1, 5, 633.
13. Hillig, W. B. (1962). *Modern Aspects of the Vitreous State* Vol. 2, p. 152; Ed. J. D. Mackenzie, Butterworths, London.
14. Van Wazer, J. R. (1962). *Inorganic Polymers*, Ed. Stone and Graham, Academic Press, New York.
15. Westman, A. E. R. (1960). *Modern Aspects of the Vitreous State* Vol. 1, Ch. 4; Ed. J. D. Mackenzie, Butterworths, London.
16. Ray, N. H. and Lewis, C. J. (1972). *J. Mat. Sci.*, **7**, 47.
17. Namikawa, H. and Munakata, M. (1965). *J. Ceram. Assoc. Japan*, **13**, 38.
18. Ray, N. H. *J. Polym. Sci.* (in press).
19. Di Marzio, E. A. (1964). *J. Res. Nat. Bur. Stds*, **68A**, 611.
20. Eisenberg, A., Farb, H. and Cool, L. G. (1966). *J. Polymer Sci.*, A2, **4**, 855.
21. Kamazawa, Ikeda, and Kawazoe, (1969). *J. Ceram. Assoc. Japan* **77**, 163.
22. Ray, N. H., Lewis, C. J. and others *Phys. Chem. Glasses.* (in press).
23. Ray, N. H., Laycock, J. N. C. and Robinson, W. D. *Phys. Chem. Glasses.* (in press).
24. Takahashi, K. (1962). *Advances in Glass Technology*: Technical papers of the VIth International Congress on Glass, Washington, p. 366, Plenum Press, New York.
25. Imaoka, M. *ibid.*, p. 149.
26. Stanworth, J. E. (1948). *J. Soc. Glass Technology*, **32**, 154T.
27. Krogh Moe, J. (1960). *Phys. Chem. Glasses*, **1**, 26. (1962). *ibid.* **3**, 1.
28. Abe, T. (1952). *J. Amer. Ceram. Soc.* **35**, 284.
29. Bray, P. J. and O'Keefe, J. G. (1963). *Phys. Chem. Glasses*, **4**, 37.
30. Gooding, E. J. and Turner, W. E. S. (1934). *J. Soc. Glass Technology*, **18**, 32.
31. Ciesla (1968). *Diss. Abs.* **29** (2), 576B.
32. Vail, J. G. (1952). *Soluble Silicates*, A.C.S. Monograph No. 116. Reinhold Corp. New York.
33. Weldes, H. H. and Lange, K. R., Amer. Soc. Chem. Eng. Symposium on Glass and related materials, part III. Philadelphia, 1968.
34. Debye, P. and Neumann, R. (1951). *J. Phys. and Colloid Chem.* **55**, 1.
35. Debye, P. and Neumann, R. (1949). *J. Chem. Phys.* **17**, 664.
36. Bacon, L. R. Unpublished work reported in Vail, J. G. ref. 32.

37. Heidtkamp, G. and Endell, K. (1936). *Glastech. Ber.* **14**, 89.
38. Endell, K. and Hellbrügge, H. (1940). *Angew. Chem.*, **53**, 271.
39. Trap, H. L. and Stevels, J. M. (1959). *Glastech. Ber.* **32K**, (VI) 51, and (1960). *Phys. Chem. Glasses* **1**, 107.
40. Clarke, J. K. and Turner, W. E. S. (1919). *J. Soc. Glass Technology*, **3**, 260.
41. Winkelmann, A. and Schott, O. (1893). *Ann. Phys. Chem.* **49**, 401.
42. Phillips, C. J. (1964). *Glass Technology*, **5**, 216.
43. Williams, M. L. and Scott, G. E. (1970). *Glass Technology* **11**, 76.
44. Bacon, J. F. International Commission on Glass (Canadian Ceramic Soc.) Toronto 1967.
45. Ray, N. H. and Stacey, M. H. (1969). *J. Materials Sci.* **4**, 73.
46. Schulze, G. (1913). *Ann. Phys.* **40**, 327.
47. Halberstadt, J. (1933). *Z. anorg. Chem.* **211**, 185.
48. Johnson, J. R. (1950). Ohio State University Thesis.
49. Johnson, J. R., Bristow, R. H. and Blau, H. H. (1959). *J. Amer. Ceram. Soc.* **42**, 271.
50. LeClerc, P. IV International Congress on Glass Paris (1957), p. 331.
51. Doremus, R. H. (1962). *Modern Aspects of the Vitreous State* Vol. 2 p. 1. (see Hillig, ref. 13).
52. Stevels, J. M. (1957). *Handbuch der Physik* Vol. 20. Springer Verlag, Berlin.
53. Pearson, A. D. (1964). *Modern Aspects of the Vitreous State* Vol. 3 p. 29.
54. Tobolsky, A. V., Owen, G. D. T. and Eisenberg, A. (1962). *J. Colloid Sci.* **17**, 717–725.
55. Di Marzio, E. A. and Gibbs, J. H. (1959). *J. Polymer Sci.* **40**, 121.
56. Marsh, D. M. (1964). *Proc. Roy. Soc.* **282A**, 33.
57. Kolomiets, B. T. and Pozdnev, V. P. (1960). *Soviet Phys. Solid St.* **2**, 23.
58. Nemilov, S. V. (1964). *J. Appl. Chem. USSR.* **37**, 1026.
59. Bacon, R. F. and Fanelli, R. (1943). *JACS.* **65**, 639.
60. Nemilov, S. V. and Petrovskii, G. T. (1963). *J. Applied Chem. USSR.* **36**, 932.
61. Arai, K. and Saito, S. (1971). *Jap. Applied Phys.* **10**, 1669–74.
62. Kurkjian, C. R., Krause, J. T. and Sigety, E. A. IXth International Congress on Glass, Versailles 1971, A1. 4, p. 503.

INDEX

Acrylic acid, 20–1, 65, 72
Acrylic soil conditioners, 242–57
  agricultural treatments, 249–51
  engineering applications, 244–9
  mode of action on soil properties, 252–7
Actinolite, 336
Activation energies for relaxation processes, 121
Adhesives, 196, 203
  tissue, 294
Alginic acid, metal salt of, 2
Aluminium
  hydroxide, 337
  isomorphous substitution of, 310, 328, 339
  phosphate, 308
Aluminium/silicon spinel, 351
Aluminosilicate gels, 322
Aluminosilicate glass, 66
Amine salts, 53
Ammonium salt, 332
Amosite, 348
Amphiboles, 18, 332–5, 346, 348–9
Amphoteric ion exchanger resins, 15–16
Analcite, 357
Anion exchange resins, 15, 16
Annealing, effect of, 152–3
Anthophyllite series, 335, 336
Antithrombogenic materials, 294
Apatite, 312
Asbestos, strength of, 348–50
ASPA cement, 226, 232, 234, 235

Barium
  sebacate, 265, 276
  terephthalate, 265, 276
Bentonites, 357
Birefringence, 144, 145
Bitumens, 243
Bolaform ions, 210
Bond switching, 46, 54, 56, 370
Boron phosphate, 308
Bound ions, 29, 30
Butadiene–acronitrile–methacrylic acid terpolymer, 57
Butadiene–acrylic acid copolymers, 55
Butadiene–lithium methacrylate copolymers, 41
Butadiene–methacrylic acid copolymers, 94, 160
Butadiene methyl(2-methyl-5-vinyl) pyridinium iodide copolymers, 41

Cadmium
  sebacate, 277
  zinc replacement by, 55
Calcium
  azelate, 270
  decanoate, 275
  dodecanedioate, 270
  montmorillonites, 357
  poly(acrylate), 242
  salts, 336
  sebacate, 262, 263, 271, 273, 274, 276, 278
  terephthalate, 276

Carbon
  fibres, 21
  polymers, 4
Carbon–carbon bond, 15
Carboxylated elastomers, 173–207
  companies producing, 173
  cross-link in oxide vulcanised, 190
  diene, 198
  history, 173–5
  mixed vulcanisation, 185
  preparation, 176–9
  properties, 179–90
  salts of, 20
  unvulcanised, 179
  uses, 194–205
  vulcanisation with metal oxides, 180
Carpets, 203
Carrageenan, 22
Cation exchange
  reaction, 13
  resins, 12
Cation valency and thermal expansion, 63
Ceramics, 360
Chain atoms, 8
Chain-extension, 17
Chrysotile, 338, 348
Clays, 359
Cleavage properties of infinite chain and 2-D sheet polymers, 346–8
Clinopyroxenes, 330
Cluster formation, 32, 34
Colloidal suspensions, stability of, 358–61
Condensed phosphates, 19
Contact lenses, 294
Covalent bond, 4, 302, 345, 348
Covalent chains, 5
  long, 17
  short, 27
Covalent network, 5, 9
Covalent segments, 5
Creep, 50, 56
  stress-relaxation, 54
Cristobalite, 351
Crocidolite, 348

Cross-linking, 17, 37–8, 47–50, 56, 93–5, 116, 159, 173, 186–7, 202, 211, 307–8, 320, 370, 374
Crystallinity, 49, 61, 70–1, 135–8, 150–5, 167
Crystallisation, 150, 154
  equation, 50
  glass, 364

Deflocculation, 360
Dehydration, 350, 356
Dehydroxylation, 350
Demineralisation of water, 13
Dentistry, 213
Depolymerisation, 355
Desalination, 293
Diaflo ultrafiltration membranes, 292
Dialysis, 293, 294
Dianions, 27
Dicarboxylic acids, metal salts of, 27
Dichroic ratio, 132
Dielectric constant, 35, 119, 120
Dielectric loss factor, 119
Dielectric relaxation, 119–20, 122–4
Differential scanning calorimetry, 138
Differential thermal analysis, 323, 350, 370
Diffusion in silicate networks, 400
Divalent metal oxides and hydroxides, 181
Divinyl benzene, 13
DMAEM, 254

Elastic recovery of polymer melts, 106
Elastomeric polymers, 173
Electrical properties, 78, 119–24
Emulsion polymerisation, 176
Energy parameter, 46
Enstatite, 352
EPDM rubbers, 175
EPR rubbers, 175
Ethylene, 69, 72, 73
  ionomers, 33, 34
Ethylene–acrylic acid copolymers
  infra-red spectra, 131

Ethylene–acrylic acid copolymers—*contd.*
  magnetic resonances, 127
  melt flow, 51–2
  properties of, 75, 116
  relaxation, 55, 116–19
  stiffness, 57
  viscosity, 94–5
  water absorption, 64
Ethylene–lithium acrylate copolymer, magnetic resonances, 127
Ethylene–metal acrylate copolymers, 42
  melting points and crystallinities, 49
Ethylene–methacrylic acid copolymers
  electrical properties, 120
  electron microscopy, 82–93
  infra-red spectra, 129
  loss modulus, 109
  melt viscosities, 95
  optical properties, 144
  properties of, 76
  proton magnetic relaxation studies, 125
  relaxation, 116–19
    spectra, 104
  spherulite, 156
  viscosity, 102, 104
  WLF plot, 100
  X-ray diffraction measurements, 134

Feldspars, 311, 343, 355
Flex modulus, 165
Flory's equation, 153
Flow
  behaviour, 51
  steady-state, 103
  viscous, 102
Fluorophlogopite, 321
Foam manufacture, 202

Galactose 2,6-disulphate, 22
Galactose 6-sulphate, 22

Gels, 20, 294–5
  aluminosilicate, 322
  polysaccharide, 21
  thermodynamic, 289
Gibbs–Di Marzio theory, 38
Glass
  network, 27
  structure, 3-D, 2
  temperature, 159
  transition, 35–46, 110
    copolymers, 41
    homopolymers, 38
    inorganic glasses, 44
    temperature, 37–8
      function of q/a parameter, 40
      versus ion concentration, 42–4
Glass–ionomer cements, 226–35
  materials, 227
  properties, 234–5
  setting mechanism, 232
  structure, 229, 232–4
Glasses, 35
  aluminosilicate, 66
  anhydrous alkali silicate, 394
  barium phosphate, 381
  borate, 386–92
    expansion coefficient, 389
    formation, 386
    properties, 387–92
    transformation temperature, 387–9
    Young's modulus, 391
  chalcogenide, 363, 366, 400–4
    constitution, 401
    mechanical properties, 404
    melt viscosity, 403
    transformation temperature, 401
    Young's modulus, 404
  crystallisation, 364
  formation, 364
  inorganic, 363–406
    glass transition, 44
    types of, 365
  invert, 27, 397
  ionic polymer, 1
  lithium silicate, 393

Glasses—*contd.*
  oxide, 363, 365–400
    cation co-ordination number and oxygen-bond strength, 372–3
    cation effects, 372
    expansion coefficient, 389
    4-connective networks, 392
    mechanical properties, 376–8
    melt viscosity, 375
    mobility of cations, 378–9
    oxygen density, 372–3, 376
    properties of, 370–1
    structure, 366–9
    tensile strength, 376
    3-connective networks, 379
    3–4 connective networks, 386
    transformation temperature, 369–72
      composition relationship, and, 373–5
    Young's modulus, 376
  phosphate, 379–86
    durability, 382–5
    formation, 380
    melt viscosity, 385
    properties, 381–6
    transformation temperature, 381
    Young's modulus, 385
  polymeric nature of, 364
  silicate, 392–400
    density, 394
    diffusion in networks, 400
    expansion coefficient, 396
    flow properties, 53
    multi-component, 397
    stiffness, 61
    transformation temperature, 396
    viscosity, 396
    Young's modulus, 398
  sodium metasilicate, 393
  soluble silicate, 392
  specific heat, 370
  strength, 62
α-L-Gulopyranosiduronate, 22

Halatopolymeric transition, 261
Halatopolymers, 261

Hexasodium tetrapolyphosphate, 326
Homopolymers, glass transition, 38
Hycar 1571, 202
Hydration, 355, 356
Hydrogels, 289, 290, 292, 294
  dry, 296
Hydrogen bonding, 93–5, 337, 342
Hydrolysis, 178, 353–5
Hydrolytic stability, 64–6

Illites, 341
Immunisation reactions, 296
Infra-red spectroscopy, 129–32
Interatomic bonding, 301
Ion
  binding, 209–12
  exchange, 3, 357
    resins, 12, 15–16, 20, 23
  pairs, 29, 31, 32, 160, 161
Ionenes, 2, 23
Ionic bonds, 2–5, 37, 44, 45, 46, 49, 261, 262, 302, 345, 348
  aggregation, 30
  chemical stability, 64
  concentration, 28–30
  effect on melting behaviour, 47
  high concentration, 38
  long covalent chains containing, 17–27
  sheets containing, 9
  short covalent chains plus, 27
Ionic domains, 33, 34
Ionic polymers, 300, 345
  cements, 208–35
    ion binding and molecular configuration, 209
  classification, 5–9
  definition, 1
  form of, 2
  hybrid organic–inorganic, 9
  inorganic, 8, 50
  organic, 8
  physical properties, 35–66
  range of, 2, 3
  representations, 36
  rigid, highly decarboxylated, 208–60

Ionic polymers—*contd.*
  structural features, 28–35
  structures, 2–5
  types, 5
  typical examples, 9–28
Ionic segments, 108
Ionic solids, 4
Ionomers, 2, 19, 69–172
  characterisation, 74–5
  clarity, 70, 79
  commercially available, 69
  definition, 69
  electrical properties, 78, 120
  electron microscopy, 79–93
  general properties, 70
  infra-red spectroscopy, 129
  laboratory preparation, 74
  mechanical properties, 77
  melt strength, 71–2
  morphology, 79, 91
  optical properties, 144–50
  physical properties, 75
  preparation, 72
  relaxation, 117–19
    spectra, 106
  rheological properties, 93–109
  structure, 154–62
    film, of, 79
  surface properties, 79
  thermal properties, 150–4
  thin film, 79
  transparency, 70–1
  vacuum-drawn film, 71
  viscoelastic properties, 104
  viscosity, 93–104
  WLF plot, 100
  X-ray diffraction measurements, 134
Ions
  aggregation, 30–5
  bound, 29, 30

Jadeite, 330

Kaolinite, 254, 337, 339, 350–2, 358
Kaolins, 341

Kossel–Lewis model, 304
Kurrol's salt, 332

Latices, 173, 177, 183, 194
  carboxylated, 200
  uses, 199–205
Lead poly(acrylate), 242
Leather reconstitution, 204
Leucite, 357
Lithium perchlorate–polypropylene glycol, 41
Lithium-7 magnetic resonance, 127
London–van der Waals attractive energy, 252
Loss modulus of ethylene–methacrylic acid copolymers, 109

Maddrell's salt, 332
Magnesium sebacate, 264, 277
β-D-Mannopyranosiduronate, 22
Maxwell–Wagner effect, 122
Mechanical properties, 54, 75, 77, 109, 278
Melt
  behaviour, 50–4
  flow, 271–6
Melting/decomposition behaviour, 47–50
Melting point, 46–50
Membrane applications, 292–4
Menschutkin reaction, 16, 24
Metakaolin, 351
Metal dicarboxylates, 261, 265–76
  crystalline structure, 265
  effect on strength of phenolic resin, 279–80
  mechanical properties, 278
  melting points, 276
  polymeric structure, 266
  preparation, 262–5
  properties, 276–8
  thermal stabilities, 276
  uses, 279–80
Metal polyacrylate salts, 237–42
Metasilicate, structure of, 12

Methacrylic acid, 21, 58, 69, 72, 73, 86
Methyl methacrylate, 21
Mica, 321, 347
Mixed ions, 24–7
Monolithic plastics, 237
Montmorillonite, 338, 340, 341, 352, 356, 358
Mooney scorch, 183, 186
Mullite, 352
Multiplets, 31, 32, 34, 160–2

Network
  atoms, 3
  bonds, 3, 29, 46, 370, 373
  former, 11
  modifier, 11
Neutralisation, 113, 116, 119, 121, 135–7, 141, 144
Non-directional bonds, 4
Non-wovens, 200
Noselite, 345
Nuclear magnetic resonance studies, 124–8

Oligomers, 324–8
Optical properties, 144–50
Orientation function, 132
Orthopyroxenes, 330

Packaging, 72
Paints, 204
PAM, 255–7
Paper manufacture, 201
Pentasodium tripolyphosphate, 326
Pentasodium tripolyphosphate-II, 327
Phosphates, 300–62
  B/A ratio, 313–16, 318, 321
  bond formation, 302
  branched chain, 332
  characterisation, 322
  cleavage properties of infinite chain and 2-D sheet polymers, 346–8
  complexes, 360
  condensation polymerisation, 313
  condensed, 19
  copolymers, 308
  covalent/ionic bond formation and connectivity, 308
  cross-linked, 332
  crystallisation from melt, 318–21
  dimers and co-dimers (1-connective), 324–6
  electronegativity, 301
  electronic structure, 301
  heat treatment, 314
  humidity effects, 315
  hydrolytic degradation, 353
  infinite chain polymers, 331–2
  interatomic bonding, 301
  linear chain polymers (1- and 2-connective), 326–7
  nature of cations, 315
  oligomers, 324–8
  polymers, 307
  precipitation from aqueous solution, 321
  reorganising melts, 318
  ring, 328
  solid-state reaction, 316–17
  structural type and B/A ratio relationship, 313
  structure, 300, 323–45
  structure/property relationships, 345–61
  synthesis, 312–22
  tetrahedral symmetry, 303
  3-D framework, 342
  2-D sheet polymers, 337–42
  see also Glasses
Phosphorus trichloride, 72
Plasticised PVC, electrical properties, 78
Plasticised vinyls, mechanical properties, 77
Plasticisers, 169
Polyacrylates, 66
  glass transition temperature, 38–9
Polyacrylic acid, 237
  metal salts, 19–20, 57
Polyanions, 9–14, 17–22, 41, 254, 284, 287

Polycations, 8, 15–17, 22–4, 41, 284, 287
Polyelectrolyte complexes, 24–6, 34, 281–99
  applications, 291–7
  formation reaction, 281
  general characteristics, 281–2
  history, 283
  in situ polymerisation, 287, 296
  isomers, comparison with, 282
  neutral, 281
    fabrication methods, 285
  non-stoichiometric, 25–6, 281, 285
  physical chemistry, 288–91
  properties, 288–91
  related materials, comparison with, 282
  solubility phase diagram, 288
  starting materials, 284
  stoichiometric, 25–6
  synthesis, 284–8
Polyelectrolytes
  anionic, 9, 14, 20
  cationic, 9, 22
    quaternary, 23
  non-ionic, 9
  snake-cage, 26
  water-soluble, 24
Polyether, structure, 3
Polyethylene, 61, 72, 91
  crystallinity, 135–8, 166, 167
  electrical properties, 78
  low angle X-ray scattering pattern, 140
  mechanical properties, 77
  relaxation, 110–15
  structure, 154
    film, of, 82
  surface properties, 79
Polyion complexes, 281
Polymethacrylic acid, metal salts, 57
Polyolefins, optical characteristics, 70
Polyphosphates, 18, 35
  glass transition temperature, 38–9
Polysaccharide gels, 21
Polysalts, 24–6, 281
Poly(sodium acrylate), 159
Polystyrene, 15

Polystyrene–sodium methacrylate, 44
Polysulphides, 28
Polyvinyl chloride, plasticised, electrical properties, 78
Polyvinylbenzenetrimethyl ammonium chloride, 24
Polyvinylbenzyltrimethyl ammonium chloride, 25, 290
Precipitation reactions, 283
Pyrophyllite, 340, 352, 356
Pyroxenes, 18, 329, 330, 346
Pyroxymanganite, 330

Radial distribution function, 138, 140
Reduced variables, method of, 99
Relaxation
  behaviour, 54–6, 109–28, 163
    ethylene–acrylic acid copolymers, 116–19
    ethylene–methacrylic acid copolymers, 116–19
    ionomers, 116–19
    nuclear magnetic resonance studies, 124–8
  processes, activation energies, 121
  spectra
    ethylene–methacrylic acid copolymer, 104
    ionomer, 107
Retarded elastic elements, 106, 107
Reverse osmosis membranes, 293
Rhodonite, 330
Rigid, highly carboxylated, ionic polymers, 208–60
Rubbers
  containing carboxyl groups, 20
  stiffness, 57
  strength, 62
  with quaternary nitrogen cross-link points, 16

SBR, 202–4
Scroll structure, 338, 348
Serpentine, 352
Shear modulus, 57–9

Shrink proofing, 198
Silica
  expansion coefficient, 396
  polymeric forms, 10
  products based, 9–12
  vitreous, 366, 392
  Young's modulus, 398
Silicates, 300–62
  anhydrous alkali, 394
  B/A ratio, 316, 318, 320, 321
  bond formation, 302
  branched chain, 332
  characterisation, 322
  cleavage properties of infinite chain and 2-D sheet polymers, 346–8
  colloidal suspensions, 358–61
  condensation polymerisation, 313
  copolymers, 308
  covalent/ionic bond formation and connectivity, 305
  cross-linked, 332
  crystallisation from melt, 318–21
  dehydroxylation, 350
  dimers and co-dimers (1-connective), 324–6
  double chain polymers, 332–6
  double stranded chain ions, 18
  electronegativity, 301
  electronic structure, 301
  flow properties, 53
  high temperature changes, 351
  hydration/dehydration properties, 356
  hydrolytic stability, 354–6
  hydrothermal synthesis methods, 317
  infinite chain polymers, 328–30
  interatomic bonding, 301
  layer lattice, 337
  linear, 17
    glass transition temperature, 38–9
  linear chain polymers (1- and 2-connective), 326–7
  naturally occurring polymers, 310
  network structures, 10–12
  oligomers, 324–8
  polymers, 305–7
  precipitation from aqueous solution, 321
  reorganising metals, 318, 320
  ring, 328
  single chain ions, 18
  single-phase amorphous, 35
  solid, 18
  solid-state reaction, 316–17
  soluble, 392
  structure, 300, 323–45
  structure/property relationships, 345–61
  synthesis, 312–22
  tetrahedral symmetry, 303
  thermal stability, 350–3
  3-D framework polymers, 342–5
  2-D sheet polymers, 337–42
  see also Glasses
Sillimanite, 334, 352
Silver salts, 332
Sodalite, 345
Sodium
  dihydrogen phosphate, 313
  metaphosphate, 326
  montmorillonites, 357
  polystyrene sulphonate, 24, 25, 287, 290
  salt, 315
  sebacate, 265
  tripolyphosphate, 360
Soil conditioners
  agricultural treatments, 249–51
  engineering applications, 244–9
  mode of action on soil properties, 252–7
Solution grafting, 177
Solution metallation reactions, 178
Solutions, 20
SPA, 255
Spherulitic structures, 79, 82, 89, 147, 154
SPMA, 254
Spodumene, 330
Stabilisers, 243
Static strain optical coefficient, 144
Stiffness, 56–61, 75
Strength, 62

Stress-relaxation, 50, 51, 56
  creep, 54
Structural features, 2–5, 28–35
Structure
  ionomers, 154–62
  polyethylene, 154
Styrene, 13, 34, 58, 72
Styrene–metal methacrylate
  copolymers, 56
Styrene–methacrylic acid
  copolymer, 59
  glass transitions, 37
Styrene–sodium methacrylate
  ionomers, 64
Styrene–vinyl pyridine block
  copolymers, 26
Sulphonic groups, 14
Surface coating, 21, 72
Surlyn, 19, 69

Talc, 339, 352
Tetramethylammonium
  tripolyphosphate, 353
Thermal expansion, 63
Thermal properties of ionomers,
  150–4
Thermodynamic gels, 289
Thermoplastic ionic polymers, 69–172
Tissue adhesives, 294
Tobermorite, 336
Tremolite series, 335

Ultrafiltration membranes, 292
Ultramarines, 345
Ultraphosphates, 336, 353, 354

Van Aarsten–Stein theory, 150
Vermiculite, 341, 350, 358
Vinyl polymers, structure, 3
Vinyl pyridine, 26
Vinyls, plasticised, mechanical
  properties, 77
Viscoelastic properties, 50, 104
Viscosity, 320
  anhydrous alkali silicates, 396

chalcogenide glasses, 403
dynamic, 104
ethylene–acrylic acid copolymer,
  94, 102, 104
internal, 97, 107, 150
ionomers, 93–104
melts, 53–4, 375
non-Newtonian, 101
phosphate glasses, 385
reduced, 99
steady-flow, 104
temperature dependence, 98–100
Voigt elements, 106
Vulcanisation, 180
  mixed, 185
  oxide, cross-link in, 190

Water
  absorption, 64, 162–9
  demineralisation, 13
Weathering, 355
WLF method, 99
β-Wollastonite, 330

Xonotlite, 336
X-ray diffraction, 134–44, 155–62,
  166, 265, 266, 322

Young's modulus
  borate glasses, 391
  oxide glasses, 385

Zeolites, 322, 344, 356–8
Zeta potential, 359, 360
Zinc
  dimerate, 267, 276–8
  oxide, 215, 237
  phosphate cement, 219
  polyacrylate, 66, 237–42
  polycarboxylate cement, 212–26
    biological properties, 225
    composition, 214

Zinc—*contd.*
  polycarboxylate cement—*contd.*
    magnesium oxide addition, 216
    materials, 213
    metal oxide powder, 215
    physical properties, 221–5
    properties, 220
    setting mechanism, 217
    structure, 220
    Types I to IV, 214–15
  replacement by cadmium, 55
  salts, 57
  sebacate, 277
  stearate, 268